Advances in
CHEMICAL ENGINEERING
PHOTOCATALYTIC TECHNOLOGIES

VOLUME **36**

Advances in
CHEMICAL ENGINEERING
PHOTOCATALYTIC TECHNOLOGIES

VOLUME **36**

Edited by

HUGO I. DE LASA

Chemical Reactor Engineering Centre (CREC)
Department of Chemical and Biochemical Engineering
University of Western Ontario
London, Ontario, Canada
N6A 5B9

BENITO SERRANO ROSALES

Programa de Ingeniería Química
Unidad Académica de Ciencias Químicas
Universidad Autónoma de Zacatecas
Km. 6 Carretera a Guadalajara
Ejido La Escondida
Zacatecas Zac
98160, Mexico

Amsterdam ● Boston ● Heidelberg ● London ● New York ● Oxford
Paris ● San Diego ● San Francisco ● Singapore ● Sydney ● Tokyo
Academic Press is an imprint of Elsevier

ELSEVIER

Academic Press is an imprint of Elsevier
Radarweg 29, PO Box 211, 1000 AE Amsterdam, The Netherlands
32 Jamestown Road, London NW1 7BY, UK
30 Corporate Drive, Suite 400, Burlington, MA 01803, USA
525 B Street, Suite 1900, San Diego, CA 92101-4495, USA

First edition 2009

Library of Congress Cataloging-in-Publication Data
A catalog record for this book is available from the Library of Congress

British Library Cataloguing in Publication Data
A catalogue record for this book is available from the British Library

ISBN: 978-0-12-374763-1
ISSN: 0065-2377

For information on all Academic Press publications
visit our website at books.elsevier.com

Printed and bound in USA
09 10 11 12   10 9 8 7 6 5 4 3 2 1

CONTENTS

Preface   ix
Contributors   xi

1. **Determination of Photoadsorption Capacity of Polychrystalline**
   TiO$_2$ Catalyst in Irradiated Slurry   **1**
   Vincenzo Augugliaro, Sedat Yurdakal, Vittorio Loddo, Giovanni Palmisano
   and Leonardo Palmisano

   1. Introduction   2
   2. Experimental   5
   3. TiO$_2$ Surface Modifications Under Irradiation   8
   4. Photoadsorption Determination   10
   5. Results   18
   6. Discussion   21
   7. Conclusions   28
       Appendices   29
       List of Symbols   33
       References   33

2. **Treatment of Chromium, Mercury, Lead, Uranium, and Arsenic**
   in Water by Heterogeneous Photocatalysis   **37**
   Marta I. Litter

   1. Introduction   37
   2. Thermodynamical Considerations and Mechanistic Pathways   41
   3. Chromium   44
   4. Mercury   49
   5. Lead   53
   6. Uranium   57
   7. Arsenic   58
   8. Conclusions   61
       Acknowledgment   62
       References   62

3. **Mineralization of Phenol in an Improved Photocatalytic Process Assisted with Ferric Ions: Reaction Network and Kinetic Modeling**   69
Aaron Ortiz-Gomez, Benito Serrano-Rosales, Jesus Moreira-del-Rio and Hugo de-Lasa

1. Introduction   70
2. Experimental Methods Used in CREC   79
3. Fe-Assisted Photocatalytic Mineralization of Phenol and its Intermediates   81
4. Kinetic Modeling: Unpromoted PC Oxidation and Fe-Assisted PC Oxidation of Phenol   92
5. Conclusions   105
   Recommendations   106
   List of Symbols   106
   References   108

4. **Photocatalytic Water Splitting Under Visible Light: Concept and Catalysts Development**   111
R.M. Navarro, F. del Valle, J.A. Villoria de la Mano, M.C. Álvarez-Galván and J.L.G. Fierro

1. Introduction   111
2. Photoelectrochemistry of Water Splitting   113
3. Photocatalysts for Water Splitting Under Visible Light   124
4. Concluding Remarks and Future Directions   140
   Acknowledgments   141
   References   141

5. **Photocatalytic Reactor Configurations for Water Purification: Experimentation and Modeling**   145
Ajay K. Ray

1. Introduction   145
2. Macrokinetic Studies   148
3. Major Challenges in the Design and Development of Large-Scale Photocatalytic Reactors for Water Purification   159
4. Conclusions   181
   Acknowledgments   183
   References   183

6. **Development and Modeling of Solar Photocatalytic Reactors**   185
Camilo A. Arancibia-Bulnes, Antonio E. Jiménez and Claudio A. Estrada

1. Introduction   186
2. Solar Photocatalytic Reactors   187
3. Radiation Transfer in Photocatalytic Reactors   206
4. The P1 Approximation   213

5. Conclusions and Perspectives   222
    Acknowledgments   223
    List of Symbols   223
    Abbreviations   225
    References   225

**7. Scaling-Up of Photoreactors: Applications to Advanced Oxidation
   Processes   229**
   Orlando M. Alfano and Alberto E. Cassano

1. Introduction   230
2. Scaling-Up of a Photocatalytic Wall Reactor with Radiation Absorption
    and Reflection   234
3. Scaling-Up of a Homogeneous Photochemical Reactor with Radiation
    Absorption   250
4. Scaling-Up of a Heterogeneous Photocatalytic Reactor with Radiation
    Absorption and Scattering   263
5. Conclusions   282
    Acknowledgments   283
    Notation   283
    References   286

8. Photocatalytic Treatment of Air: From Basic Aspects to Reactors   289
   Yaron Paz

1. Introduction   290
2. Types of Air-Treatment Applications   293
3. Basic Aspects of Photocatalysis for Air Treatment   296
4. Types of Target Pollutants   303
5. Photocatalytic Reactors for Air Treatment: Modes of Operation   310
6. Types of Photocatalytic Reactors for Air Treatment   312
7. Current Problems and Future Trends   329
8. Concluding Remarks   331
    Acknowledgment   331
    List of Symbols   331
    Abbreviations   332
    References   333

Index   337

Contents of Volumes in this Serial   353

PREFACE

In recent years, the international community has been increasingly concerned about the stresses imposed on the natural environment by many chemical and energy-generating processes. As a result, the world is witnessing an accelerated development and implementation of new green technologies. These green technologies are called to provide ecologically responsible solutions for the much needed supply of drinking water, clean air, and various forms of energy.

Photocatalysis holds great promise for delivering these ground-breaking technologies. Photocatalysis is a truly environmentally friendly process where irradiation, either near UV or solar light, promotes photoexcitation of semi-conductor solid surfaces. As a result, mobile electrons and positive surface charges are generated. These excited sites and electrons accelerate oxidation and reduction reactions, which are essential steps for pollutant degradation and other photoinduced chemical transformations such as water splitting.

Photocatalysis and its related technological issues have been strongly influenced by recent publications. The present Volume 36-*Photocatalytic Technologies* of the Elsevier's *Advances in Chemical Engineering Series* aims at offering a comprehensive overview of the state-of-the-art photocatalytic technology. In order to accomplish this, several prominent researchers were invited to contribute a chapter for the Volume 36.

Chapter 1 examines the phenomenological principles involved in the modeling of photocatalytic reactions including the photo-adsorption of chemical species. This chapter proposes a method to quantify photo-adsorbed species onto irradiated TiO_2. The technique is applied to the oxidation of phenol and benzyl alcohol.

Chapter 2 considers the removal of inorganic water contaminants using photocatalysis. Metal cations react via one-electron steps first leading to unstable chemical intermediates, and later to stable species. Three possible mechanisms are identified: (a) direct reduction via photo-generated conduction band electrons, (b) indirect reduction by intermediates generated from electron donors, and (c) oxidative removal by electron holes or hydroxyl radicals. The provided examples show the significance of these mechanisms for the removal of water contaminants such as chromium, mercury, lead, uranium, and arsenic.

Chapter 3 addresses the photocatalytic mineralization of organic species in water and its enhancement by using ferric ions. This methodology uses Photo CREC reactors with Fe-promoted TiO_2. It is shown that 5 ppm of Fe in water provides an optimum iron concentration able to maximize the rates of oxidation and mineralization for both phenol and its aromatic intermediates. This chapter also describes a parallel–series kinetic reaction network. This reaction network and the derived kinetic parameters are most suitable for describing the improved phenol photocatalytic oxidation with ferric ions.

Chapter 4 reports research progress on hydrogen production via water splitting using photocatalysis. It is stated that while water splitting with UV light shows good prospects, water splitting under visible light requires a significant efficiency improvement provided by an enhanced utilization of irradiated photons per molecule of hydrogen produced. In order to accomplish this, new nanomaterials manufactured under close control of crystallinity, electronic structure, and morphology are proposed.

Chapter 5 addresses the scaling-up in photocatalytic reactors with catalyst irradiation being identified as a most important engineering design parameter. It is stated that the photocatalytic reactor design involves a skilful combination of a highly and uniformly irradiated photocatalyst, and an intensive mixing of the TiO_2 suspension. In order to attain these design objectives, several reactor designs are reviewed such as a multiple tube reactor, a tube light reactor, a rotating tube reactor, and a Taylor vortex reactor.

Chapter 6 describes solar-powered photocatalytic reactors for the conversion of organic water pollutants. Nonconcentrating reactors are identified as some of the most energetically efficient units. It is reported that the absorption of radiation is a critical parameter in the efficiency reactor evaluation. The radiative transfer equation (RTE) solution under the simplified conditions given by the P1 approximation is proposed for these assessments.

Chapter 7 reports a scaling-up procedure for photocatalytic reactors. The described methodology uses a model which involves absorption of radiation and photocatalyst reflection coefficients. The needed kinetics is obtained in a small flat plate unit and extrapolated to a larger reactor made of three concentric photocatalyst-coated cylindrical tubes. This procedure is applied to the photocatalytic conversion of perchloroethylene in air and to the degradation of formic acid and 4-chlorophenol in water.

Chapter 8 addresses the treatment of contaminated air streams using photocatalysis. Special attention is given to the distinction between reaction kinetics and mass transport processes. The reviewed studies show the evolution from the early days of TiO_2 photocatalysis, where the aim was to understand the basic process parameters, to today's development of phenomenological models assisting in the scaling-up of units.

In summary, the present issue of *Advances in Chemical Engineering* Volume 36 offers an up-to-date overview and discussion of principles and applications of photo catalytic reaction engineering. Altogether, Volume 36 is an invitation to reflect on the possibilities of photocatalysis as a promising technology for green reaction engineering.

Hugo I. de Lasa and Benito Serrano Rosales,
December 2008.

CONTRIBUTORS

Orlando M. Alfano, *INTEC (Universidad Nacional del Litoral and CONICET), 3000 Santa Fe, Argentina*

M.C. Álvarez-Galván, *Instituto de Catálisis y Petroleoquímica (CSIC), C/ Marie Curie 2, 28049, Madrid, Spain*

Camilo A. Arancibia-Bulnes, *Centro de Investigación en Energía, Universidad Nacional Autónoma de México, Privada Xochicalco s/n, Col. Centro, A. P. 34, Temixco, 62580 Morelos, México*

Vincenzo Augugliaro, *"Schiavello-Grillone" Photocatalysis Group, Dipartimento di Ingegneria Chimica dei Processi e dei Materiali, Università di Palermo, Viale delle Scienze, 90128 Palermo, Italy*

Alberto E. Cassano, *INTEC (Universidad Nacional del Litoral and CONICET), 3000 Santa Fe, Argentina*

Hugo de-Lasa, *Chemical Reactor Engineering Centre-CREC, Department of Chemical and Biochemical Engineering, The University of Western Ontario, London, Ontario, Canada N6A5B9*

F. del Valle, *Instituto de Catálisis y Petroleoquímica (CSIC), C/ Marie Curie 2, 28049, Madrid, Spain*

Claudio A. Estrada, *Centro de Investigación en Energía, Universidad Nacional Autónoma de México, Privada Xochicalco s/n, Col. Centro, A. P. 34, Temixco, 62580 Morelos, México*

J.L.G. Fierro, *Instituto de Catálisis y Petroleoquímica (CSIC), C/ Marie Curie 2, 28049, Madrid, Spain*

Antonio E. Jiménez, *Centro de Investigación en Energía, Universidad Nacional Autónoma de México, Privada Xochicalco s/n, Col. Centro, A. P. 34, Temixco, 62580 Morelos, México*

Marta I. Litter, *Gerencia Química, Centro Atómico Constituyentes, Comisión Nacional de Energía Atómica, Av. Gral. Paz 1499, 1650 San Martín, Prov. de Buenos Aires, Argentina*

Vittorio Loddo, *"Schiavello-Grillone" Photocatalysis Group, Dipartimento di Ingegneria Chimica dei Processi e dei Materiali, Università di Palermo, Viale delle Scienze, 90128 Palermo, Italy*

Jesus Moreira-del-Rio, *Chemical Reactor Engineering Centre-CREC, Department of Chemical and Biochemical Engineering, The University of Western Ontario, London, Ontario, Canada N6A5B9*

R.M. Navarro, *Instituto de Catálisis y Petroleoquímica (CSIC), C/Marie Curie 2, 28049, Madrid, Spain*

Aaron Ortiz-Gomez, *Chemical Reactor Engineering Centre-CREC, Department of Chemical and Biochemical Engineering, The University of Western Ontario, London, Ontario, Canada N6A5B9*

Giovanni Palmisano, *"Schiavello-Grillone" Photocatalysis Group, Dipartimento di Ingegneria Chimica dei Processi e dei Materiali, Università di Palermo, Viale delle Scienze, 90128 Palermo, Italy*

Leonardo Palmisano, *"Schiavello-Grillone" Photocatalysis Group, Dipartimento di Ingegneria Chimica dei Processi e dei Materiali, Università di Palermo, Viale delle Scienze, 90128 Palermo, Italy*

Yaron Paz, *Department of Chemical Engineering, Technion-Israel Institute of Technology, Haifa, Israel*

Ajay K. Ray, *Department of Chemical and Biochemical Engineering, University of Western Ontario, London, ON N6A 5B9, Canada*

Benito Serrano-Rosales, *Programa de Ingeniería Química, Unidad Académica de Ciencias Químicas, Universidad Autonoma de Zacatecas, Km.6 Carretera a Guadalajara, Ejido La Escondida, Zacatecas, Zac., 98160, Mexico*

J.A. Villoria de la Mano, *Instituto de Catálisis y Petroleoquímica (CSIC), C/ Marie Curie 2, 28049, Madrid, Spain*

Sedat Yurdakal, *"Schiavello-Grillone" Photocatalysis Group, Dipartimento di Ingegneria Chimica dei Processi e dei Materiali, Università di Palermo, Viale delle Scienze, 90128 Palermo, Italy; Kimya Bölümü, Fen Fakültesi, Anadolu Üniversitesi, Yunus Emre Kampüsü, 26470 Eskişehir, Turkey*

Determination of Photoadsorption Capacity of Polychrystalline TiO$_2$ Catalyst in Irradiated Slurry

Vincenzo Augugliaro[1,*], **Sedat Yurdakal**[1,2], **Vittorio Loddo**[1], **Giovanni Palmisano**[1], and **Leonardo Palmisano**[1]

Contents   1. Introduction 2
2. Experimental 5
3. TiO$_2$ Surface Modifications Under Irradiation 8
4. Photoadsorption Determination 10
    4.1 Langmuir isotherm 13
    4.2 Freundlich isotherm 14
    4.3 Redlich–Peterson isotherm 16
5. Results 18
6. Discussion 21
    6.1 Benzyl alcohol photoadsorption 22
    6.2 Phenol photoadsorption 25
    6.3 Reaction mechanism 26
7. Conclusions 28
Appendices 29
A1. Asymptotic Cases of Langmuir Photoadsorption Isotherm 29
A2. Temkin Isotherm 31
List of Symbols 33
References 33

[1] "Schiavello-Grillone" Photocatalysis Group, Dipartimento di Ingegneria Chimica dei Processi e dei Materiali, Università di Palermo, Viale delle Scienze, 90128 Palermo, Italy

[2] Kimya Bölümü, Fen Fakültesi, Anadolu Üniversitesi, Yunus Emre Kampüsü, 26470 Eskişehir, Turkey

[*] Corresponding author.
E-mail address: augugliaro@dicpm.unipa.it

Advances in Chemical Engineering, Volume 36
ISSN 0065-2377, DOI: 10.1016/S0065-2377(09)00401-3

1. INTRODUCTION

In the field of heterogeneous catalysis the need of kinetic investigation is strictly connected to the main task of a chemical engineer, that is, designing properly a chemical reactor. A successful reactor design should thus start from reliable kinetic models that describe the rate of catalytic reactions and, therefore, from the reaction mechanisms, which means understanding reactions at a molecular level. In catalysis, due to the complex nature of this phenomenon, adsorption and desorption of reactants as well as several steps for surface reactions must be taken into account. For heterogeneous photocatalysis, which may be considered a special case of heterogeneous catalysis, the previous considerations hold true with the added difficulty that the light absorbed by the photocatalyst affects both adsorption (photoadsorption) and surface reactions.

The use of irradiation to initiate chemical reactions is the principle on which heterogeneous photocatalysis is based. When a wide band gap semiconductor like titanium dioxide (Carp et al., 2004) is irradiated with suitable light, excited electron–hole pairs result that can be applied in chemical processes to modify specific compounds. If recombination or lattice reaction does not involve all the photogenerated pairs, the conduction band electrons participate in reduction reactions on the catalyst surface while positive holes are involved in oxidation reactions. Suitable substrates must be adsorbed on the catalyst surface for the occurrence of a photoreaction process which always starts with the substrate(s) adsorption and eventually ends with the product(s) desorption. On these grounds heterogeneous photocatalysis is defined as follows (Braslavsky, 2007): "*Change in the rate of a chemical reaction or its initiation under the action of ultraviolet, visible, or infrared radiation in the presence of a substance, the photocatalyst, that absorbs light and is involved in the chemical transformation of the reaction partners.*" Symbolically overall photocatalytic reaction is expressed by the equation:

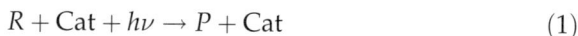

$$R + Cat + h\nu \rightarrow P + Cat \tag{1}$$

where R and P are reactants and reaction products, respectively, present in the gas or liquid phase, Cat is the solid photoadsorbent (photocatalyst), and $h\nu$ is the symbol of photons able to be absorbed by the photocatalyst.

The knowledge of heterogeneous photocatalytic systems has grown very much since the pioneering work on water photolysis carried out with a semiconductor electrode (Fujishima and Honda, 1972). The basic principles of heterogeneous photocatalysis are now well established (Fujishima et al., 1999; Kaneko and Okura, 2002; Schiavello, 1997) and also the applicative aspects of this technology are being investigated in the fields not only of environment remediation (Augugliaro et al., 2006; Fujishima et al., 2000; Mills and Le Hunte, 1997) but also of green chemistry (Gonzalez et al., 1999; Mohamed et al., 2002; Yurdakal et al., 2008a). There are, however, many important aspects waiting to be investigated. One of these is the correct approach for the determination of the photoadsorption capacity under photoprocess occurrence, that is, of the amount of substrate adsorbed on the surface of a photocatalyst which is being irradiated.

Photon absorption by photocatalyst is regarded as the first stage of photo-excitation of heterogeneous system; the photoexcitation pathways of wide band gap solids may involve photogeneration of excitons and/or free charge carriers, depending on photocatalyst features such as fundamental absorption band, extrinsic/intrinsic defect absorption bands, or UV-induced color center bands. Independently of photoexcitation type, photon absorption has two main effects: (i) it changes the characteristics of photocatalyst surface and (ii) it generates active photoadsorption centers. A typical case of the first effect is that band gap irradiation induces superhydrophilicity (photoinduced superhydrophilicity, PSH) on the TiO_2 surface, which shows hydrophobic features under dark conditions (Fujishima and Zhang, 2006; Fujishima et al., 2000; Wang et al., 1997). This PSH is accompanied by photocatalytic activity, as both phenomena have a common ground, so that the surface-adsorbed compounds may be either photooxidized or washed away by water.

The second important effect is that irradiation absorption generates active states of the photoadsorption centers with trapped electrons and holes. By definition (Serpone and Emeline, 2002) *"the photoadsorption center is a surface site which reaches an active state after photoexcitation and then it is able to form photoadsorbed species by chemical interaction with substrate (molecules, or atoms, or ions) at solid/fluid interface."* In turn, the active state of a surface photo-adsorption center is *"an electronically excited surface center, i.e. surface defect with trapped photogenerated charge carrier that interacts with atoms, molecules or ions at the solid/gas or solid/liquid interfaces with formation of chemisorbed species."*

Adsorption initiated by light absorbed by the solid surface (photoadsorption) can be expressed by the following simple mechanism (Ryabchuk, 2004):

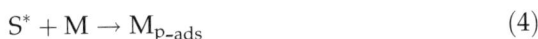

$$S + h\nu \rightarrow S^* \tag{2}$$

$$S^* \rightarrow S \tag{3}$$

$$S^* + M \rightarrow M_{p-ads} \tag{4}$$

where S is the photoadsorption center, S^* the active state of photoadsorption center, M the substrate in the fluid phase, and M_{p-ads} the photoadsorbed substrate. Equation (2) describes the photoexcitation of adsorbent with formation of active S^* centers and Equation (4) the adsorption of molecule M or "chemical decay" of the active states S^*, while Equation (3) depicts the "physical decay" of S^* state.

For a liquid–solid catalytic reaction the common technique for determining the adsorbed amount of a species dissolved in the solution is that of performing experiments in a batch not-reacting system and of measuring: (i) the volume of liquid solution; (ii) the concentration of the adsorbing species in the starting solution; and (iii), after that a known mass of catalyst is added to the liquid and steady-state conditions are reached, the concentration decrease determined in the starting solution due to the added catalyst. This procedure is based on the adequate assumptions that the catalyst's superficial features are not affected by the composition of the surrounding fluid phase and that the measured decrease of the species amount in the solution is equal to the amount of species adsorbed on the catalyst. The same procedure cannot be applied for the photoadsorption determination; in fact, photoadsorption occurs under the simultaneous presence of irradiation and of reducing and oxidizing species needed for charge carriers to be trapped on the semiconductor surface. Under these conditions the photoreaction also starts so that the measured decrease of species in the batch-irradiated slurry is determined both by photoadsorption and by reaction, these contributions being indistinguishable from the solution side.

This complexity determines that investigations on heterogeneous photocatalytic processes sometimes report information only on dark adsorption and use this information for discussing the results obtained under irradiation. This extrapolation is not adequate as the characteristics of photocatalyst surface change under irradiation and, moreover, active photoadsorption centers are generated. Nowadays very effective methods allow a sound characterization of bulk properties of catalysts, and powerful spectroscopies give valuable information on surface properties. Unfortunately information on the photoadsorption extent under real reaction conditions, that is, at the same operative conditions at which the photoreactivity tests are performed, are not available. For the cases in which photoreaction events only occur on the catalyst surface, a critical step to affect the effectiveness of the transformation of a given compound is to understand the adsorption process of that compound on the catalyst surface. The study of the adsorbability of the substrate allows one to predict the mechanism and kinetics that promote its photoreaction and also to correctly compare the performance of different photocatalytic systems.

This chapter presents a quantitative method to determine the photoadsorption capacity of a polycrystalline semiconductor oxide irradiated in liquid–solid system. The determination is performed under reaction conditions so that it is really indicative of the photoadsorption capacity. The method uses the experimental results obtained in typical batch photoreactivity runs; on this ground it has been applied to the following photocatalytic processes carried out in aqueous suspensions: (i) oxidation of phenol in the presence of a commercial TiO_2 catalyst (Degussa P25) and (ii) oxidation of benzyl alcohol in the presence of a home-prepared TiO_2

catalyst. The influence on photoactivity of substrate concentration, catalyst amount, and irradiation power is investigated. The kinetic modeling of the photooxidation processes is carried out by taking into account the photo-adsorption phenomenon by means of three types of isotherm equations, that is, Langmuir, Freundlich, and Redlich–Peterson. Nonlinear regression analysis applied to all the photoreactivity results allows establishing the most appropriate correlation for the photoadsorption isotherm and also to determine the values of the model parameters. The best fitting model is evaluated by choosing the Marquardt's percent standard deviation (MPSD) as error estimation tool.

2. EXPERIMENTAL

Photoreactivity runs of phenol degradation were carried out in aqueous suspensions of a commercial TiO_2 (Degussa P25) while for benzyl alcohol degradation a home-prepared nanostructured TiO_2 specimen was used. The preparation method of home-prepared catalyst is summarized here; the details are elsewhere reported (Addamo et al., 2004). The precursor solution was obtained by slowly adding 5 mL of $TiCl_4$ drop by drop into a 200-mL beaker containing 50 mL of water; during the addition, which lasted 5 min, the solution was magnetically stirred by a cylindrical bar (length, 3 cm; diameter, 0.5 cm) at 600 rpm. After that the beaker was closed and mixing was prolonged for 12 h at room temperature, eventually obtaining a clear solution. This solution was transferred to a round-bottom flask having on its top a Graham condenser (Palmisano et al., 2007a). The flask was put in boiling water, thus determining the boiling of the solution; the duration of the boiling was of 0.5 h, obtaining a white suspension at the end of the treatment. The suspension was then dried at 323 K by means of a rotovapor machine (model Buchi Rotovapor M) working at 150 rpm, in order to obtain the final powdered, poorly crystalline anatase-phase catalyst.

The flow-sheet of the experimental setup is shown in Figure 1. The details are reported elsewhere (Palmisano et al., 2007b). A cylindrical batch photo-reactor of Pyrex glass with immersed lamp was used for the photocatalytic runs of benzyl alcohol and phenol oxidation. On the top of the reactor three ports allowed the inflow and outflow of gases, the pH and temperature measurements, and the withdrawal of samples for analysis.

The catalyst was used in aqueous suspension well mixed by means of a magnetic stirrer. The reacting mixture was illuminated by a mercury medium-pressure lamp (type B, Helios Italquartz, Milan, Italy) coaxial with the photoreactor. A Pyrex thimble surrounding the lamp allowed the circulation of distilled water in order to cool the lamp and cutoff infrared radiation; in these conditions the reactor temperature was of 295 ± 2 K. Lamps of 125, 500, or 1,000 W electric power were used; average irradiances

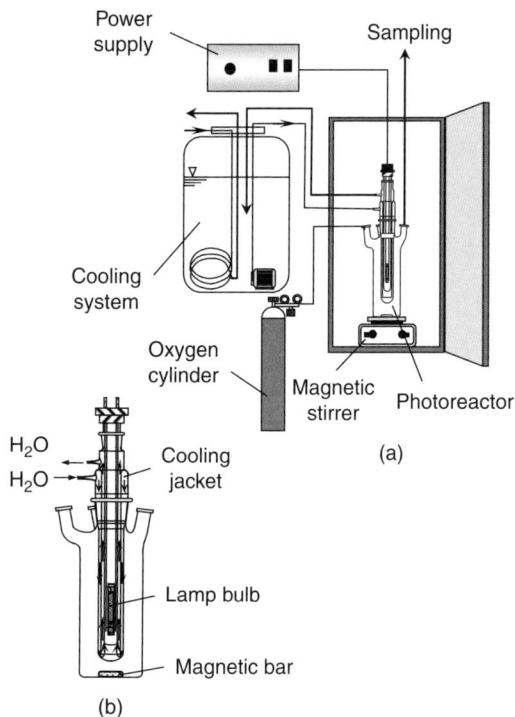

Figure 1   (a) Set up of experimental apparatus and (b) Photoreactor (Palmisano et al., 2007b).

impinging onto and leaving the suspension were measured by using a radiometer (UVX, Digital) at $\lambda = 360$ nm; these measurements were carried out for each lamp and each catalyst concentration. By considering that the lamp thimble does not transmit radiation with wavelength less than 300 nm and that the lamps mainly emit at 310 and 360 nm, the previous measurements were allowed to have an estimate of the photon flow absorbed by the suspension. These values, even if they are approximate, are however useful in quantitatively ordering the different suspensions with respect to their ability in absorbing photons. The benzyl alcohol and phenol initial concentrations varied in the 0.5–5.0 and 0.25–4.0 mM ranges, respectively, and the catalyst amounts were in the 0.05–0.8 g L^{-1} range.

The standardized procedure of the runs was the following one and it was strictly observed; the procedure stages are depicted in Figure 2. The photoreactor *without the immersed lamp* was filled with a fixed volume of substrate aqueous solution of known concentration and pure oxygen (or air for a few runs) was bubbled for 30 min under magnetic stirring (bubbling and stirring were never interrupted in the course of runs); then the desired amount of catalyst was added and samples (5 cm^3) of the suspension were taken at

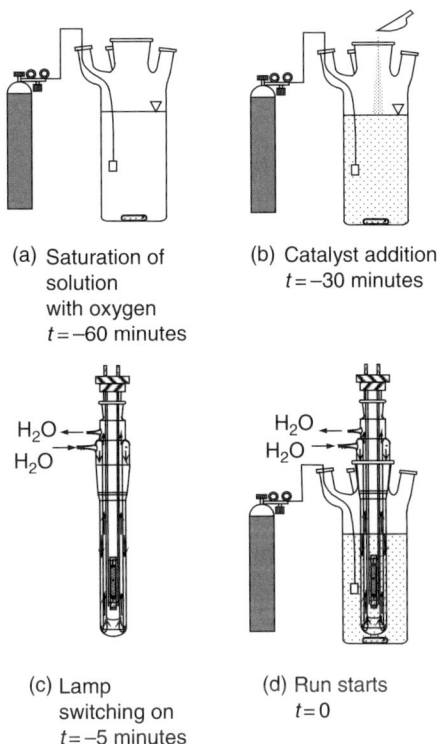

(a) Saturation of solution with oxygen
$t = -60$ minutes

(b) Catalyst addition
$t = -30$ minutes

(c) Lamp switching on
$t = -5$ minutes

(d) Run starts
$t = 0$

Figure 2 Experimental run procedure from (a) to (d).

fixed intervals of time until the dark equilibrium conditions were reached (about 30 min after catalyst addition). At that time the lamp *outside the photoreactor* was switched on and time was allowed for reaching steady-state conditions of irradiation (about 5 min). Once the irradiation did not change, as checked by the radiometer, the lamp was fast immersed into the suspension and that time was taken as the zero time of the photoreactivity run; samples (5 cm^3) for analysis were withdrawn by means of a syringe every 15 min in the first 2 h of the run and then every 0.5 or 1 h. The samples were immediately filtered by inserting a cellulose acetate filter (Millipore, Billerica, MA USA, pore diameter of 0.45 μm) on the exit hole of the sampling syringe.

For benzyl alcohol degradation runs the quantitative determination and identification of the species present in the reacting suspension was performed by means of a high-performance liquid chromatograph (HPLC) (Beckman Coulter, Fullerton, CA USA, System Gold 126 Solvent Module and 168 Diode Array Detector), equipped with a Luna 5 μ Phenyl–Hexyl column (250 mm long × 2 mm i.d.) using standards (Sigma-Aldrich, St. Louis, MO USA). The eluent consisted of 17.5% acetonitrile, 17.5% methanol, 65% 40 mM KH_2PO_4 aqueous solution. For phenol degradation runs the quantitative determination and identification was performed by using the same HPLC and the eluent consisted of 35% methanol and 65% 40 mM KH_2PO_4 aqueous solution.

For TOC determinations $2\,cm^3$ of the irradiated slurry was withdrawn from the photoreactor every 30 min and, after filtration over the Millipore filter, analyzed by a TOC Shimadzu analyzer.

All the used chemicals (Sigma-Aldrich) had a purity $> 99.0\%$.

3. TiO₂ SURFACE MODIFICATIONS UNDER IRRADIATION

For gas–solid and liquid–solid systems, the interaction of a species with the solid surface depends on the chemical nature of the species and on the chemical and physical nature of the solid. For nonilluminated surface of semiconductor oxides, a thermodynamic equilibrium between a species and the solid is established only when the electrochemical potential of the electrons in the entire system is uniform. When the adsorption–desorption equilibrium is established, an aliquot of the species is located in an adsorbed layer, held at the surface by either weak or strong bonding forces.

When a photocatalyst goes from dark conditions into irradiated ones, the radiation absorption determines that its surface undergoes a series of changes needed for eventually allowing the occurrence of photoprocesses (Bickley, 1985a, b). Essentially the interaction of a photon with a solid semiconductor gives rise to an increase of the vibrational state of the lattice or the number of quasi-free charge carriers, that is, it generates an excited state of the solid. The illumination with band gap (or greater) energy creates a perturbation to this adsorption–desorption equilibrium established in the dark; under irradiation the previous equilibrium is displaced determining a net photoadsorption or photodesorption of species. A new equilibrium is achieved when the species and the solid acquire the same electrochemical potential (Bickley, 1988a, b).

Direct observation of photosorption phenomena at the TiO₂ surface has been reported in several earlier studies, focussing on the role played by surface OH groups in the photoadsorption of oxygen and therefore in regulating the TiO₂ photoactivity. For example, UV illumination stimulates desorption of oxygen and water molecules adsorbed at the TiO₂ surface (Meriaudeau and Vedrine, 1976; Murphy et al., 1976; Sakai et al., 1998; Wang et al., 1999; Wu et al., 2000) while studies by IR and XPS reveal that the amount of OH groups on the TiO₂ surface increases by UV illumination (Asakuma et al., 2003; Hoffmann et al., 1995; Ignatchenko et al., 2006; Linsebigler et al., 1995; Mori et al., 2007). These changes are generally fast and reversible, that is, once irradiation is stopped the surface recovers its previous features, at equal initial conditions. A well-recognized indication of the fact that irradiation modifies the surface of wide band gap solids is the post-sorption or memory effect, that is, the adsorption in the dark caused by the pre-irradiation of solid surface (Solonitzyn and Terenin, 1959; Solonitzyn et al., 1982).

A more recent and clear evidence of this behavior is the phenomenon of induced superhydrophilicity, that is, the generation of a highly hydrophilic TiO_2 surface by UV illumination (Lee et al., 2003; Miyauchi et al., 2002; Nakajima et al., 2001; Sakai et al., 1998; Takeuchi et al., 2005; Wang et al., 1997; Watanabe et al., 1999). Superhydrophilic property of the TiO_2 surface allows water to spread completely across the surface rather than remaining as droplets. In the case of a film which consists of only TiO_2, the contact angle of water almost becomes zero during UV irradiation. However, it is found that the contact angle goes up and is restored comparatively quickly in a dark place. One of the most interesting aspects of TiO_2 is that photocatalysis and induced hydrophilicity can take place simultaneously on the same surface even though the mechanisms are different. In the case of photocatalysis, UV light excites the catalyst and pairs of electrons and holes are generated; the photogenerated electrons then react with molecular oxygen to produce superoxide radical anions and the photogenerated holes react with water to produce hydroxyl radicals. These two types of rather reactive radicals then work together to carry out redox reactions with species adsorbed on the TiO_2 surface. On the other hand, surface hydroxyl groups can trap more photogenerated holes and improve the separation of electrons and holes which results in the enhancement of photocatalysis.

In the case of superhydrophilicity electrons and holes are still produced, but they react in a different way. The electrons tend to reduce the Ti(IV) cations to the Ti(III) state, and the holes oxidize the oxygen anions. In the process, oxygen atoms are ejected, creating oxygen vacancies. Water molecules can then occupy these oxygen vacancies, producing adsorbed OH groups, which tend to make the surface hydrophilic. Depending upon the composition and the processing, the surface can have more photocatalytic character and less superhydrophilic character, or vice versa. In spite of the different mechanism of photocatalytic and hydrophilic effect, they behave synergetically. Because more OH groups can be adsorbed on the surface due to hydrophilicity, the photocatalytic activity is enhanced; so hydrophilicity can improve photocatalysis. On the other hand, the surface can adsorb contaminated compounds which tend to turn the hydrophilic surface to hydrophobic surface. Photocatalysis can decompose the organic compounds on the surface resulting in the restoration of hydrophilicity. From this point, photocatalysis can improve hydrophilicity and maintain this characteristic for a long time. The proposed mechanism means that the surface structure changes during the reaction. In other words, the rate of photo-induced hydrophilicity may depend on the history of the sample.

Specifically for liquid–solid systems, adsorption phenomena involving ions occur with a transfer of electric charge, causing a significant variation of the electronic band structure of the surface. The charge transfer is responsible for the formation of the so-called space-charge layer where the potential difference between surface and bulk creates an electric field that can play a beneficial role in the further adsorption steps. Most photons are absorbed in the superficial layer of the solid where the space-charge exists and the photoelectrons straightforwardly interact with the electric field present here. The consequence is a more efficient separation of the electrons and holes within the space-charge layer, that is, the increase of their mean lifetime.

Photoadsorbed species can act (i) as surface-hole trapping and photoelectrons can be trapped in the bulk of the solid or (ii) as surface-electron trapping and holes can react with OH surface groups and/or H_2O. Both the alternatives depend on the chemical nature of the molecule to be adsorbed and on the type of the solid adsorbent. It is worth noting that in the gas–solid regime only gaseous species or lattice ions can be involved, whereas in the liquid phase also the interaction with the solvent (often H_2O) should also be considered.

All the above considerations allow to conclude that the electronic surface modifications created under irradiation eventually induce photoadsorption both in gas–solid and in liquid–solid regimes, being the photoadsorption phenomenon strictly related with the photoactivity of the solid photocatalysts. When light is switched off, a reversal of the process could be observed, although rarely the reversibility is complete and the achievement of a new equilibrium depends mainly on kinetics rather than thermodynamic factors.

4. PHOTOADSORPTION DETERMINATION

For heterogeneous photocatalytic processes, it is generally agreed that the expression for the degradation rate of organic substrates on TiO_2 surfaces in the presence of oxygen follows (with minor variations) the Langmuir–Hinshelwood (LH) model, which is widely used in liquid- and gas-phase systems (de Lasa et al., 2005; Gora et al., 2006; Ibrahim and de Lasa, 2004; Murzin and Salmi, 2005; Palmisano et al., 2007c; Turchi and Ollis, 1989; Vorontsov et al., 1999). This model successfully explains the kinetics of reactions that occur between two adsorbed species, a free radical and an adsorbed substrate, or a surface-bound radical and a free substrate. In the LH model, adsorption equilibrium is assumed to be established at all times so that the rate of reaction is taken to be much less than the rate of adsorption or desorption. Reaction is assumed to occur between adsorbed species whose coverage on the catalyst surface is always in equilibrium with the species concentration in the fluid phase so that the rate-determining step of the photocatalytic process is the surface reaction. The concentrations of adsorbed species are therefore determined by adsorption equilibrium as given by the suitable isotherm. It is useful to report that the simple rate form of the LH approach may have origins which take into account different photoreaction mechanisms (Demeestere et al., 2004; Krýsa et al., 2006; Minero and Vione, 2006).

In batch photocatalytic reactors working in liquid–solid regime, the depletion of a species is the combined result of photoadsorption and photo-conversion processes. To describe this depletion, a mole balance applied to the species at whatever time (de Lasa et al., 2005) can be represented as

$$n_T = n_L + n_S \tag{5}$$

where n_T is the total number of moles present in the photoreactor, n_L the number of moles in the fluid phase, and n_S the number of moles photo-adsorbed on the solid. When Equation (5) is divided by the volume of the liquid phase, V, one obtains:

$$C_T = C_L + \frac{n_S}{V} \tag{6}$$

where C_T is the total concentration of the species and C_L the concentration in the liquid phase.

By considering that both substrate and oxygen must be present in the system for the occurrence of photoreaction, it is assumed that the total disappearance rate of substrate per unit surface area, r_T, follows a second-order kinetics of first order with respect to the substrate coverage and of first order with respect to the oxygen coverage:

$$r_T \equiv -\frac{1}{S}\frac{dn_T}{dt} = k''\theta_{Sub}\theta_{Ox} \tag{7}$$

in which S is the catalyst surface area, t the time, k'' the second-order rate constant, and θ_{Sub} and θ_{Ox} are the substrate and oxygen fractional coverages of the surface, respectively. θ_{Sub} and θ_{Ox} are defined in the following way:

$$\theta_{Sub} \equiv \frac{n_S}{WN_S^*} \tag{8}$$

$$\theta_{Ox} \equiv \frac{n_{S,Ox}}{WN_{S,Ox}^*} \tag{9}$$

where $n_{S,Ox}$ is the number of oxygen moles photoadsorbed on the solid, N_S^* and $N_{S,Ox}^*$ are the maximum capacity of photoadsorbed moles of substrate and oxygen, respectively, on the unit mass of irradiated solid, and W the mass of catalyst.

By considering that in the course of the runs the oxygen bubbling in the dispersion is not stopped, its concentration in the liquid phase does not change during the occurrence of substrate degradation and, moreover, it is always in excess. Then the θ_{Ox} term of Equation (7) does not depend on time, that is, it is a constant. By putting $k = k'' \cdot \theta_{Ox}$, Equation (7) turns a pseudo-first-order rate equation:

$$r_T \equiv -\frac{1}{S}\frac{dn_T}{dt} = k\theta_{Sub} \tag{10}$$

in which k is the pseudo-first-order rate constant. Introducing in Equation (10) the liquid volume, V, and the surface area per unit mass of catalyst, S_S, one obtains

$$r_T \equiv -\frac{1}{S}\frac{dn_T}{dt} = -\frac{1}{S_S W}V\frac{dC_T}{dt} = k\theta_{Sub} \tag{11}$$

By considering that the kinetic information on the photoprocess consists on the knowledge of substrate concentration values in the liquid phase, C_L, as a function of irradiation time, the C_T and θ_{Sub} variables of Equation (11) must to be transformed as a function of C_L. With concern to θ_{Sub}, it is necessary to choose an adsorption isotherm, that is, a relationship between θ_{Sub} [and therefore n_S, see Equation (8)] and C_L, while for C_T Equation (6) coupled with the chosen isotherm may be used. Therefore, Equation (11) may be formally written in the following way:

$$-\frac{1}{S_S W}V\frac{d}{dt}C_T(C_L) = k\theta_{Sub}(C_L) \tag{12}$$

It is important to remind that the kinetic modeling is carried out by assuming that the main assumptions of the LH model hold; it means that Equation (12) gives the evolution with irradiation time of the concentration in the liquid phase, C_L, of a species which is in photoadsorption equilibrium on the catalyst surface over which the species undergoes a slow transformation process. The main implication of the previous statement is that for a batch photocatalytic run the substrate concentration values measured in the liquid phase at a certain time represent the substrate concentration in equilibrium with an (unknown) substrate amount photoadsorbed on the catalyst surface. This feature belongs to all the measured values of substrate concentration except to the initial one. The substrate concentration measured at the start of a photoreactivity run is characteristic of a system without irradiation. As a consequence, when a kinetic model is fitted to the experimental data, the regression analysis must be done with the concentration values measured in the course of the run excluding the value of initial substrate concentration because the condition of the initial concentration value (dark) is not that of the other ones (irradiated) (Yurdakal et al., 2008b).

4.1. Langmuir isotherm

The Langmuir isotherm theory (Satterfield, 1980) assumes a monolayer coverage of adsorbate over a homogenous adsorbent surface. A plateau characterizes the Langmuir isotherm; therefore, a saturation point is reached where no further adsorption can occur. Adsorption is assumed to take place at specific homogeneous sites within the adsorbent. Once a molecule occupies a site, no further adsorption can take place at that site. The Langmuir model assumes that the catalyst surface is completely uniform, that is, (i) the adsorption onto the surface of each molecule has equal activation energy and (ii) no transmigration of adsorbate in the plane of the surface occurs. Those assumptions are hypothesized here to be suitable to describe also photoadsorption. The Langmuir adsorption isotherm is described by the following relationship:

$$\theta_{Sub} \equiv \frac{n_S}{WN_S^*} = \frac{K_L^* C_L}{1 + K_L^* C_L} \tag{13}$$

where K_L^* is the photoadsorption equilibrium constant, which is related to the free energy of photoadsorption, and N_S^* the monolayer adsorption capacity. K_L^* may be considered a measure of the intrinsic photoreactivity of the catalyst surface. By solving Equation (13) with respect to n_S and substituting in Equation (6), the result is

$$C_T = C_L + \frac{WN_S^*}{V} \frac{K_L^* C_L}{1 + K_L^* C_L} \tag{14}$$

Substituting in Equation (11), the Langmuir relationship [Equation (13)] produces

$$-\frac{V}{WS_S} \frac{dC_T}{dt} = k \frac{K_L^* C_L}{1 + K_L^* C_L} \tag{15}$$

Taking the derivative of Equation (14) with respect to time, it yields

$$\frac{dC_T}{dt} = \left[1 + \frac{W N_S^* K_L^*}{V} \frac{1}{(1 + K_L^* C_L)^2} \right] \frac{dC_L}{dt} \tag{16}$$

Substituting Equation (16) in the left-hand side term of Equation (15), one obtains

$$-\frac{V}{WS_S}\left[1+\frac{WN_S^*K_L^*}{V}\frac{1}{(1+K_L^*C_L)^2}\right]\frac{dC_L}{dt} = k\frac{K_L^*C_L}{1+K_L^*C_L} \quad (17)$$

and, rearranging and separating the variables, the following differential equation is obtained:

$$-\frac{V}{WS_S}\frac{1}{kK_L^*}\frac{dC_L}{C_L} - \frac{V}{WS_S}\frac{1}{k}dC_L - \frac{1}{S_S}\frac{N_S^*}{k}\frac{dC_L}{C_L(1+K_L^*C_L)} = dt \quad (18)$$

As outlined before, Equation (18) gives the evolution with irradiation time of the concentration in the liquid phase of a species which is in photoadsorption equilibrium on the catalyst surface over which the species undergoes a slow transformation process. On this ground, the integration of Equation (18) must be performed with the condition that at $t=0$ the substrate concentration in the liquid phase is that in equilibrium with the initial photoadsorbed amount, $C_{L,0}$; this initial concentration is unknown but it may be determined by the regression analysis carried out with the experimental data obtained after the start of irradiation. The integration yields

$$\frac{V}{WS_S}\frac{1}{kK_L^*}\ln\frac{C_{L,0}}{C_L} + \frac{V}{WS_S}\frac{1}{k}(C_{L,0}-C_L) + \frac{1}{S_S}\frac{N_S^*}{k}\ln\left(\frac{1+K_L^*C_L}{1+K_L^*C_{L,0}}\frac{C_{L,0}}{C_L}\right) = t \quad (19)$$

Equation (19) contains four unknown parameters, K_L^*, N_S^*, k, and $C_{L,0}$, whose determination may be carried out by a best fitting procedure.

4.2. Freundlich isotherm

The Langmuir theory of adsorption is an approximation of the real surfaces as surface reconstruction frequently occurs. An assumption generally used for the description of the physical chemistry of real adsorbed layers is that surface sites are different, the so-called biographical non-uniformity (Murzin and Salmi, 2005). By applying the Langmuir adsorption isotherm to a distribution of energies among the sites such that the heat of adsorption decreases logarithmically with coverage, the Freundlich isotherm is derived (Satterfield, 1980); it is described by the following relationship:

$$\theta_{Sub} \equiv \frac{n_S}{WN_S^*} = K_F^*C_L^{1/n} \quad (20)$$

where K_F^* is the Freundlich isotherm constant indicative of the relative adsorption capacity of the adsorbent and n a dimensionless parameter indicative of the intensity of the adsorption. The n parameter represents the mutual interaction of adsorbed species; a value of n greater than unity is interpreted to mean that adsorbed molecules repulse one another. The Freundlich expression is an exponential equation and therefore it assumes that as the adsorbate concentration in the liquid phase increases, the concentration of adsorbate on the adsorbent surface also increases. Using the empirical form of Freundlich isotherm, an infinite amount of adsorption can occur, what is contrary to chemisorption. The statistical derivation of this isotherm, however, sets a value of the Freundlich maximum adsorption capacity, N_S^*. By solving Equation (20) with respect to n_S and substituting in Equation (6), the result is

$$C_T = C_L + \frac{WN_S^*}{V} K_F^* C_L^{1/n} \tag{21}$$

Substituting in Equation (11) the Freundlich relationship [Equation (20)] produces

$$-\frac{V}{WS_S} \frac{dC_T}{dt} = kK_F^* C_L^{1/n} \tag{22}$$

Taking the derivative of Equation (21) with respect to time, it yields

$$\frac{dC_T}{dt} = \left[1 + \frac{WN_S^* K_F^*}{V} \frac{1}{n} C_L^{(1-n)/n}\right] \frac{dC_L}{dt} \tag{23}$$

Substituting Equation (23) into the left-hand side term of Equation (22), one obtains

$$-\frac{V}{WS_S} \left[1 + \frac{WN_S^* K_F^*}{V} \frac{1}{n} C_L^{\,1-n/n}\right] \frac{dC_L}{dt} = kK_F^* C_L^{1/n} \tag{24}$$

and, rearranging and separating the variables, the following differential equation is obtained:

$$-\frac{V}{WS_S} \frac{1}{kK_F^*} \frac{dC_L}{C_L^{1/n}} - \frac{1}{S_S} \frac{N_S^*}{kn} \frac{dC_L}{C_L} = dt \tag{25}$$

As done in the case of Langmuir adsorption isotherm, integration of Equation (25) is performed with the condition that at $t = 0$ the substrate concentration in the liquid phase is that in equilibrium with the initial photoadsorbed amount, $C_{L,0}$; this initial concentration is unknown, but it may be determined by the regression analysis carried out with the experimental data obtained after the start of irradiation. The integration yields

$$\frac{V}{WS_S}\frac{1}{kK_F^*}\frac{n}{n-1}\left(C_{L,0}^{(n-1)/n}-C_L^{(n-1)/n}\right)+\frac{1}{S_S}\frac{N_S^*}{kn}\ln\frac{C_{L,0}}{C_L}=t \qquad (26)$$

Equation (26) contains five unknown parameters, K_F^*, N_S^*, n, k, and $C_{L,0}$, whose determination may be carried out by a best fitting procedure.

4.3. Redlich–Peterson isotherm

Since in heterogenous catalysis most of the adsorbents follow Langmuir or Freundlich models, much better than Temkin model, an attempt is made here to test the photoadsorption validity with respect to Redlich–Peterson (R–P) model (Redlich and Peterson, 1959). The R–P isotherm model combines elements from both the Langmuir and the Freundlich equations and the mechanism of adsorption is a hybrid one and it does not follow ideal monolayer adsorption. The R–P isotherm is an empirical equation, designated the "three parameter equation," which may be used to represent adsorption equilibria over a wide concentration range (Ahmaruzzaman and Sharma, 2005; Allen et al., 2004; Kumar and Sivanesan, 2005; Matthews and Weber, 1976). The R–P adsorption isotherm is described by the following relationship:

$$\frac{n_S}{W}=\frac{K_{R-P}^*C_L}{1+\alpha_{R-P}^*C_L^{\beta}} \qquad (27)$$

where K_{R-P}^* and α_{R-P}^* are R–P isotherm constants which refer to the adsorption capacity and the surface energy, respectively. The exponent β is the heterogeneity factor which lies between 1 and 0. The equation can be reduced to the Langmuir equation as β approaches 1 while it becomes the Henry's law equation when $\beta=0$. By solving Equation (27) with respect to n_S and substituting in Equation (6), the result is

$$C_T=C_L+\frac{W}{V}\frac{K_{R-P}^*C_L}{1+\alpha_{R-P}^*C_L^{\beta}} \qquad (28)$$

In the cases of Langmuir and Freundlich isotherms, it has been assumed that the total disappearance rate of substrate per unit surface area, r_T, follows a pseudo-first-order kinetics with respect to the substrate concentration which is expressed by its fractional coverage. The same assumption is made for R–P isotherm; however, as the R–P isotherm relates an adsorbed amount (and not a fractional coverage) with the equilibrium concentration in the liquid phase, the r_T term is written as

$$r_T \equiv -\frac{V}{WS_S}\frac{dC_T}{dt} = kC_{Surface} = k\frac{n_S}{WS_S} \qquad (29)$$

in which $C_{Surface}$ is the surface concentration of adsorbed species (adsorbed moles/catalyst surface area) (Minero and Vione, 2006). Substituting in Equation (29) the R–P relationship [Equation (27)] produces

$$-\frac{V}{W}\frac{dC_T}{dt} = k\frac{K_{R-P}^* C_L}{1 + \alpha_{R-P}^* C_L^\beta} \qquad (30)$$

Taking the derivative of Equation (28) with respect to time, it yields

$$\frac{dC_T}{dt} = \left[1 + \frac{WK_{R-P}^*}{V}\frac{1 + \alpha_{R-P}^* C_L^\beta(1 - \beta K_{R-P}^*)}{(1 + \alpha_{R-P}^* C_L^\beta)^2}\right]\frac{dC_L}{dt} \qquad (31)$$

Substituting Equation (31) in the left-hand side term of Equation (30), one obtains

$$-\frac{V}{W}\left[1 + \frac{WK_{R-P}^*}{V}\frac{1 + \alpha_{R-P}^* C_L^\beta(1 - \beta K_{R-P}^*)}{(1 + \alpha_{R-P}^* C_L^\beta)^2}\right]\frac{dC_L}{dt} = k\frac{K_{R-P}^* C_L}{1 + \alpha_{R-P}^* C_L^\beta} \qquad (32)$$

and, rearranging and separating the variables, the following differential equation is obtained:

$$-\frac{V}{WkK_{R-P}^*}\frac{dC_L}{C_L} - \frac{\alpha_{R-P}^* V}{WkK_{R-P}^*}C_L^{\beta-1}dC_L - \frac{1}{k}\frac{dC_L}{C_L(1 + \alpha_{R-P}^* C_L^\beta)}$$
$$-\frac{V\alpha_{R-P}^*(1 - \beta K_{R-P}^*)}{WkK_{R-P}^*}\frac{C_L^{\beta-1}dC_L}{1 + \alpha_{R-P}^* C_L^\beta} = dt \qquad (33)$$

As done in the case of previous adsorption isotherms, integration of Equation (33) is performed with the condition that at $t=0$ the substrate concentration in the liquid phase is that in equilibrium with the initial photoadsorbed amount, $C_{L,0}$; this initial concentration is unknown, but it may be determined by the regression analysis carried out with the experimental data obtained after the start of irradiation. The integration yields

$$\frac{V}{WkK_{R-P}^*}\ln\left(\frac{C_{L,0}}{C_L}\right) + \frac{\alpha_{R-P}^* V}{Wk\beta K_{R-P}^*}(C_{L,0}^\beta - C_L^\beta) + \frac{1}{k\beta}\ln\left(\frac{1 + \alpha_{R-P}^* C_L^\beta}{1 + \alpha_{R-P}^* C_{L,0}^\beta}\frac{C_{L,0}^\beta}{C_L^\beta}\right)$$
$$+ \frac{V(1 - \beta K_{R-P}^*)}{Wk\beta K_{R-P}^*}\ln\left(\frac{1 + \alpha_{R-P}^* C_{L,0}^\beta}{1 + \alpha_{R-P}^* C_L^\beta}\right) = t \qquad (34)$$

Equation (34) contains five unknown parameters, K^*_{R-P}, α^*_{R-P}, β, k, and $C_{L,0}$, whose determination may be carried out by a best fitting procedure.

5. RESULTS

As expected, no photodegradation of benzyl alcohol or phenol was observed in oxygenated solution under irradiation but without catalyst. In the presence of catalyst but without irradiation no oxidation of benzyl alcohol or phenol was detected but a small adsorption of benzyl alcohol was measured while for phenol it was negligible.

The photoreactivity runs carried out by bubbling pure oxygen or air gave the same results so that it may be concluded that the oxygen concentration is not limiting the substrate degradation rate at the used experimental conditions.

Figures 3 and 4 report the photoreactivity results obtained from representative runs carried out at different initial benzyl alcohol concentrations and two lamp powers; for all of these runs, the same amount of catalyst was used.

In Figures 3–5, the concentration values reported for the zero time correspond to those of the starting solution, that is, without catalyst and irradiation. It must be reported that at the end of dark period (lasting 30 min) the benzyl alcohol concentration in the suspension is a little lesser than that of the starting solution. The calculations of photoadsorbed amount, however, have been performed by hypothesizing that all the molecules present on the catalyst surface participate in the

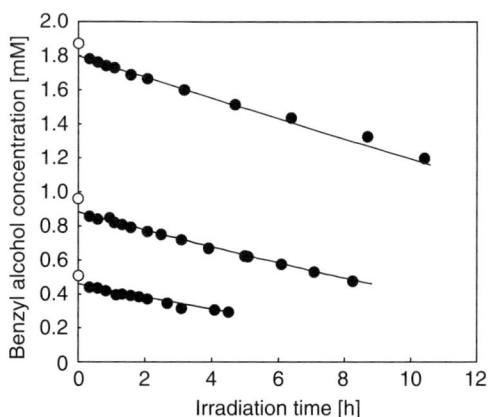

Figure 3 Experimental values of benzyl alcohol concentration vs. irradiation time. Home-prepared catalyst amount: 0.4 g L^{-1}, lamp power, 125 W. The empty symbols indicate the concentration of starting solution. The solid lines represent the Langmuir photoadsorption model [Equation (19)] (Yurdakal et al., 2008b).

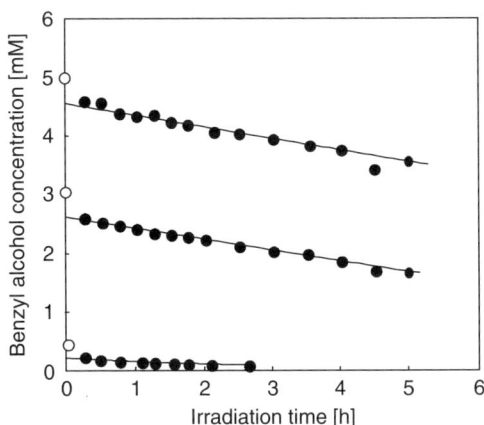

Figure 4 Experimental values of benzyl alcohol concentration vs. irradiation time. Home-prepared catalyst amount: 0.4 g L^{-1}, lamp power, 500 W. The empty symbols indicate the concentration of starting solution. The solid lines represent the Langmuir photoadsorption model [Equation (19)] (Yurdakal et al., 2008b).

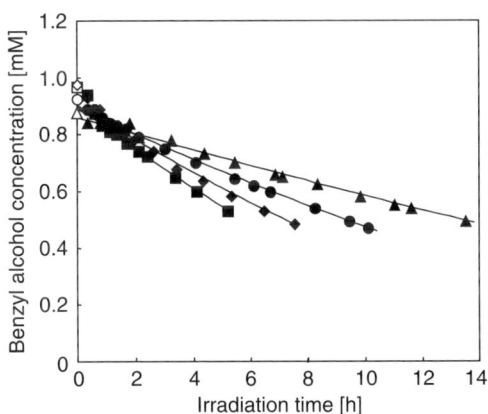

Figure 5 Experimental values of benzyl alcohol concentration vs. irradiation time. Home prepared catalyst amount: (▲) 0.1 g L^{-1}; (•) 0.2 g L^{-1}; (□) 0.6 g L^{-1}; (■) 0.8 g L^{-1}. Lamp power, 125 W. The empty symbols indicate the concentration of starting solution. The solid lines represent the photoadsorption model [Equation (19)] (Yurdakal et al., 2008b).

photoprocess and therefore the concentration of the starting solution has been taken into account.

Benzaldehyde was the intermediate product detected in the course of benzyl alcohol photocatalytic oxidation, CO_2 being the other oxidation product. No other intermediates were detected in all the course of the runs indicating that at the used experimental conditions the produced aldehyde does not compete with alcohol for photoadsorption and oxidation. This feature implies that the benzyl alcohol degradation kinetics did not change in the course of the runs due to the accumulation of intermediate products in the reaction ambient.

Figure 5 reports the results obtained from runs carried out with different amounts of catalyst; for all these runs the same initial benzyl alcohol concentration and lamp power were used.

Figures 6 and 7 report the photoreactivity results obtained from representative runs carried out at different initial phenol concentrations (with the same amount of catalyst) and at different amounts of catalyst (with the same initial phenol concentration), respectively.

In Figures 6 and 7, the phenol concentration values reported for the zero time correspond to those of the starting solution, that is, without catalyst and irradiation. It is useful to report that the adsorption of phenol in the dark is negligible so that the phenol concentration in the solution did not change after the addition of the catalyst.

In the course of phenol photocatalytic oxidation, the main products determined through TOC and HPLC analyses were CO_2 and hydroquinone and catechol, as expected having in mind that the photocatalytic oxidation of organic compounds on illuminated TiO_2 proceeds via $^{\bullet}OH$ attack on the substrate. The experimental results indicate that since the start of irradiation two parallel reaction pathways take place: direct mineralization of phenol to CO_2 (occurring through a series of reactions taking place over the catalyst surface and producing intermediates not desorbing to the bulk of solution) and partial oxidation to hydroxylated compounds (Salaices et al., 2004).

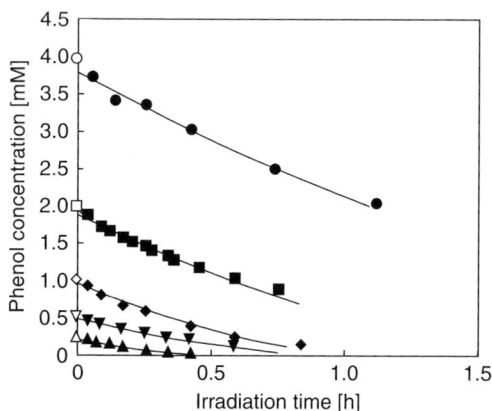

Figure 6 Experimental values of phenol concentration vs. irradiation time. TiO_2 P25 catalyst amount: 0.32 g L^{-1}, lamp power, 500 W. The empty symbols indicate the concentration of starting solution. The solid lines represent the Freundlich photoadsorption model [Equation (26)].

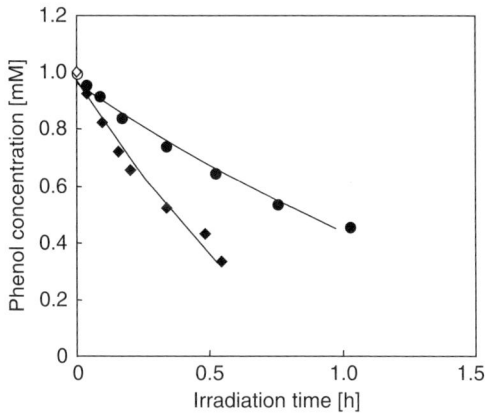

Figure 7 Experimental values of phenol concentration vs. irradiation time. TiO_2 P25 catalyst amount: (•) 0.05 g L^{-1} and (♦) 0.64 g L^{-1}. Lamp power, 500 W. The empty symbols indicate the concentration of starting solution. The solid lines represent the Freundlich photoadsorption model [Equation (26)].

The carbon balance between reacted phenol and produced CO_2 and dihydroxylated compounds was satisfactory ($> 98\%$) for phenol conversion less than 60%. On this basis it may be assumed that at the used experimental conditions the stable intermediate products of phenol photodegradation do not compete with phenol for photoadsorption and oxidation, at least in the time needed for 60% conversion.

This feature implies that the phenol degradation kinetics did not change in the course of the runs due to the accumulation of intermediate products in the reaction ambient. As a consequence, only the phenol concentration values relative to conversion below 60% have been taken into account in the course of photoadsorption modeling.

6. DISCUSSION

Nonlinear optimization techniques have been applied to determine isotherm parameters. It is well known (Ncibi, 2008) that the use of linear expressions, obtained by transformation of nonlinear one, distorts the experimental error by creating an inherent error estimation problem. In fact, the linear analysis method assumes that (i) the scatter of points follows a Gaussian distribution and (ii) the error distribution is the same at every value of the equilibrium liquid-phase concentration. Such behavior is not exhibited by equilibrium isotherm models since they have nonlinear shape; for this reason the error distribution gets altered after transforming the data to linear. Nonlinear regression method avoids such errors, making this analyzing technique the most appropriate to obtain more realistic isotherm parameters.

In this work, among different error functions the MPSD was used to enable the optimization process to determine and evaluate the fit of the isotherm equation to the experimental data. This error function is similar to a geometric mean error distribution which is modified to allow for the number of degrees of freedom of the system. For each case the isotherm parameters were determined by minimizing the MPSD function across the liquid-phase concentration range using the Data-Plot solver (Heckert and Filliben, 2003a, b). It was assumed that both the liquid-phase and the solid-phase concentrations contribute equally to weighting the error criterion chosen for the model solution procedure.

6.1. Benzyl alcohol photoadsorption

The fitting of the Langmuir, Freundlich, and R–P models to the data has been firstly applied to the photoreactivity results obtained from runs carried out at equal mass of catalyst and lamp power. For the Langmuir model, the following procedure has been followed. In order to have an estimate of parameters values, the data at high initial concentration of benzyl alcohol have been fitted to Equation (A6) (see Appendix A1) and those at low initial concentration to Equation (A13). The parameters obtained by these fitting procedures have been used to determine K_L^* and N_S^* by means of Equation (A18). As inequalities A1 and A7 were not strongly satisfied, the fitting procedure has been repeated by using the general equation of the proposed model [Equation (19)].

For the runs carried out at different catalyst amounts, the general equations for the liquid concentration profile [Equations (19), (26), and (34)] have been used. For Freundlich and Redlich–Peterson isotherms, the fitting obtained for each run was quite satisfactory ($R^2 > 0.98$); however, for the runs carried out at equal amount of catalyst, the Freundlich and Redlich–Peterson parameter values showed a dependence on initial benzyl alcohol concentration; on the contrary, same values were obtained for Langmuir parameters. On this basis the Langmuir isotherm seems to best describe the photoadsorption phenomenon of benzyl alcohol on home-prepared TiO_2 catalyst. The continuous lines drawn through the data of Figures 3–5 represent Equation (19), and a very satisfactory fitting ($R^2 > 0.99$) may be noted. The values of kinetic and Langmuir parameters (k, K_L^*, and N_S^*), obtained by the best fitting procedure, are reported later (see Figure 10).

The suitability of Langmuir model to describe the photoadsorption phenomenon has been further checked by using the linear form of Langmuir equation [Equation (A18)]. Figure 8 reports the values of the $C_{L,0}/(C_{T,0} - C_{L,0})$

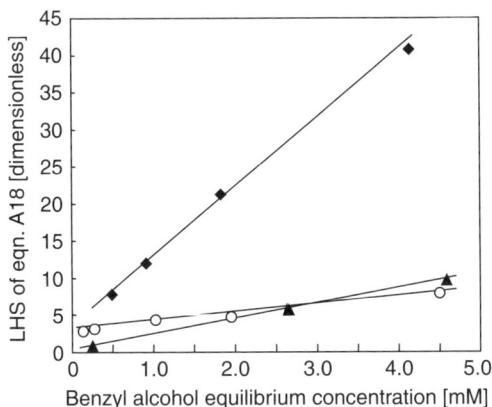

Figure 8 Linear form of Langmuir model. The left-hand side of Equation (A18), $C_{L,0}/(C_{T,0} - C_{L,0})$, vs. $C_{L,0}$ for runs carried out under irradiation with lamp power of 125 W (♦), 500 W (▲), and in the dark (○). The dark data are multiplied for 10^{-3}. The solid lines represent the Langmuir photoadsorption model [Equation (A18)] (Yurdakal et al., 2008b).

group obtained from runs carried out at equal mass of catalyst and lamp power vs. the benzyl alcohol equilibrium concentration. The straight lines drawn through the data represent Equation (A18), and a very good fitting ($R^2 > 0.99$) may be noted.

The slopes and linear coefficients of lines allow to determine the N_S^* and K_L^* values corresponding to these runs; these values are almost the same as those determined by means of Equation (19) (error percentage, $\pm 4\%$). Figure 8 also reports the adsorption results obtained in the absence of irradiation, that is, in the dark. A least-squares best fitting procedure allows to determine the values ($R^2 > 0.99$) of the Langmuir equilibrium constant and the maximum adsorption capacity in the absence of irradiation, that is, $K_L = 350$ M^{-1} and $N_S = 4.57 \times 10^{-6}$ mol/g of catalyst.

The effect of catalyst amount on photoadsorption capacity is shown in Figure 9. This figure reports the benzyl alcohol moles photoadsorbed per unit mass of catalyst vs. the catalyst amount; the reported data refer to runs carried out at equal initial benzyl alcohol concentration and lamp power. From the observation of data of Figure 9, a decrease of specific photoadsorption capacity by increasing the catalyst amount, differently from that expected on thermodynamic basis for which an increase of catalyst amount determines a corresponding increase of adsorbed substrate, may be noted.

This finding may be explained by considering that an increase of catalyst amount determines an increase of photons absorbed by the suspension but the photon flow absorbed by the unit mass of catalyst decreases. As photoadsorption is likely to be strongly dependent on absorbed photon flow, the increase of catalyst amount is eventually detrimental for specific photoadsorption.

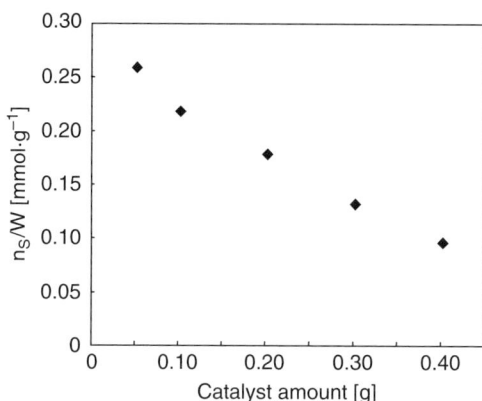

Figure 9 Photoadsorbed amount of benzyl alcohol per unit mass of catalyst, n_S/W, vs. the mass of catalyst. Initial benzyl alcohol concentration, 1 mM, lamp power, 125 W (Yurdakal et al., 2008b).

The consideration that the photon flow absorbed by the unit mass of catalyst is the parameter mainly affecting the photoadsorption phenomenon is strengthened by the results reported in Figure 10. This figure reports the values of the model parameters, K_L^*, N_S^*, and k, vs. the absorbed photon flow per unit mass of catalyst. The K_L and N_S values obtained from dark experiments are also reported. It may be noted that all parameters increase by increasing the specific photon absorption. While it is known that in thermal catalysis a temperature

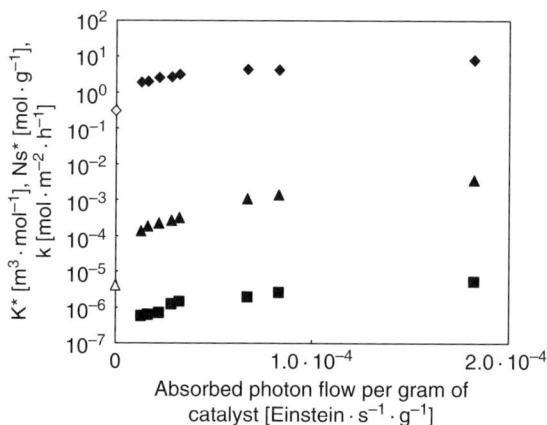

Figure 10 Values of kinetic and Langmuir model parameters [see Equation (19)] vs. the absorbed photon flow per unit mass of catalyst. ♦, K_L^*; ▲, N_S^*; ■, k. The empty symbols refer to dark conditions (Yurdakal et al., 2008b).

increase determines an increase of kinetic and equilibrium adsorption constants, the results of Figure 10 also show a beneficial effect of absorbed photons on the photoadsorbed amount of solute. This effect is very noticeable if a comparison is done between irradiated and dark conditions. The values of K_L^* and N_S^* are one and two orders of magnitude, respectively, higher than those obtained in the absence of irradiation.

6.2. Phenol photoadsorption

The fitting of photoreactivity data obtained from runs carried out at equal mass of catalyst and at different catalyst amounts has been carried out by applying the general equations of Langmuir, Freundlich, and Redlich–Peterson isotherms [Equations (19), (26), and (34)]. For Langmuir and Redlich–Peterson isotherms, the fitting obtained for each run was quite satisfactory ($R^2 > 0.98$); however, for the runs carried out at equal amount of catalyst, the Langmuir and Redlich–Peterson parameters values showed a dependence on initial phenol concentration; on the contrary, same values were obtained for Freundlich parameters. On this basis the Freundlich isotherm seems to best describe the photoadsorption phenomenon of phenol on commercial TiO_2 catalyst (Degussa P25). The continuous lines drawn through the data of Figures 6 and 7 represent Equation (26) with $n = 2$, and a satisfactory fitting ($R^2 > 0.98$) may be noted.

The suitability of Freundlich model to describe the photoadsorption phenomenon has been further checked by using the Freundlich equation [Equation (20)] in its linear form. Figure 11 reports the values of the n_S/W group obtained from runs carried out at equal mass of catalyst vs. the square

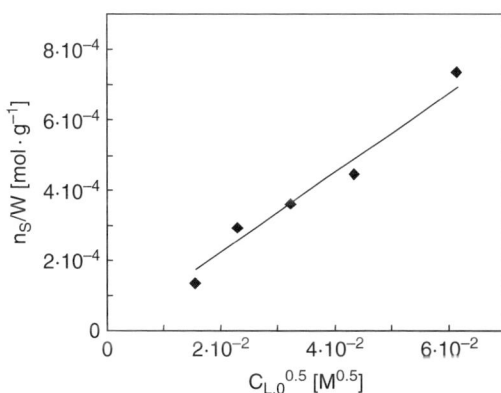

Figure 11   Linear form of Freundlich model. The left-hand side of Equation (19), n_S/W, vs. $C_{L,0}^{0.5}$ for runs carried out under irradiation with lamp power of 500 W. The solid line represents the Freundlich photoadsorption model [Equation (20)].

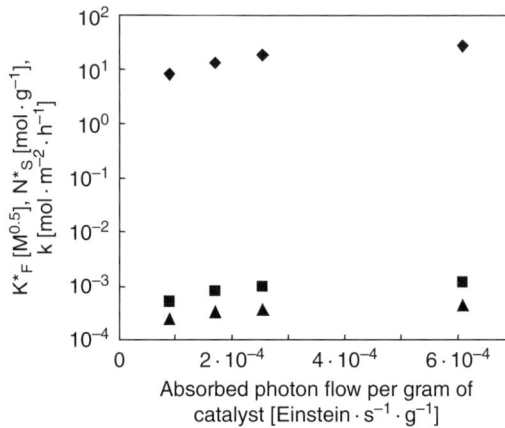

Figure 12 Values of kinetic and Freundlich model parameters [see Equation (26)] vs. the absorbed photon flow per unit mass of catalyst. ♦, K_F^*; ▲, N_S^*; ■, k.

root of phenol equilibrium concentration. The straight line drawn through the data represents Equation (20); the slope value of this line, equal to the $(K_F^* * N_S^*)$ product, is 0.0112, very near to the 0.0107 value obtained from the best fitting procedure.

The values of kinetic and Freundlich parameters (k, K_F^*, and N_S^*), obtained from runs carried out with different amounts of catalyst, are reported in Figure 12 vs. the absorbed photon flow per unit mass of catalyst. These values show the same feature of Langmuir parameters obtained for benzyl alcohol oxidation, that is, they decrease by decreasing the photon flow absorbed *by the unit mass of catalyst*. As in the case of benzyl alcohol, the consideration that the photon flow absorbed by the unit mass of catalyst is the parameter mainly affecting the photoadsorption phenomenon is strengthened by the results reported in Figure 12.

6.3. Reaction mechanism

According to the preceding discussion, the primary events occurring at the photocatalyst surface may be described by the following mechanism: The primary photochemical act, following the light absorption by the semiconductor, is the generation of electron/hole pairs. The light absorbed by the unit surface of catalyst and unit time produces equivalent bulk concentrations of charge carriers, e^- and h^+:

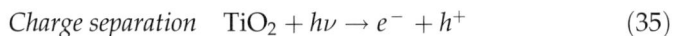

$$\text{Charge separation} \quad TiO_2 + h\nu \rightarrow e^- + h^+ \tag{35}$$

The charge carriers can recombine either in the bulk or in the surface:

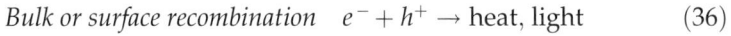

$$\text{Bulk or surface recombination} \quad e^- + h^+ \rightarrow \text{heat, light} \tag{36}$$

or migrate to the surface where they are trapped by suitable electron and hole traps forming the corresponding active centers:

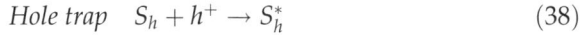

$$\text{Electron trap} \quad S_e + e^- \rightarrow S_e^* \tag{37}$$

$$\text{Hole trap} \quad S_h + h^+ \rightarrow S_h^* \tag{38}$$

If electron acceptors (Ox_2, like oxygen) or electron donors (Red_1, as organic substrates) are present at the surface, they can be photoadsorbed on the active centers:

$$\text{Photoadsorption on } S_e^* \quad S_e^* + Ox_2 \rightarrow Ox_{2(\text{photoads})} \tag{39}$$

$$\text{Photoadsorption on } S_h^* \quad S_h^* + Red_1 \rightarrow Red_{1(\text{photoads})} \tag{40}$$

The photoadsorption steps are followed by oxidation of the organic substrate:

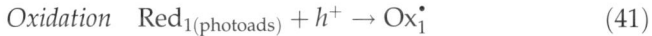

$$\text{Oxidation} \quad Red_{1(\text{photoads})} + h^+ \rightarrow Ox_1^{\bullet} \tag{41}$$

and by reduction of the oxidant species:

$$\text{Reduction} \quad Ox_{2(\text{photoads})} + e^- \rightarrow Red_2^{\bullet} \tag{42}$$

Competitive with reactions (37–40) are the back reactions:

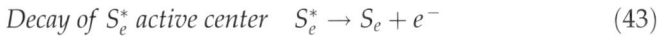

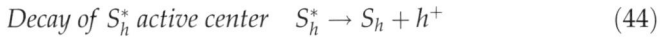

$$\text{Decay of } S_e^* \text{ active center} \quad S_e^* \rightarrow S_e + e^- \tag{43}$$

$$\text{Decay of } S_h^* \text{ active center} \quad S_h^* \rightarrow S_h + h^+ \tag{44}$$

$$\text{Photodesorption from } S_e^* \quad Ox_{2(\text{photoads})} \rightarrow S_e^* + Ox_2 \tag{45}$$

$$\text{Photodesorption from } S_h^* \quad Red_{1(\text{photoads})} \rightarrow S_h^* + Red_1 \tag{46}$$

The combined reactions (35) and (36) or (37), (38), (43) and (44) or (39), (40), (45), and (46) yield a net cycle, where there is no use of the absorbed photons. The radicals formed through reactions 41 and 42 may be further transformed by subsequent reaction with photogenerated active species or through reaction with solvent, other species present in solution (such as O_2, H_2O_2, O_2^-), elimination of molecular groups or ions (Calza et al., 1997; Piccinini et al., 1997), or dimerization (Minero et al., 1995). The additional transformations may lead to the complete degradation of the organic compound to CO_2 and inorganic anions. In addition to all the preceding reactions, which hold for a substrate that can only be oxidized, there are possible concurrent oxidative and reductive reactions, as observed for halomethanes degradation (Calza et al., 1997).

7. CONCLUSIONS

The method here proposed for the determination of photoadsorption capacity under reaction conditions gives valuable information on the influence of absorbed photons on kinetics and thermodynamics of a photocatalytic reaction. The consideration that the photon flow absorbed by the unit mass of catalyst is the parameter mainly affecting the photoadsorption phenomenon is strengthened by all the results. Under irradiation, the catalyst surface undergoes noticeable changes (for example, from hydrophobic to hydrophilic character), these changes depending on photon absorption. The photoadsorbed amount of solute shows firstly a sharp increase with absorbed photons, this increase being less important at high photon absorption. This behavior appears to indicate that the number of photoactive sites increases with photon absorption until a value is reached for which an increase of absorbed photon does not affect the amount of photoadsorbed substrate. This limiting amount should be characteristic of the catalyst surface and the physicochemical features of solute and medium (pH, ionic strength, etc.).

The basic assumptions of model are that (i) the TiO_2 surface, once firstly irradiated, needs a certain time to reach a thermodynamic equilibrium with the surrounding medium and (ii) photoadsorption is fast with respect to the reaction. The best fitting procedure suggests that the Langmuir isotherm is the most suitable one for describing benzyl alcohol photoadsorption while the Freundlich isotherm better describes phenol photoadsorption. In both cases the other tested isotherms (Freundlich and Redlich–Peterson ones for benzyl alcohol and Langmuir and Redlich–Peterson ones for phenol) did not give a satisfactory fitting. It must be outlined, however, that it is likely that the chemical features of solute determine the most suitable isotherm for modeling the photoadsorption phenomenon.

The modeling results indicate that the parameters of the Langmuir and Freundlich models depend on the absorbed photon flow. In order to take into account the dependence on photon flows, it would be necessary to develop a kinetic model based on the proposed reaction mechanism in which the photons appear as reactants. Work on this field is in progress.

APPENDICES

A1. ASYMPTOTIC CASES OF LANGMUIR PHOTOADSORPTION ISOTHERM

A straightforward determination of K_L^*, N_S^*, k, and $C_{L,0}$ parameters can be done in the following way. k and $C_{L,0}$ may be easily determined by considering two asymptotic situations of Langmuir adsorption isotherm, that is, very high or very low substrate concentration. At very high concentrations of substrate the following inequality may be assumed to hold:

$$K_L^* C_L \gg 1 \tag{A1}$$

In this case from Equation (13) one obtains

$$\theta_{Sub} \equiv \frac{n_S}{WN_S^*} = \frac{K_L^* C_L}{1 + K_L^* C_L} \approx 1 \tag{A2}$$

and Equation (14) transforms in

$$C_T = C_L + \frac{WN_S^*}{V} \tag{A3}$$

The total disappearance rate of substrate per unit surface area, r_T, gets of zero order so that Equation (15) can be written as:

$$-\frac{V}{WS_S} \frac{dC_T}{dt} = k \tag{A4}$$

Taking the derivative of Equation (A3) with respect to time, it yields that $dC_T/dt = dC_L/dt$. Substituting this equality in Equation (A4) and rearranging, one obtains

$$-\frac{dC_L}{dt} = \frac{WS_S}{V} k \tag{A5}$$

Integration of Equation (A5) with the condition that $C_L = C_{L,0}$ for $t = 0$ yields the following relationship:

$$C_L = C_{L,0} - \frac{WS_S}{V} kt \tag{A6}$$

which is a linear relationship between the concentration values measured in the solution and the irradiation time. The intercept and slope values allow to determine the $C_{L,0}$ and k values.

At very low concentrations of substrate, the following inequality may be assumed to hold:

$$K_L^* C_L \ll 1 \tag{A7}$$

In this case from Equation (13) one obtains

$$\theta_{Sub} \equiv \frac{n_S}{WN_S^*} = \frac{K_L^* C_L}{1 + K_L^* C_L} \approx K_L^* C_L \tag{A8}$$

and from Equation (14) one obtains

$$C_T = C_L + \frac{WN_S^* K_L^*}{V} C_L \tag{A9}$$

The total disappearance rate of substrate per unit surface area, r_T, turns of first order with respect to the substrate concentration and Equation (15) can be written as

$$-\frac{V}{WS_S} \frac{dC_T}{dt} = k K_L^* C_L \tag{A10}$$

Taking the derivative of Equation (A9) with respect to time, it yields

$$\frac{dC_T}{dt} = \left(1 + \frac{WN_S^* K_L^*}{V}\right) \frac{dC_L}{dt} \tag{A11}$$

Substituting this equality in Equation (A10) and rearranging, one obtains

$$-\frac{dC_L}{dt} = \frac{WS_S k K_L^*}{V + N_S^* K_L^*} C_L \tag{A12}$$

By integrating with the condition that $C_L = C_{L,0}$ for $t = 0$, one obtains

$$\ln C_L = \ln C_{L,0} - \frac{WS_S k K_L^*}{V + WN_S^* K_L^*} t \tag{A13}$$

which is a linear relationship in a semilogarithmic plot of concentration values vs. the irradiation time. The intercept value allows to determine the $C_{L,0}$ value while the slope contains two parameters, K_L^* and N_S^*, that may be determined in the following way:

From the photoreactivity runs for which the inequalities (A1) and (A7) hold, the values of $C_{L,0}$ may be determined. Applying to the substrate the molar balance expressed by Equation (6) at the start of irradiation, that is, the zero time, yields

$$C_{T,0} = C_{L,0} + \frac{n_{S,0}}{V} \tag{A14}$$

in which $C_{T,0}$ is the total concentration of substrate and coincides with the concentration of initial solution and $n_{S,0}$ is the number of photoadsorbed moles in equilibrium with the solution at $C_{L,0}$ concentration. In the statement of Equation (A14), it is assumed that only photoadsorption can occur on the surface of irradiated particles of catalyst and that the whole particles are irradiated. Equation (A14) allows to determine the $n_{S,0}$ value corresponding to a certain $C_{L,0}$:

$$n_{S,0} = V(C_{T,0} - C_{L,0}) \tag{A15}$$

All the $n_{S,0} - C_{L,0}$ couples of values are related by the Langmuir model [Equation (13)]:

$$\frac{n_{S,0}}{WN_S^*} = \frac{K_L^* C_{L,0}}{1 + K_L^* C_{L,0}} \tag{A16}$$

Inverting the terms of Equation (A16) and rearranging, one has

$$\frac{1}{n_{S,0}} = \frac{1}{WN_S^* K_L^* C_{L,0}} + \frac{1}{WN_S^*} \tag{A17}$$

Introducing Equation (A15) into Equation (A17) and multiplying Equation (A17) by $C_{L,0}$ yield Equation (A18), which is the linear form of Equation (A16):

$$\frac{C_{L,0}}{C_{T,0} - C_{L,0}} = \frac{V}{WN_S^* K_L^*} + \frac{V}{WN_S^*} C_{L,0} \tag{A18}$$

where $C_{L,0}/(C_{T,0} - C_{L,0})$ is a dependent variable, $C_{L,0}$ an independent variable, $V/(WK_L^* N_S^*)$ the linear coefficient, and $V/(WN_S^*)$ the angular coefficient of the straight line. Thus, by plotting $C_{L,0}/(C_{T,0} - C_{L,0})$ vs. $C_{L,0}$ one can determine the maximum adsorption capacity, N_S^*, and photoadsorption equilibrium constant, K_L^*, respectively, through the slope and linear coefficient of the straight line.

The determination of the value of K_L^* allows checking the inequalities expressed by Equations (A1) and (A7). For the runs for which the previous inequalities do not hold, Equation (19) must be used for determining the only unknown parameter contained in it, that is, the $C_{L,0}$ value.

A2. TEMKIN ISOTHERM

The Temkin isotherm model assumes that the heat of adsorption of all the molecules in the layer decreases linearly with coverage due to adsorbent–adsorbate interactions, and that the adsorption is characterized by a uniform distribution of the binding energies, up to some maximum binding energy. By applying the Langmuir adsorption isotherm to this distribution of energies, the Temkin isotherm equation is derived (Satterfield, 1980):

$$\theta_{Sub} \equiv \frac{n_S}{WN_S^*} = \frac{1}{f} \ln\left(K_T^* C_L\right) \tag{A19}$$

where f is a constant related to the differential heat of adsorption at zero surface coverage, and K_T^* the Temkin equilibrium adsorption constant. By solving Equation (A19) with respect to n_S and substituting in Equation (6), the result is

$$C_T = C_L + \frac{WN_S^*}{V}\frac{1}{f} \ln\left(K_T^* C_L\right) \tag{A20}$$

Substituting in Equation (11), the Temkin relationship (Equation (A19)) produces

$$-\frac{V}{WS_S}\frac{dC_T}{dt} = k\frac{1}{f} \ln\left(K_T^* C_L\right) \tag{A21}$$

Taking the derivative of Equation (A20) with respect to time, it yields

$$\frac{dC_T}{dt} = \left[1 + \frac{WN_S^*}{V}\frac{1}{f}\frac{1}{C_L}\right]\frac{dC_L}{dt} \tag{A22}$$

Substituting Equation (A22) in the left-hand side term of Equation (A21), one obtains

$$-\frac{V}{WS_S}\left[1 + \frac{WN_S^*}{V}\frac{1}{f}\frac{1}{C_L}\right]\frac{dC_L}{dt} = k\frac{1}{f} \ln\left(K_T^* C_L\right) \tag{A23}$$

and, rearranging and separating the variables, the following differential equation is obtained:

$$-\frac{V}{WS_S}\frac{f}{k}\frac{dC_L}{\ln(K_T^* C_L)} - \frac{N_S^*}{S_S}\frac{1}{k}\frac{dC_L}{C_L \ln(K_T^* C_L)} = dt \tag{A24}$$

Equation (A24) contains five unknown parameters, K_T^*, N_S^*, k, f, and $C_{L,0}$, whose determination may be carried out by a best fitting procedure. As done in the cases previously described, integration of Equation (A24) must be performed with the condition that at $t = 0$ the substrate concentration in the liquid phase is that in equilibrium with the initial photoadsorbed amount, $C_{L,0}$; this initial concentration is unknown, but it may be determined by the regression analysis carried out with the experimental data obtained after the start of irradiation.

LIST OF SYMBOLS

C_L concentration in the liquid phase (M)

$C_{L,0}$ substrate initial concentration in photoadsorption equilibrium (M)

C_T total concentration of the species (M)

f Temkin constant (dimensionless)

h Planck's constant (6.626×10^{-34} J·s)

k pseudo-first-order rate constant (mol m^{-2} h^{-1})

k'' second-order rate constant (mol m^{-2} h^{-1})

K_F^* Freundlich isotherm constant (M^n)

K_L^* Langmuir photoadsorption equilibrium constant (M^{-1})

K_{R-P}^* Redlich–Peterson isotherm constant (dm^3·g^{-1})

K_T^* Temkin equilibrium adsorption constant (M^{-1})

n Freundlich parameter (dimensionless)

n_L number of moles in the fluid phase (moles)

n_S number of moles photoadsorbed on the solid phase (moles)

N_S^* maximum capacity of photoadsorbed moles of substrate (mol g^{-1})

$n_{S,Ox}$ oxygen moles photoadsorbed on the solid (moles)

$N_{S,Ox}^*$ maximum capacity of photoadsorbed moles of oxygen (mol g^{-1})

n_T total number of moles present in the photoreactor (moles)

r_T total disappearance rate of substrate per unit surface area (mol m^{-2} h^{-1})

S catalyst surface area (m^2)

S_S catalyst specific surface area (m^2 g^{-1})

t time (h)

V volume of the liquid phase (dm^3)

W mass of catalyst (g)

α_{R-P}^* Redlich–Peterson isotherm constant (M$^{-\beta}$)

β Redlich–Peterson heterogeneity factor (dimensionless)

v radiation wavenumber (nm^{-1})

θ_{Ox} oxygen fractional coverage of the surface (dimensionless)

θ_{Sub} substrate fractional coverage of the surface (dimensionless)

REFERENCES

Addamo, M., Augugliaro, V., Di Paola, A., García-López, E., Loddo, V., Marcì, G., Molinari, R., Palmisano, L., and Schiavello, M. *J. Phys. Chem. B* **108**, 3303 (2004).

Ahmaruzzaman, M., and Sharma, D.K. *J. Colloid Interface Sci.* **287**, 14 (2005).

Allen, S.J., Mckay, G., and Porter, J.F. *J. Colloid Interface Sci.* **280**, 322 (2004).

Asakuma, N., Fukui, T., Toki, M., Awazu, K., and Imai, H. *Thin Solid Films* **445**, 284 (2003).

Augugliaro, V., Litter, M., Palmisano, L., and Soria, J. *J. Photochem. Photobiol. C* **7**, 127 (2006).

Bickley, R.I. "Fundamental Aspects of the Adsorption and the Desorption of Gases at SolidSurfaces under Illumination" *in* M. Schiavello (Ed.), "Photoelectrochemistry,

Photocatalysis and Photoreactors, Fundamentals and Developments". Reidel, Dordrecht (1985a), pp. 379–388.

Bickley, R.I. "Some Experimental Investigations of Photosorption Phenomena at the Gas-Solid Interface" *in* M. Schiavello (Ed.), "Photoelectrochemistry, Photocatalysis and Photoreactors, Fundamentals and Developments". Reidel, Dordrecht (1985b), pp. 491–502.

Bickley, R.I. "Photoadsorption and Photodesorption at the Gas-Solid Interface Part I Fundamental Concepts" *in* M. Schiavello (Ed.) "Photocatalysis and Environment Trends and Applications". Kluwer, Dordrecht (1988a), pp. 223–232.

Bickley, R.I. "Photoadsorption and Photodesorption at the Gas-Solid Interface Part II Photoelectronic Effects Relating to Photochromic Changes and to Photosorption" *in* M. Schiavello (Ed.), "Photocatalysis and Environment Trends and Applications". Kluwer, Dordrecht (1988b), pp. 233–239.

Braslavsky, S.E. *Pure Appl. Chem.* **79**, 293 (2007).

Calza, P., Minero, C., and Pelizzetti, E. *Environ. Sci. Technol.* **31**, 2198 (1997).

Carp, O., Huisman, C.L., and Reller, A. *Prog. Solid State Chem.* **32**, 33 (2004).

de Lasa, H., Serrano, B., and Salaices, M. "Photocatalytic Reaction Engineering". Springer, New York (2005).

Demeestere, K., De Visscher, A., Dewulf, J., Van Leeuwen, M., and Van Langenhove, H. *Appl. Catal. B* **54**, 261 (2004).

Fujishima, A., and Honda, K. *Nature* **238**, 37 (1972).

Fujishima, A., Hashimoto, K., and Watanabe, T. "TiO$_2$ Photocatalysis: Fundamentals and Applications". Bkc Inc., , Tokyo (1999).

Fujishima, A., Rao, T.N., and Tryk, D.A. *J. Photochem. Photobiol. C* **1**, 1 (2000).

Fujishima, A., and Zhang, X. *Crit. Rev. Chim.* **9**, 7 (2006).

Gonzalez, M.A., Howell, S.G., and Sikdar, S.K. *J. Catal.* **183**, 159 (1999).

Gora, A., Toepfer, B., Puddu, V., and Li Puma, G. *Appl. Catal. B* **65**, 1 (2006).

Heckert, N.A., and Filliben, J.J. "NIST Handbook 148: DATAPLOT Reference Manual, Volume I: Commands". National Institute of Standards and Technology Handbook Series, Gaithersburg (2003a).

Heckert, N.A., and Filliben, J.J. "NIST Handbook 148: Dataplot Reference Manual, Volume II: Let Subcommands and Library Functions". National Institute of Standards and Technology Handbook Series, Gaithersburg (2003b).

Hoffmann, M.R., Martin, S.T., Choi, W.Y., and Bahnemann, D.W. *Chem. Rev.* **95**, 69 (1995).

Ibrahim, H., and De Lasa, H. *AIChE J.* **50**, 1017 (2004).

Ignatchenko, A., Nealon, D.G., Dushane, R., and Humphries, K. *J. Mol. Cat. A* **256**, 57 (2006).

Kaneko, M., and Okura, I. (Eds.), "Photocatalysis: Science and Technology". Springer-Verlag, Heildelberg, New York (2002).

Krýsa, J., Waldner, G., Měšťánková, H., Jirkovský, J., and Grabner, G. *Appl. Catal. B* **64**, 290 (2006).

Kumar, K.V., and Sivanesan, S. *J. Hazard. Mater.* **126**, 198 (2005).

Lee, Y.C., Hong, Y.P., Lee, H.Y., Kim, H., Jung, Y.J., Ko, K.H., Jung, H.S., and Hong, K.S. *J. Colloid Interface Sci.* **267**, 127 (2003).

Linsebigler, A.L., Lu, G.G., and Yates, J.T. *Chem. Rev.* **95**, 735 (1995).

Matthews, A.P., and Weber Jr., W.J. *AIChE Symp. Ser.* **73**, 91 (1976).

Meriaudeau, P., and Vedrine, J.C. *J. Chem. Soc. Faraday Trans. II* **72**, 472 (1976).

Mills, A., and Le Hunte, S. *J. Photochem. Photobiol. A* **108**, 1 (1997).

Minero, C., Pelizzetti, E., Pichat, P., Sega, M., and Vincenti, M. *Environ. Sci. Technol.* **29**, 2226 (1995).

Minero, C., and Vione, D. *Appl. Catal. B* **67**, 257 (2006).

Miyauchi, M., Kieda, N., Hishita, S., Mitsuhashi, T., Nakajima, A., Watanabe, T., and Hashimoto, K. *Surf. Sci.* **511**, 401 (2002).

Mohamed, O.S., Gaber, A.E.M., and Abdel-Wahab, A.A. *J. Photochem. Photobiol. A* **148**, 205 (2002).

Mori, K., Imaoka, S., Nishio, S., Nishiyama, Y., Nishiyama, N., and Yamashita, H. *Microporous Mesoporous Mater.* **101**, 288 (2007).

Murphy, W.R., Veerkamp, T.H., and Leland, T.W. *J. Catal.* **43**, 304 (1976).

Murzin, D.Y., and Salmi, T. "Catalytic Kinetics". Elsevier, Amsterdam (2005).

Nakajima, A., Koizumi, S., Watanabe, T., and Hashimoto, K. *J. Photochem. Photobiol. A* **146**, 129 (2001).

Ncibi, M.C. *J. Hazard. Mater.* **153**, 207 (2008).

Palmisano, G., Addamo, M., Augugliaro, V., Caronna, T., Di Paola, A., García-López, E., Loddo, V., Marcì, G., Palmisano, L., and Schiavello, M. *Catal. Today* **122**, 118 (2007b).

Palmisano, G., Loddo, V., Yurdakal, S., Augugliaro, V., and Palmisano, L. *AIChE J.* **53**, 961 (2007c).

Palmisano, G., Yurdakal, S., Augugliaro, V., Loddo, V., and Palmisano, L. *Adv. Synth. Catal.* **349**, 964 (2007a).

Piccinini, P., Minero, C., Vincenti, M., and Pelizzetti, E. *J. Chem. Soc. Faraday Trans.* **93**, 1993 (1997).

Redlich, O., and Peterson, D.L. *J. Phys. Chem.* **63**, 1024 (1959).

Ryabchuk, V. *Int. J. Photoenergy* **6**, 95 (2004).

Salaices, M., Serrano, B., and de Lasa, H.I. *Chem. Eng. Sci.* **59**, 3 (2004).

Sakai, N., Wang, R., Fujishima, A., Watanabe, T., and Hashimoto, K. *Langmuir* **14**, 5918 (1998).

Satterfield, C.N. "Heterogeneous Catalysis in Practice". McGraw-Hill, New York (1980).

Schiavello, M. (Ed.), "Heterogeneous Photocatalysis". Wiley, New York (1997).

Serpone, N., and Emeline, A.V. *Int. J. Photoenergy* **4**, 91 (2002).

Solonitzyn, Yu.P., Prudnikov, I.M., and Yurkin, V.M. *Russ. J. Phys. Chem.* **57**, 2028 (1982).

Solonitzyn, Yu.P., and Terenin, A.N. *Discuss Faraday. Soc.* **28**, 28 (1959).

Takeuchi, M., Sakamoto, K., Martra, G., Coluccia, S., and Anpo, M.*J. Phys. Chem. B* **109**, 15422 (2005).

Turchi, C.S., and Ollis, D.F. *J. Catal.* **119**, 483 (1989).

Vorontsov, A.V., Kurkin, E.N., and Savinov, E.N. *J. Catal.* **186**, 318 (1999).

Yurdakal, S., Palmisano, G., Loddo, V., Augugliaro, V., and Palmisano, L. *J. Am. Chem. Soc.* **130**, 1568 (2008a).

Yurdakal, S., Loddo, V., Palmisano, G., Augugliaro, V., and Palmisano, L. *Catal. Today*, DOI:10.1016/j.cattod.2008.06.032, (2008b).

Wang, R., Hashimoto, K., Fujishima, A., Chikuni, M., Kojima, E., Kitamura, A., Shimohigoshi, M., and Watanabe, T. *Nature* **388**, 431 (1997).

Wang, R., Sakai, N., Fujishima, A., Watanabe, T., and Hashimoto, K. *J. Phys. Chem. B* **103**, 2188 (1999).

Watanabe, T., Nakajima, A., Wang, R., Minabe, M., Koizumi, S., Fujishima, A., and Hashimoto, K. *Thin Solid Films* **351**, 260 (1999).

Wu, W.C., Liao, L.F., Shiu, J.S., and Lin, J.L. *Phys. Chem. Chem. Phys.* **2**, 4441 (2000).

CHAPTER 2

Treatment of Chromium, Mercury, Lead, Uranium, and Arsenic in Water by Heterogeneous Photocatalysis

Marta I. Litter

Contents		
	1. Introduction	37
	2. Thermodynamical Considerations and Mechanistic Pathways	41
	3. Chromium	44
	4. Mercury	49
	5. Lead	53
	6. Uranium	57
	7. Arsenic	58
	8. Conclusions	61
	Acknowledgment	62
	References	62

1. INTRODUCTION

Presence of heavy metals and metalloids in water nowadays represents one of the most important environmental problems because the yearly total toxicity of mobilized species overcomes the total toxicity of anthropogenically generated organic and radioactive wastes. In addition to natural sources, anthropic activities introduce hundreds of billions of tons per year of heavy metals in the terrestrial ambient. The accumulation of metals and metalloids in effluents and in industrial wastes represents significant losses in raw materials and causes perturbation of the ecological equilibrium. Precious and common metals enter waters through washing, rinsing, pickling, and surface treatment procedures of industrial processes such as

Gerencia Química, Centro Atomico Constituyentes, Comisión Nacional de Energía Atómica, Av Gral. Paz 1499, 1650 San Martín, Prov. de Buenos Aires, Argentina; Instituto de Investigación e Ingeniería Ambiental, Universidad de General San Martín, Peatonal Belgrano 3563, 18 piso, 1650 San Martín, Prov. de Buenos Aires, Argentina; Consejo Nacional de Investigaciones Científicas y Técnicas, Argentina.

E-mail address: litter@cnea.gov.ar

Advances in Chemical Engineering, Volume 36
ISSN 0065-2377, DOI: 10.1016/S0065-2377(09)00402-5

hydrometallurgy, plating, or photography. Metal and metalloid species such as chromium, mercury, uranium, and arsenic are considered priority pollutants by several environmental agencies, and the allowed amount of these species in drinking water is becoming more and more stringent.

Although removal of organic and microbiological pollutants from waters has been thoroughly studied, less attention has been paid to the transformation of metal or metalloid ions in species of lower toxicity or more easily isolated. Metals in their various oxidation states have infinite lifetimes, and chemical or biological treatments present severe restrictions or are economically prohibitive. Removal of these species is carried out, generally, by precipitation, electrolysis, chemical oxidation, adsorption, or chelation, all of them presenting drawbacks.

Heterogeneous photocatalysis is a very well-known technology, valuable for purification and remediation of water and air. Several excellent revisions exist on the subject, with different approaches (Bahnemann et al., 1994; Emeline et al., 2005; Fujishima and Zhang, 2006; Hoffmann et al., 1995; Legrini et al., 1993; Linsebigler et al., 1995; Mills and Le Hunte, 1997; Rajeshwar, 1995; Serpone, 1997).

In heterogeneous photocatalysis, after excitation of semiconductors with light of energy equal to or higher than the bandgap (E_g), conduction band electrons (e_{cb}^-) and valence band holes (h_{vb}^+) are created. TiO_2 is so far the most useful semiconductor material for photocatalytic purposes because of its exceptional optical and electronic properties, chemical stability, nontoxicity, and low cost. The energy bandgaps of the photocatalytic forms of TiO_2, anatase and rutile, are 3.23 eV (corresponding to 384 nm) and 3.02 eV (corresponding to 411 nm) (Rajeshwar, 1995). The German company Degussa produces the most popular commercial form of TiO_2 under the name P-25. For Degussa P-25, the values of the edges of conduction and valence bands at pH 0 have been calculated as −0.3 and +2.9 V[1], respectively (Martin et al., 1994). Photogenerated holes and electrons can recombine or migrate to the surface where they can react with donor (D) or acceptor (A) species (Figure 1). The energy level at the bottom of the conduction band is actually the reduction potential of photoelectrons and the energy level at the top of the valence band determines the oxidizing ability of photoholes, each value reflecting the ability of the system to promote reductions and oxidations. Valence band holes are strong oxidants that may attack directly oxidizable species D or form hydroxyl radicals ($HO^{\bullet}$) from water or surface hydroxide ions, while conduction band electrons are mild reducing acceptors. From a thermodynamic point of view, couples can be photocatalytically reduced by conduction band electrons if they have redox potentials more positive than the flatband potential (V_{fb}) of the conduction band and can be

---

[1] All standard reduction potentials given in this work are vs. NHE.

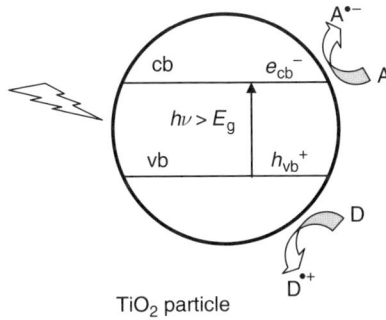

Figure 1 Simplified diagram of the heterogeneous photocatalytic processes occurring at an illuminated TiO_2 particle.

oxidized by valence band holes if they have redox potentials more negative than the V_{fb} (Bahnemann et al., 1994; Hoffmann et al., 1995; Linsebigler et al., 1995; Mills and Le Hunte, 1997; Serpone, 1997).

The above-described scheme is completed by the following basic equations where e_{cb}^-, h_{vb}^+, and $HO^\bullet$ are involved as follows:

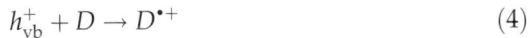

$$TiO_2 + h\nu \rightarrow e_{cb}^- + h_{vb}^+ \tag{1}$$

$$e_{cb}^- + A \rightarrow A^{\bullet-} \tag{2}$$

$$h_{vb}^- + H_2O \rightarrow HO^\bullet + H^+ \tag{3}$$

$$h_{vb}^+ + D \rightarrow D^{\bullet+} \tag{4}$$

In particular, O_2 adsorbed to TiO_2 can be reduced by e_{cb}^-, generating $O_2^{\bullet-}$ in a thermodynamically feasible but rather slow electron transfer reaction (Hoffmann et al., 1995). Values of $E^0(O_2/O_2^{\bullet-}) = -0.3$ V and $E^0(O_2/HO_2^\bullet) = -0.05$ V have been reported for homogeneous solutions; the reduction potentials onto the TiO_2 surface are probably less negative. As the following set of equations indicates, this cathodic pathway is an additional source of hydroxyl radicals:

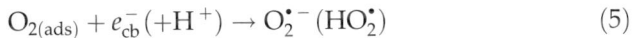

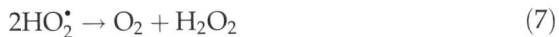

$$O_{2(ads)} + e_{cb}^- (+H^+) \rightarrow O_2^{\bullet-} (HO_2^\bullet) \tag{5}$$

$$2O_2^{\bullet-} + 2H_2O \rightarrow 2HO_2^\bullet + 2OH \tag{6}$$

$$2HO_2^\bullet \rightarrow O_2 + H_2O_2 \tag{7}$$

$$H_2O_2 + O_2^{\bullet -} \rightarrow HO^- + HO^\bullet + O_2 \qquad (8)$$

In anoxic conditions, protons are the strongest electron acceptors, being reduced to hydrogen atoms, which combine then in hydrogen molecules (Bahnemann et al., 1987):

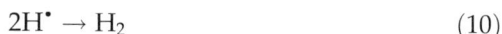

$$H^+ + e_{cb}^- \rightarrow H^\bullet \qquad (9)$$

$$2H^\bullet \rightarrow H_2 \qquad (10)$$

If in solution there is a metal ion of convenient redox potential, conduction band electrons can reduce the species to a lower oxidation state:

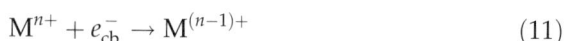

$$M^{n+} + e_{cb}^- \rightarrow M^{(n-1)+} \qquad (11)$$

Alternatively, the metal can be oxidized by holes or hydroxyl radicals:

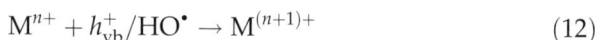

$$M^{n+} + h_{vb}^+/HO^\bullet \rightarrow M^{(n+1)+} \qquad (12)$$

Photocatalytic treatments can convert the ionic species into their metallic solid forms and deposit them over the semiconductor surface or transform them in less toxic soluble species. When a transformation to the zerovalent state is possible, this allows the recovery of the metal by mechanical or chemical procedures, with an important economical return. From the beginning of the development of heterogeneous photocatalysis, transformation and deposit of metals – principally the most noble, expensive, and toxic ones – was visualized as one of the potential applications of the technology in view of the involved economical and environmental aspects. Various semiconductors have been applied in the photocatalytic transformation or deposition of metals such as chromium, gold, silver, platinum, palladium, rhodium, mercury, lead, manganese, thallium, and copper, among others (Serpone et al., 1988). Other applications of heterogeneous photocatalysis related to metal ions are light energy storage, photographic imaging, prevention of semiconductor corrosion, and preparation of modified semiconductors, but these topics will not be treated here.

In view of the enormous literature published on the subject, only the cases of chromium, mercury, lead, uranium, and arsenic are reviewed here. In 1999, we published an extensive review on metal treatment by heterogeneous photocatalysis in which the early literature is mentioned (Litter, 1999). In this chapter, we will remind the most important issues and update the most recent information.

2. THERMODYNAMICAL CONSIDERATIONS AND MECHANISTIC PATHWAYS

The redox level of the metallic couples related to the levels of the conduction and valence bands can be considered as the most important parameter to predict the feasibility of transformation of the species, and it is possible to take these values for a first approach. Figure 2 shows the redox levels of

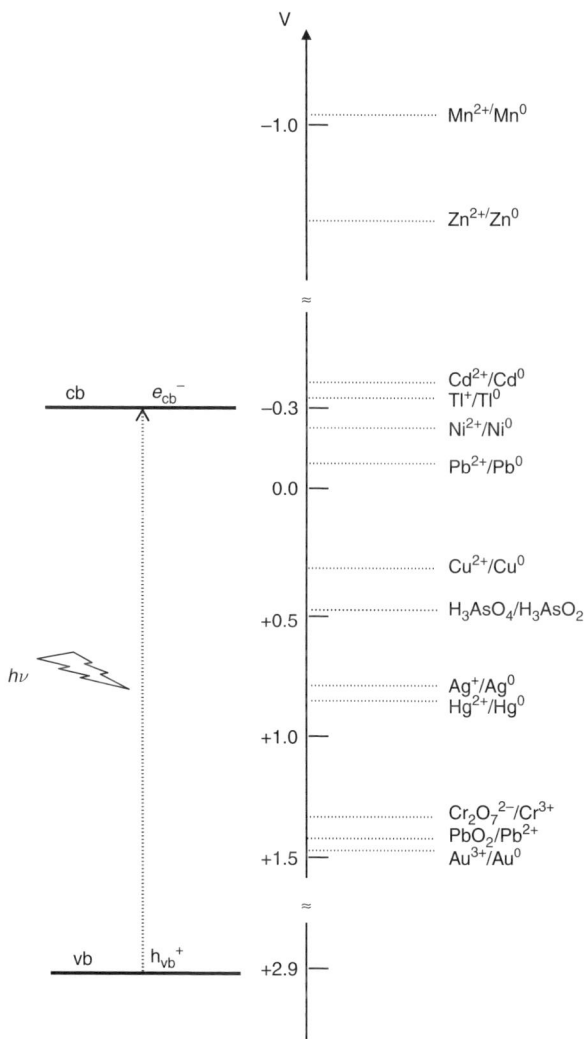

Figure 2 Position of the reduction potentials of various metallic couples (vs. NHE) related to the energy levels of the conduction and valence bands of Degussa P-25 TiO$_2$ at pH 0. Reduction potentials are taken from Bard et al. (1985).

couples and the thermodynamic ability of the TiO_2 photocatalytic system to reduce or oxidize the corresponding species. According to this, metals have been separated into two groups: those as Cu^{2+}, As^{5+}, Ag^+, Hg^{2+}, Cr^{6+}, and Au^{3+}, which show a strong tendency to undergo photocatalytic reduction, and those as Mn^{2+}, Zn^{2+}, Cd^{2+}, Tl^+, Ni^{2+} and Pb^{2+}, which show essentially no or a very weak tendency to accept photogenerated electrons from TiO_2 (Rajeshwar et al., 2002).

However, in this chapter, it will be emphasized that the above concepts – generally used when describing heterogeneous photocatalysis applied to metal systems – only express the feasibility of the overall process to occur, implicitly assuming a global multielectronic transfer reaction in many of the cases. Nevertheless, under the working conditions of ordinary photocatalytic reactions, that is, under nonintense irradiation, multielectronic reactions are rather unlikely, considering the frequency of photon absorption (Grela and Colussi, 1996). Recent experiments of our research group (see sections 3, 4 and 5) led to the conclusion that most photocatalytic processes on metal ions occur through successive one-electron pathways that produce unstable intermediates until the most stable species is formed.

Three types of mechanisms can be considered for the photocatalytic removal of metal ions: (a) direct reduction by photogenerated electrons, (b) indirect reduction by intermediates generated by hole or hydroxyl radical oxidation of electron donors present in the media, and (c) oxidative removal by holes or hydroxyl radicals (Lin and Rajeshwar, 1997), all of them represented in Figure 3.

In the direct reduction (a), the initial electron transfer step, reaction (11), is usually considered as the rate determining one (Mills and Valenzuela, 2004). For predicting the feasibility of the transformation, the reduction potential of the first step related to the energy of the conduction band has to be considered. The conjugate anodic reaction of Equation (11) is oxidation of water by holes or $HO^\bullet$, initiated by reaction (3) and ending in protons and oxygen:

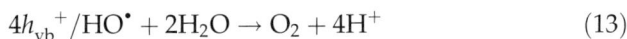

$$4h_{vb}^+/HO^\bullet + 2H_2O \rightarrow O_2 + 4H^+ \tag{13}$$

The photogenerated holes or $HO^\bullet$ radicals can reoxidize also the species to the original one, causing a nonproductive short-circuiting of the overall process:

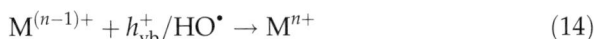

$$M^{(n-1)+} + h_{vb}^+/HO^\bullet \rightarrow M^{n+} \tag{14}$$

As the water oxidation reaction (13) is a very slow four-electron process, metal reductive transformation can be generally improved by addition of sacrificial organic agents (Prairie and Stange, 1993; Prairie et al., 1993a). The process can be improved even more if a strong reducing species is formed

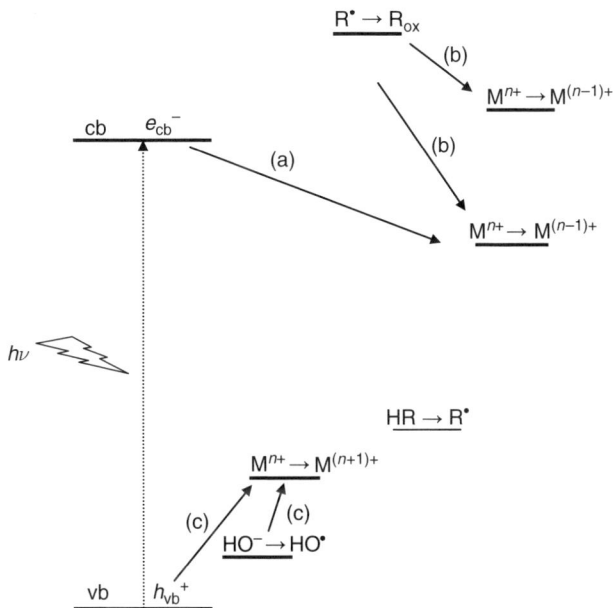

Figure 3 Schematic diagram for the photocatalytic transformation of metal ions on TiO_2. Different pathways (a), (b), and (c) are indicated (see text). The diagram of energy levels is only qualitative.

from the sacrificial agent, giving rise to an indirect pathway (type (b) reaction), as we will see later. The process will be very dependent on the nature of the added agent: low-molecular-weight acids, alcohols, and aldehydes do not cause any effect, while easily oxidizable organics such as EDTA, salicylic acid, and citric acid provide very fast reduction rates. These organic compounds are oxidized by holes or hydroxyl radicals in irreversible reactions, avoiding recombination of electron–hole pairs and enhancing reduction of metal ions:

$$RCOO^- + h_{vb}^+/HO^{\bullet} \rightarrow R^{\bullet} + CO_2 \qquad (15)$$

In addition, this process hinders the short-circuiting depicted by reaction (14).

Indirect reduction (mechanism (b)) was suggested by Baba et al. (1986), who analyzed the photodeposition of Pt, Ag, and Au on an anodically biased TiO_2 electrode in a solution containing alcohols (MeOH, EtOH, or 2-PrOH). As no metal deposition was observed in the absence of alcohols, these authors proposed that photogenerated conduction band TiO_2 electrons do not take part directly in the deposition. As a result, the reduction is then assumed to be driven by intermediates generated by attack of h_{vb}^+ or

HO$^{\bullet}$ to alcohols or carboxylic acids (HR), with the generation of highly energetic species (R$^{\bullet}$) (Figure 3):

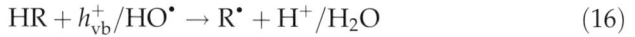

$$HR + h_{vb}^+/HO^{\bullet} \rightarrow R^{\bullet} + H^+/H_2O \tag{16}$$

For the cases of methanol, ethanol, or 2-propanol, R$^{\bullet}$ are 1-hydroxyalkyl radicals, while for formic or oxalic acids the strong reducing $CO_2^{\bullet-}$ is formed, with other carboxylic acid generating similar reducing species. Concurrently, reaction (14) is hindered if HR is added at relatively high concentrations because reaction (16) will predominate. The conjugated cathodic reaction in anoxic conditions will be H$^+$ reduction by conduction band electrons, with H$_2$ formation (reactions (9) and (10)). Once produced, R$^{\bullet}$ would be the effective reducing species of M^{n+}:

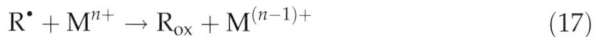

$$R^{\bullet} + M^{n+} \rightarrow R_{ox} + M^{(n-1)+} \tag{17}$$

where R$_{ox}$ is an aldehyde, a ketone, or CO$_2$ depending on the compound. In Figure 3, a simplified diagram of this process is presented.

Obviously, if the potential for the one-electron reduction of the metal is adequate, as described for M^{n+} in Figure 3, the direct reaction (a) can take place. This will not be the case for Pb(II), Ni(II), or Tl(I), which will only react through the indirect mechanism (b), reaction (12). After this first stage, other reducing steps driven by R$^{\bullet}$ or e_{cb}^- may take place until a stable metal species is formed. It has been pointed out that direct reduction of metal ions to the zerovalent state by reducing radicals is rather slow, but once some metal nuclei are formed, they serve as cathodic site to facilitate further reduction (Baba et al., 1986).

In mechanism (c), oxidative transformation of the metal species takes place by holes or hydroxyl radicals (or other reactive oxygen species, ROS) attack (Figure 3). This occurs according to reaction (12) when the oxidation of the metal or metalloid to a higher oxidation state is thermodynamically possible (cases of Pb(II), Mn(II), Tl(I), and As(III)).

3. CHROMIUM

Chromium (VI) is a frequent contaminant in wastewaters arising from industrial processes such as electroplating, leather tanning, or paints due to its carcinogenic properties, its concentration in drinking waters has been regulated in many countries. The World Health Organization (WHO) recommends a limit of 0.05 mg L^{-1} in drinking waters (World Health Organization, 2006), value adopted by several national environmental agencies. The preferred treatment is reduction to the less harmful Cr(III), non-toxic and less mobile. This process is performed generally with chemical reagents such as sodium thiosulfate, ferrous sulfate, sodium meta-bisulfite, or sulfur dioxide; in this way, the ion can be precipitated in neutral or alkaline solutions as $Cr(OH)_3$ and removed from the aqueous phase.

The net reaction for Cr(VI) reduction in acid aqueous solutions is:

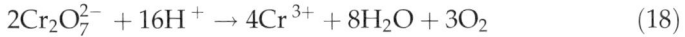

$$2Cr_2O_7^{2-} + 16H^+ \rightarrow 4Cr^{3+} + 8H_2O + 3O_2 \tag{18}$$

and at neutral pH:

$$4CrO_4^{2-} + 20H^+ \rightarrow 4Cr^{3+} + 10H_2O + 3O_2 \tag{19}$$

The process is thermodynamically feasible ($\Delta G_{298}^0 = -115.8\,kJ$), but dichromate is stable in aqueous acidic solutions due to the high overpotential of the reaction of water oxidation. Cr(VI) photocatalytic reduction has been found effective over several semiconductors (in form of powders or electrodes) such as TiO_2, ZnO, CdS, ZnS, and WO_3. Several examples, all of them investigated before 1999, have already been described in our previous review (Litter, 1999), including reactors for technological applications (Aguado et al., 1991, 1992; Domènech, 1993; Domènech and Muñoz, 1987a, b, 1990; Domènech et al., 1986; Giménez et al., 1996; Lin et al., 1993; Miyake et al., 1977; Muñoz and Domènech, 1990; Prairie and Stange, 1993; Prairie et al., 1992, 1993a; Sabaté et al., 1992; Selli et al., 1996; Wang et al., 1992; Xu and Chen, 1990; Yoneyama et al., 1979). Other papers appeared later (Aarthi and Madras, 2008; Cappelletti et al., 2008; Chenthamarakshan and Rajeshwar, 2000; Chen and Ray, 2001; Colón et al., 2001a, b; Das et al., 2006; Fu et al., 1998; Hidalgo et al., 2007; Horváth et al., 2005; Jiang et al., 2006; Kajitvichyanukul et al., 2005; Kanki et al., 2004; Khalil et al., 1998; Ku and Jung, 2001; Mohapatra et al., 2005; Papadam et al., 2007; Rengaraj et al., 2007; Ryu and Choi, 2008; Schrank et al., 2002; Shim et al., 2008; Tuprakay and Liengcharernsit, 2005; Tzou et al., 2005; Wang et al., 2008; Xu et al., 2006; Yang and Lee, 2006; Zheng et al., 2000). The list includes microparticles used in slurries or conveniently supported, mixed, and modified semiconductors and even nanomaterials (nanoparticles, nanotubes).

The photocatalytic Cr(VI) reduction is more feasible at low pH because the net reaction consumes protons (Equations (18) and (19)), but use of neutral or alkaline conditions can be more convenient because Cr(III) can be precipitated as the hydroxide and immobilized, avoiding expensive separation steps; after the photocatalytic process, an adequate acid or strong basic treatment easily separates Cr(III) from the catalyst (Lin et al., 1993).

Cr(VI) photocatalytic reaction is very slow in the absence of scavengers of holes or $HO^•$, because, as said, reaction (13) is kinetically sluggish. The reaction can be accelerated by the addition of organic compounds (Colón et al., 2001a, b; Fu et al., 1998; Ku and Jung, 2001; Miyake et al., 1977; Papadam et al., 2007; Prairie and Stange, 1993; Rajeshwar et al., 2002; Shrank et al., 2002; Tzou et al., 2005; Wang et al., 2008); even humic acids (Selli et al., 1996; Yang and

Lee, 2006) and MTBE (Xu et al., 2006) were found to be efficient donors. Prairie and Stange (1993) and Wang et al. (2008) found that Cr(VI) removal rate in the presence of different organics increases in the order: p-hydroxybenzoic acid $<$ phenol $<$ formic acid $<$ salicylic acid $<$ citric acid. In many cases, these organic compounds can be present simultaneously with Cr(VI) in the waste-waters of different industrial processes. Sometimes, the agent is not able to enhance markedly Cr(VI) reduction but the opposite is true: Cr(VI) acts as a very strong oxidant for organic species, as in the cases of the photoinduced oxidative degradation of anionic (sodium dodecyl sulfate, SDS) and cationic (cetyltrimethylammonium bromide, CTAB) surfactants (Horváth et al., 2005).

Cappelletti et al. (2008), using nanocrystalline TiO_2 samples and different sacrificial molecules (formic acid, isopropyl alcohol and sodium sulfite), found Cr(0) formation at the TiO_2 surface. Since the reduction potential for Cr(III) to Cr(II) is $-0.42\,V$ and the global reduction potential from Cr(III) to Cr(0) is $-0.74\,V$ (Bard et al., 1985), these processes cannot be driven by TiO_2 conduction band electrons. In contrast, Cr(0) should have been formed by reducing radicals coming from the oxidation of the sacrificial molecules, according to reactions (16) and (17).

A different type of synergy was found in a binary Cu(II)/Cr(VI) system: under N_2 and UV light, acceleration of both Cr(VI) and Cu(II) reduction was observed (Goeringer et al., 2001). While no explanations were provided concerning this behavior, it can be proposed that the enhancement may be due to the formation of a layer or spots of metallic copper that promote photocatalysis.

Our research group (Botta et al., 1999; Di Iorio et al., 2008; Meichtry et al., 2007, 2008b; Navío et al., 1998, 1999; San Román et al., 1998; Siemon et al., 2002; Testa et al., 2001, 2004) reported several studies on Cr(VI) photocatalytic reduction using TiO_2, ZrO_2, and Fe-doped samples and gained insight about the processes taking place in the system. The standard reduction potentials of Cr(VI)/Cr(V), Cr(V)/Cr(IV), and Cr(IV)/Cr(III) couples are $+0.55$, $+1.34$, and $+2.10\,V$, respectively (Bard et al., 1985). All of them are positive enough to be reduced by TiO_2 conduction band electrons (Figure 3). Therefore, working with Cr(VI) in pure water TiO_2 suspensions or in the presence of donor agents such as EDTA, oxalic acid, and citric acid, our team proposed for the first time that Cr(VI) photocatalytic reduction over TiO_2 takes place by successive one-electron transfer reducing steps leading to Cr(III), the stable final product (Meichtry et al., 2007; Testa et al., 2001, 2004):

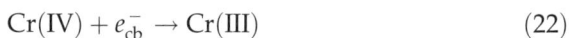

$$Cr(VI) + e_{cb}^- \rightarrow Cr(V) \tag{20}$$

$$Cr(V) + e_{cb}^- \rightarrow Cr(IV) \tag{21}$$

$$Cr(IV) + e_{cb}^- \rightarrow Cr(III) \tag{22}$$

These assumptions were validated, at least for the initial one-electron process (20), by EPR experiments in which Cr(V) species were detected. While the fate of Cr(V) and Cr(IV) is not clear, it is hypothesized that they are probably reduced by conduction band electrons or can suffer disproportionation to other species. As the anodic reaction in the absence of donors is sluggish (reaction (13)), an arrest of Cr(VI) reduction is observed after some time of reaction, because of the reoxidation of Cr(V)/(IV)/(III) species by holes or HO$^\bullet$. On the contrary, Cr(VI) removal in the presence of the donors is total in a very short time, as the short-circuiting process is hindered.

Our research group also reached to very important conclusions concerning the role of dissolved molecular oxygen in the photocatalytic reduction of Cr(VI), which was the object of controversy for many years. In fact, it can be possible to think that O_2 inhibits Cr(VI) reduction, given the likelihood that conduction band electrons are consumed via reaction (5). However, research of our team and some other investigations show no particular effect of O_2 (Domènech and Muñoz, 1987a; Domènech et al., 1986; Navío et al., 1998; Siemon et al., 2002; Testa et al., 2004). There is in this respect important evidence supporting this lack of effect of oxygen:

1. There is no difference in the Cr(VI) photocatalytic reduction efficiency while working under nitrogen or under oxygen (or air) (Siemon et al., 2002; Testa et al., 2004).
2. There is no variation in Cr(VI) photocatalytic reduction efficiency while measured over pure or platinized TiO_2 (Siemon et al., 2002). If O_2 competes with Cr(VI) for conduction band electrons, a faster Cr(VI) reduction should have been obtained over Pt/TiO_2, as platinum reduces the overpotential for oxygen reduction. Experimental results suggest that electron transfer to metallic ion from the conduction band or from Pt is rapid and there is no oxygen mediation requirement.
3. Spectroscopic evidences (an absorption band at 375 nm) of the formation of a charge-transfer complex between Cr(VI) and TiO_2 nanoparticles, showing a strong interaction of the metal ion with the semiconductor surface, were reported in a recent paper (Di Iorio et al., 2008).

The behavior of Cr(VI) contrasts strongly with that of other metal ions such as Hg(II) (see section 4), whose reduction is greatly inhibited by dissolved oxygen. In this sense, it is a unique system and the fact that its photocatalytic reduction can be made in air represents an important technological advantage.

The photocatalytic Cr(VI) reduction has been reported to take place also under visible irradiation. Kyung et al. (2005) found simultaneous and synergistic conversion of Cr(VI) and Acid Orange 7 over TiO_2 using light of $\lambda > 420$ nm, explained by internal light-induced electron transfer in a complex formed between both Cr(VI) and the dye. A similar behavior was observed in the presence of the nonionic surfactant Brij (Cho et al., 2004). Sun et al. (2005) found the simultaneous visible-light Cr(VI) reduction and oxidation of 4-chlorophenol (4-CP) or salicylic acid by TiO_2 irradiation at $\lambda > 400$ nm; the effect was attributed to visible-light excitation of electrons from the valence band to oxygen vacancy states located around 2 eV above the valence band, followed by electron transfer to Cr(VI); alternatively, the transfer could be to surface defects or trap states present at ca. 0.5 eV below the conduction band, from where Cr(VI) may capture electrons.

In the presence of dyes, it is widely accepted that the photocatalytic process under visible light is different from that under UV light. As it can be seen in Figure 4, the dye is excited to a singlet state from which it can inject electrons to the TiO_2 conduction band, leaving behind a radical cation that replaces h_{vb}^+ or $HO^\bullet$ as usual oxidant entities, lowering the oxidizing ability of the system in comparison with that under UV irradiation (Hodak et al., 1996; Meichtry et al., 2008b). However, the reducing power remains intact, making possible Cr(VI) reduction.

In a recent paper, our research group showed that alizarin red chelated to TiO_2 nanoparticles promotes reduction of Cr(VI) to Cr(V) under visible light; Cr(V) formation was confirmed by EPR measurements (Di Iorio et al., 2008).

Another dye, hydroxoaluminiumtricarboxymonoamide phthalocyanine (AlTCPc), adsorbed on TiO_2 particles at different loadings was tested for Cr(VI) photocatalytic reduction under visible irradiation in the presence of 4-CP as sacrificial donor. Direct evidence of the one-electron reduction of Cr(VI) to Cr(V) was also obtained by EPR experiments. The inhibition

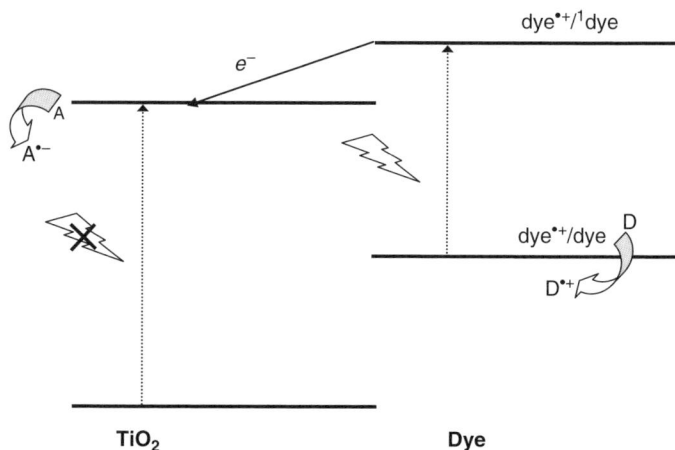

Figure 4 Simplified diagram of the heterogeneous photocatalytic processes occurring under visible irradiation when a dye is attached to a semiconductor.

of formation of ROS in a photocatalytic reductive pathway by the fast trapping of electrons by Cr(VI) and the easier oxidability of 4-CP compared to AlTCPc protects the dye and avoids its photobleaching, making feasible Cr(VI) reduction by the use of solar radiation (Meichtry et al., 2008b).

In conclusion, much experimental work has been done on Cr(VI) photocatalytic reduction since 1999. The advances are related to the elucidation of mechanistic pathways, detection of intermediary species, kinetic calculations, role of dissolved oxygen, and potential use of visible light. However, several interesting points are still worthy of investigation with the aim of optimizing the technology for real use in wastewaters; the presence of synergetic organic compounds such as carboxylic acids or phenols, common constituents of real wastes, makes the process even more attractive.

4. MERCURY

Mercury (II) is a frequent component of industrial wastewaters, remarkably toxic at concentrations higher than $0.005\,mg\,L^{-1}$. The World Health Organization (2006) and national environmental agencies recommend a limit of $0.006\,mg\,L^{-1}$ of inorganic mercury in drinking water. The health hazards due to the toxic effect of mercury at Minamata, Japan, and Iraq are very well known (Bockris, 1997).

The major use of mercury compounds is as agricultural pesticides. It is also used in the chlorine-alkali industry, in paints, as a catalyst in chemical and petrochemical industries, in electrical apparatus, cosmetics, thermometers, gauges, batteries, and dental materials. For this reason, it is a very common pollutant in wastewaters.

Removal of mercuric species in aqueous solutions is difficult because they are hard to be bio- or chemically degraded. At high concentrations, mercury can be removed from the solution by membrane filtration, precipitation with chemicals, ion exchange, adsorption, and reduction (Botta et al., 2002). However, these methods are much less efficient and very expensive for concentrations lower than $100\,mg\,L^{-1}$ (Manohar et al., 2002).

Mercury transformation by heterogeneous photocatalysis with semiconductors (including electrodes, micro- and nanoparticles) such as ZnO, TiO$_2$, WO$_3$ under UV, visible irradiation, and even solar light has been reported in a series of papers. Some of them have been detailed in our previous review (Aguado, et al., 1995; Chen et al., 1997; Clechet et al., 1978; Domènech and Andrés, 1987a, b; Kaluza and Boehm, 1971; Lin et al., 1993; Litter, 1999; Prairie and Stange, 1993; Prairie et al., 1993a, b; Rajeshwar et al., 2002; Serpone et al., 1987, 1988; Tanaka et al., 1986; Tennakone, 1984; Tennakone and Ketipcaralichi, 1995; Tennakone and Wickramanayake, 1986; Tennakone et al., 1993; Wang and Zhuang, 1993), and some others appeared later (Bussi et al., 2002; Horváth and Hegyi, 2001, 2004; Khalil et al., 2002;

Lau et al., 1998; Rader et al., 1994; Skubal and Meshkov, 2002; Wang et al., 2004). It was concluded that the removal efficiency depends strongly on pH, that the reaction is inhibited by oxygen and that there is an enhancement by organic donors.

Recent studies of our research team on TiO_2-photocatalysis of three inorganic mercuric salts, $Hg(NO_3)_2$, $HgCl_2$ and $Hg(ClO_4)_2$, shed light on this complex system (Botta et al., 2002). Depending on the conditions, different products are formed on the photocatalyst surface: Hg(0), HgO, or calomel. Time profiles of Hg(II) concentration with time were characterized by a relatively rapid initial conversion followed by a decrease or an arrest of the rate. Three pH values (3, 7, and 11) were tested, finding that the faster transformation takes place at pH 11 for all salts. Inhibition by oxygen was observed in acid and neutral media but not at basic pH.

The global reaction for metallic mercury deposition is

$$Hg(II) + H_2O \rightarrow 2H^+ + {}^1\!/_2 O_2 + Hg(0) \tag{23}$$

Accordingly, the photocatalytic reaction was found to occur with proton release (Serpone et al., 1987).

In line with the fact that photocatalytic reactions under nonintense photon fluxes take place through monoelectronic steps, a direct reductive mechanism (pathway (a) in Figure 3) involving successive one-electron charge-transfer reactions was postulated (Botta et al., 2002; Custo et al., 2006):

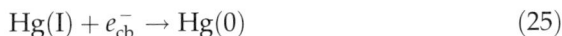

$$Hg(II) + e_{cb}^- \rightarrow Hg(I) \tag{24}$$

$$Hg(I) + e_{cb}^- \rightarrow Hg(0) \tag{25}$$

This reaction sequence explains the formation of calomel when the photocatalytic reaction starts from mercuric chloride. At pH 11, however, calomel was not observed, as it disproportionates to Hg(0) and HgO.

On the other hand, reduced mercury species can be reoxidized by holes or HO$^\bullet$ as follows (Botta et al., 2002; Custo et al., 2006):

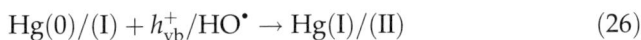

$$Hg(0)/(I) + h_{vb}^+/HO^\bullet \rightarrow Hg(I)/(II) \tag{26}$$

This reoxidation can be hindered by synergistic electron donors. In these cases, mechanism (a) or (b) can take place, according to Figure 3. EDTA was found to be one of the best reductants (Chen and Ray, 2001; Custo et al., 2006), probably because it forms stable complexes with Hg(II). Complexation with citrate also proved to be a very good alternative for Hg(II) removal from water through metallic mercury deposition (Tennakone and Ketipearachchi, 1995). Ethanol enhanced Hg(II) removal dramatically (Horváth and Hegyi, 2001, 2004); SDS was an even better hole scavenger (Horváth and Hegyi, 2004; Horváth et al., 2005). This could have important environmental results for treating domestic and industrial laundry wastewaters containing both Hg(II) and SDS. On the contrary, the cationic CTAB surfactant promoted only slight photoreduction of Hg(II) because repulsion due to its positive charge hinders an efficient reaction (Horváth and Hegyi, 2004; Horváth et al., 2005). Formic acid also enhanced Hg(II) removal, the effect increasing with the organic donor concentration until a limiting value (Wang et al., 2004).

An interesting application of Hg(II) photocatalysis is the use of an activated carbon developed from municipal sewage sludge using $ZnCl_2$ as chemical activation reagent combined with TiO_2. Hg(II) was first photoreduced to Hg(0) and then adsorbed on the carbon and TiO_2 surfaces, with a final recovery of the metal on a silver trap by heating (Zhang et al., 2004).

Another remarkable example is the separation of Cr(VI)/Hg(II) mixtures, which are present in dental office wastewaters (Wang et al., 2004). When Cr(VI) is added to TiO_2 suspensions at pH 4, there is an inhibition of the initial Hg(II) reduction rate. This phenomenon was attributed to catalyst deactivation by Cr(III) species deposited on the TiO_2 surface.

Another interesting application of the photocatalytic technology is the removal (more than 99%) of mercury from hazardous wastes of a chlor-alkali plant. The process begins by acid attack of the solid wastes and treatment of the acid solution under UV irradiation in the presence of TiO_2 and citric acid. The selective precipitation of reduced mercury took place, while the other metal compounds remained in the solution. It was claimed that the final effluents reached a quality close to that of the standards imposed by international environmental agencies (Bussi et al., 2002).

The toxicity of organic mercury compounds, for example, methyl- or phenylmercury, is considerably higher than that of the inorganic species. For example, the massive case of poisoning in Japan, the Minamata Bay incident, was attributed to industrial discharge of organomercurials, and declining bird populations in Sweden was blamed on the use of phenyl- and methylmercurial pesticides as seed dressings (Baughman et al., 1973). Phenylmercury chloride (PMC) and phenylmercury acetate (PMA) were commonly used as pesticides; in particular, PMA was widely used in Argentina until 1971, when it was forbidden.

TiO_2-photocatalytic treatment of methylmercury was tested. It was found that metallic mercury can be deposited only in the presence of methanol and absence of oxygen, according to an indirect photocatalytic reduction (Serpone et al., 1987, 1988).

Interesting applications for treatment of dicyanomercury (II) and tetracyanomercurate (II) ion, high-toxic pollutants coming from precious metal cyanidation processes, were also studied (Rader et al., 1994). Both compounds underwent over 99% removal by TiO_2-photocatalysis in alkaline solutions with deposition of Hg and HgO and oxidation of cyanide to nitrate.

Complete mineralization of the dye mercurochrome (merbromin) by TiO_2 photocatalytic oxidation was found to occur in oxygenated solutions in the presence of citrate, with the corresponding deposition of metallic mercury (Tennakone et al., 1993).

UV/TiO_2 photocatalysis of PMC and PMA in acid aqueous solutions was studied recently by our research group (de la Fournière et al., 2007). Previous work of Prairie et al. (1993b) reports fast PMC mineralization with simultaneous Hg immobilization onto the catalyst. Removal of Hg(II) took place in both cases, with Hg(0) deposition when starting from PMA, and mixtures of Hg(0) and Hg_2Cl_2 when starting from PMC. The reaction was faster at pH 11, with formation of mixtures of Hg and HgO. Oxygen inhibited the reaction. Phenol was detected as a product of both PMA and PMC. It was found that, fortunately, no dangerous methyl- or ethylmercury species were formed in the case of PMA. Calomel formation from PMC under nitrogen reinforces the two successive one-electron transfer reactions, as in the case of inorganic salts. As a result, it can be proposed that the organic moiety plays the role of an electron donor, according to the following sequence of simplified reactions:

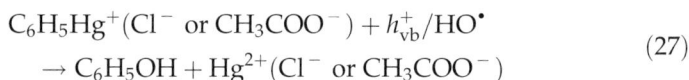

$$C_6H_5Hg^+(Cl^- \text{ or } CH_3COO^-) + h_{vb}^+/HO^{\bullet}$$
$$\rightarrow C_6H_5OH + Hg^{2+}(Cl^- \text{ or } CH_3COO^-) \tag{27}$$

followed by reactions (24) and (25). The anodic pathways would continue through:

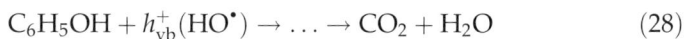

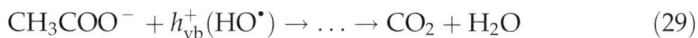

$$C_6H_5OH + h_{vb}^+(HO^{\bullet}) \rightarrow \ldots \rightarrow CO_2 + H_2O \tag{28}$$

$$CH_3COO^- + h_{vb}^+(HO^{\bullet}) \rightarrow \ldots \rightarrow CO_2 + H_2O \tag{29}$$

It is also possible to consider a simultaneous Hg(II) reduction and oxidation of the organic moiety, taking place on the initial organomercurial. However, it is not possible to distinguish between simultaneous or consecutive steps.

In closing, important advances have been performed after 1999 on Hg(II) photocatalysis, especially concerning the highly toxic organomercuric compounds. However, and as said in a previous paper (Botta et al., 2002), it is worthwhile to remark that remediation of Hg(II) in aqueous solutions is hard to attain completely, because of the low levels needed to avoid toxicity (in the order of $\mu g\ L^{-1}$). Thus, very sensitive analytical tools must be used to control the concentration of species in the solution. The physicochemical properties of the products derived from the treatment also introduce serious difficulties. Although zerovalent mercury can be carefully distilled off by mild heating, trapped and recondensed, or it can be dissolved with nitric acid or *aqua regia* for confinement or further treatment of smaller volumes of the effluent, metallic Hg is volatile and somewhat water soluble, HgO is also fairly water soluble, and Hg(I) and Hg(II) nitrates and perchlorates are water soluble.

In spite of these potential complexities, it shall be emphasized that it is always better to have the pollutant immobilized as metallic deposit, treating it later on the solid residue as a hazardous species. It must be also reminded that calomel, if formed, is a less toxic species than $HgCl_2$ with all this leading to a less hazardous chemical system.

5. LEAD

Lead (II) is a toxic metal ion frequently found in wastewater coming from industrial effluents. Lead pollution is mainly anthropogenic and originates in municipal sewages, mining, refining of Pb-bearing ores, chemical manufacture, and other sources. It is a component of insecticides, batteries, water pipes, paints, alloys, food containers, and so on. It has been extensively used as a gasoline additive (tetramethyl- or other alkyl-lead compounds). Although this application has been fortunately forbidden or reduced in most countries, some dangerous wastes could still be present, and they have to be treated. Lead may also be present naturally in groundwater, soils, plants, and animal tissues (Vohra and Davis, 2000). The World Health Organization (2006) and national agencies recommend a maximum of $0.01 \, mg \, L^{-1}$ in drinking water.

Removal of lead from water is performed generally by precipitation as carbonate or hydroxide with or without coagulation. Chelation with EDTA, nitrilotriacetic acid, or other agents is another usual treatment followed by recovery using precipitation, electrolysis, or chemical oxidation (Borrell-Damián and Ollis, 1999). However, most of these treatments are expensive, and some other ways of lead elimination from wastewater are necessary to be developed.

Heterogeneous photocatalysis of Pb(II) systems has received scarce attention. In our previous review (Litter, 1999), we cited a few early papers (Inoue et al., 1978, 1980a, b; Kobayashi et al., 1983; Lawless et al., 1990; Maillard-Dupuy et al., 1994; Rajh et al., 1996a, b; Tennakone, 1984; Tennakone and Wijayantha, 1998; Thurnauer et al., 1997; Torres and Cervera-March, 1992), and although some new papers appeared later (Aarthi and Madras, 2008; Chen and Ray, 2001; Chenthamarakshan et al., 1999; Kabra et al., 2007, 2008; Kobayashi et al., 1983; Mishra et al., 2007; Rajeshwar et al., 2002), information was and continues to be scant. The mechanisms of transformation of lead (II) in water by UV-TiO$_2$ are especially attractive because they depend very much on the reaction conditions, related to the nature of the photocatalyst, the effect of oxygen, and the presence of electron donors.

Recent results of our group (Murruni et al., 2007, 2008) using TiO_2 and Pt–TiO_2 under different conditions tried to shed light on the mechanisms related to this complex system. As in previous reported cases (Chen and Ray, 2001; Kobayashi et al., 1983; Tanaka et al., 1986; Torres and Cervera-March, 1992), a low Pb(II) removal was obtained over pure TiO_2 either in the absence or in the presence of oxygen. The efficiency could be enhanced (1) by using platinized TiO_2 in oxygenated suspensions, (2) by bubbling ozone, and (3) by addition of hole/$HO^•$ scavengers.

All these results can be explained by the occurrence of oxidative or reductive pathways. The oxidative route to Pb(IV) species through hole or $HO^•$ attack (pathway (c), Figure 3) is one possible process. The global reaction is

$$2Pb^{2+} + 2H_2O + O_2 \rightarrow 2PbO_2 + 4H^+ \tag{30}$$

In this route, PbO_2 is formed as the final product, obtained as a dark brown deposit on TiO_2 (Inoue et al., 1980a, b; Murruni et al., 2007; Tennakone, 1984). Considering two consecutive one-electron charge-transfer steps, the first one will be h_{vb}^+ or $HO^•$ attack leading to the trivalent state (Murruni et al., 2007):

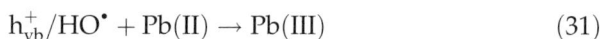

$$h_{vb}^+/HO^• + Pb(II) \rightarrow Pb(III) \tag{31}$$

The one-electron reduction potential for the Pb(III)/Pb(II) couple is not reported, but it is not unreasonable to assume that Pb(II) can be easily oxidized. Pb(III) is unstable, forming Pb(IV) by simple oxidation by O_2 or by stronger oxidants (h_{vb}^+, ROS, etc.); disproportionation is also possible:

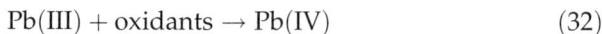

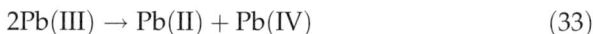

$$Pb(III) + oxidants \rightarrow Pb(IV) \tag{32}$$

$$2Pb(III) \rightarrow Pb(II) + Pb(IV) \tag{33}$$

The conjugate cathodic rate-limiting reaction is O_2 reduction to $O_2^{•-}$, reaction (5). Removal is poor over pure TiO_2 in O_2 or air due to the high overpotential of this reaction. In contrast, the reaction is remarkably rapid over platinized TiO_2, because Pt facilitates the reaction (Chenthamarakshan et al., 1999; Lawless et al., 1990; Murruni et al., 2007; Tanaka et al., 1986; Torres and Cervera-March, 1992). In agreement, the concentration of dissolved O_2 was found to decrease on illumination (Tanaka et al., 1986) and a dependence of the removal rate with O_2 concentration was found (Torres and Cervera-March, 1992). Thus, the oxidative route seems to be the preferred photocatalytic pathway in the absence of electron donors.

Bubbling of ozone improves Pb(II) removal over Pt–TiO_2 and allows the oxidation even on pure TiO_2 in comparison with the same reaction under oxygen (Murruni et al., 2007). It is important to note that there is no Pb(II) removal by O_3 in the absence of photocatalyst, even under UV irradiation. The enhancement by ozone is attributed to the production of a higher concentration of ROS (H_2O_2, $HO^\bullet$, $O_2^{\bullet-}$, $O_3^{\bullet-}$) through reactions (34) to (41), including reaction (7), where the ozonide radical ($O_3^{\bullet-}$) is a relevant species:

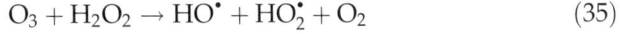

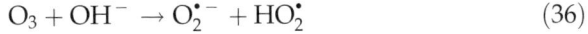

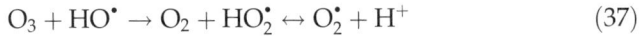

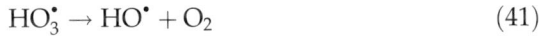

$$O_3 + H_2O + h\nu(\lambda > 300 \text{ nm}) \rightarrow H_2O_2 + O_2 \tag{34}$$

$$O_3 + H_2O_2 \rightarrow HO^\bullet + HO_2^\bullet + O_2 \tag{35}$$

$$O_3 + OH^- \rightarrow O_2^{\bullet-} + HO_2^\bullet \tag{36}$$

$$O_3 + HO^\bullet \rightarrow O_2 + HO_2^\bullet \leftrightarrow O_2^\bullet + H^+ \tag{37}$$

$$O_3 + HO_2^\bullet \rightarrow 2O_2 + HO^\bullet \tag{38}$$

$$2HO_2^\bullet \rightarrow O_2 + H_2O_2 \tag{7}$$

$$O_{3(ads)} + e_{cb}^- \rightarrow O_3^{\bullet-} \tag{39}$$

$$O_3^{\bullet-} + H^+ \rightarrow HO_3^\bullet \tag{40}$$

$$HO_3^\bullet \rightarrow HO^\bullet + O_2 \tag{41}$$

In the ozone systems, $PbO_{1.37}$ and PbO_2 are identified as deposits on the catalyst.

It is also possible to propose a direct reductive route driven by TiO_2-conduction band electrons through successive one-electron reduction steps finishing in metallic lead (pathway (a), Figure 3), with oxidation of water by holes or $HO^\bullet$, ending in protons and oxygen, reaction (13):

$$Pb(II) + e_{cb}^- \rightarrow Pb(I) \tag{42}$$

$$Pb(I) + e_{cb}^- \rightarrow Pb(0) \tag{43}$$

$$4h_{vb}^+/HO^\bullet + 2H_2O \rightarrow O_2 + 4H^+ \tag{13}$$

The global reaction releases protons as follows:

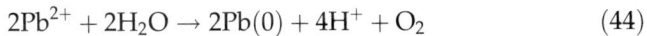

$$2Pb^{2+} + 2H_2O \rightarrow 2Pb(0) + 4H^+ + O_2 \tag{44}$$

However, the direct reductive route by TiO_2 conduction band electrons is not feasible, considering that the one-electron reduction potential of the couple Pb(II)/Pb(I) is very negative ($E^0 = -1.0$ V, Breitenkamp et al., 1976). Accordingly, Pb shows essentially no or a very weak tendency to accept photogenerated electrons from TiO_2, as said in Section 2. This reduction pathway was not observed even using platinized TiO_2 under nitrogen. (Chenthamarakshan et al., 1999; Lawless et al., 1990; Murruni et al., 2007).

In fact, direct reduction of Pb(II) to Pb(0) by a bielectronic process has been reported under laser irradiation, where due to the high photonic frequency, accumulation of electrons may allow multielectronic injection (Rajh et al., 1996a, b; Thurnauer et al., 1997).

The reductive mechanism may take place, however, through the indirect pathway (b) (Figure 3), in the presence of hole/HO$^\bullet$ scavengers, that is, alcohols or carboxylic acids that form powerful reducing intermediates. Methanol, ethanol, 2-propanol, 1-butanol, t-butanol, and citric acid and formic acid were tested successfully under nitrogen and over pure TiO_2 (Chenthamarakshan et al., 1999; Lawless et al., 1990; Murruni et al., 2007, 2008). Reactions (16) and (11) took place, where M^{n+} and $M^{(n-1)+}$ are Pb^{2+} and Pb^+, respectively, while R$^\bullet$ are the reducing species formed from the organics. Once a small amount of Pb^+ is formed, easy reduction (E^0(Pb(I)/ Pb(0)) = +0.75 V, Breitenkamp et al., 1976) or disproportionation (reaction (45)) takes place:

$$2Pb(I) \rightarrow Pb(II) + Pb(0) \qquad (45)$$

Formic acid and 2-propanol promoted the highest efficiency. However, formic acid is considered a better additive because it can be used at lower concentrations, and it does not introduce toxic degradation products in the system (Murruni et al., 2008). In those cases, no lead deposits on the photocatalyst are obtained, but stains were observed on the lamp surface, composed mainly by colloidal zerovalent Pb, as demonstrated by X-ray diffraction (XRD) and X-ray photoelectron spectroscopy (XPS) analysis.

Reactions in the presence of electron donors are inhibited by oxygen or air, because of the competition of O_2 with Pb(II) for the reducing species (Kabra et al., 2007, 2008). In these conditions, the oxidative mechanism leading to Pb(IV) is not feasible, considering that the donor competes with Pb(II) for the oxidant, leading to a decrease more than to an enhancement of Pb(II) removal.

In conclusion, although an oxidative Pb(II) removal from water could be viable, leading to immobilization by formation of lead oxides, the use of hole/HO$^\bullet$ scavengers can be considered the most effective and economic method. This approach avoids expensive platinization of the catalyst or use of ozone. In addition, reactions can be conducted under air, and lead (II) and organic scavengers (alcohols, carboxylates) may be present together in industrial wastes, with this approach having the potential of providing economical methods for the removal of the very toxic lead-related water pollutants.

6. URANIUM

Uranium occurs naturally in the III, IV, V, and VI oxidation states, but the most common natural form is the hexavalent one, as uranyl ion (UO_2^{2+}). Natural uranium is a mixture of three radionuclides, ^{234}U, ^{235}U, and ^{238}U, with higher proportion of the last one. Uranium is present in the nature in granite and other mineral deposits. In some groundwaters, levels of uranium reach 50 mg L^{-1} (Noubactep et al., 2006) and considerable concentrations of natural uranium and its radionuclide daughters (radium, polonium, and lead) can be present in well waters used for drinking purposes. In oxic waters, U is present as UO_2^{2+} or hydroxycomplexes, although it easily forms complexes with carbonates, phosphates, and sulfates. The main applications of uranium are as fuel for nuclear power plants, in catalysts, and in pigments. Pollution sources are lixiviation of deposits, emissions from the nuclear industry, combustion of coal or other fuels, and U-containing phosphate fertilizers (WHO, 2004).

In human beings, uranium can provoke nephritis, and it is considered carcinogen, causing bone cancer. These consequences are even more noxious than radiological risks (Katsoyiannis, 2007). The World Health Organization (2006) and national regulatory agencies recommend no more than 0.015 mg L^{-1} in drinking water.

Removal methods of uranium from water are ionic exchange, ultrafiltration, adsorption on granular ferric hydroxide, iron oxides (ferrihydrite, hematite, magnetite, and goethite), activated carbon or TiO_2; evaporation, biorreduction, and zerovalent iron have been also tested (Behrends and van Cappellen, 2005; Cantrell et al., 1995, Chen et al. 1999; Charlet et al., 1998; Gu et al., 1998; Kryvoruchko et al., 2004; Liger et al., 1999; Mellah et al., 2006; Naftz et al., 2002; Noubactep et al., 2006; O'Loughlin et al., 2003; Vaaramaa and Lehto, 2003).

There are very few reports in the literature concerning heterogeneous photocatalysis for uranium treatment in water. In our previous review, only one case of photocatalytic reaction on uranium salts was reported (Amadelli et al., 1991). Taking into account the standard reduction potentials, U(VI) can be photocatalytically reduced by TiO_2 conduction band electrons to U(V) and then to U(IV) ($E^0 - | 0.16$ V and +0.58 V, respectively, Bard et al., 1985). However, more reduced U(III) and U(0) forms cannot be generated because of very negative redox potentials (Bard et al., 1985). In addition, U(V) rapidly disproportionates to U(VI) and U(IV), and its chemistry is very complex (Selbin and Ortego, 1969). For example, uranyl carbonate, which can be easily formed in natural or wastewaters, has a more negative standard redox potential than the aqueous uranyl ($E^0 = -0.69$ V, Grenthe et al., 2004), making unfeasible its reduction to the pentavalent state via conduction band electrons. Accordingly, Amadelli et al. (1991) obtained photoreduction of uranyl solutions on illuminated TiO_2 suspensions and electrodes only in the presence of hole scavengers (2-propanol, acetate, or formate), obtaining an uranium oxide of stoichiometry close to U_3O_8. Complexation of uranyl with the hole scavengers was found to play an essential role in the photoredox process, and, in addition, this indicates the predominance of pathway (b) of Figure 3. Similar results are presented in by Chen et al. (1999). No uranium deposition in the presence of air was observed; however, in the presence of EDTA and absence of oxygen, a reductive deposition took place. Further exposure to air reoxidizes and redissolves the uranium species. Platinization of TiO_2 enhanced the reaction only slightly, confirming the predominance of a reductive process. This method was proposed for recovering uranyl from aqueous solutions of dilute uranium(VI)-EDTA species, which are usually present in wastewaters of nuclear power stations.

Another work indicates the possibility of photocatalytic treatment of the wastes of nuclear fuels with separation of Np, Pu, and U (Boxall et al., 2005).

While mechanistic studies on the photocatalytic removal of uranium-related species are scarce and merit further research, additional investigation is mandatory to clarify major reaction engineering issues following this approach.

7. ARSENIC

The toxicity of the metalloid arsenic is well known. Ingestion of more than 100 mg of the element causes acute poisoning, but ingestion of small amounts during a long period causes the occurrence of chronic regional endemic hydroarsenicism ("hidroarsenicismo crónico regional endémico", HACRE, in Spanish), which produces skin alterations and cancer (Manahan, 1990; Rajeshwar, 1995). The World Health Organization (2006) and most national regulatory agencies recommend $10 \mu g$ L^{-1} as the maximum allowable arsenic concentration in drinking water.

Arsenic pollution can be originated in anthropic activities (mining, use of biocides, wood preservers). However, most pollution is natural, coming from mineral dissolution in surface or groundwaters (Bundschuh et al., 2000, 2004; Litter, 2002). Predominant As forms in natural ground and surfacewaters (neutral pH) are arsenate (As(V), as $H_2AsO_4^-$ and $HAsO_4^{2-}$) and arsenite (As(III), as neutral $H_3AsO_3^0$). The mobility of arsenical forms in waters is very dependent on pH, Eh conditions, and presence of different chemical species (Smedley et al., 2002). Consequently, removal methods must take into account these physicochemical properties. While As(V) can be easily removed by conventional ion exchange and adsorption techniques, As(III) removal is difficult due to its nonionic form in aqueous solutions at pH < 9. For this reason, treatment includes generally an oxidation step with agents such as chlorine, permanganate, H_2O_2, solid MnO_2, iron chloride, and even microorganisms. Conventional treatments are oxidation–coagulation–adsorption processes on flocs of iron or aluminum hydroxides, ionic interchange, activated alumina, lime softening, and reverse electrodialysis and osmosis (Emett and Khoe, 2001; Ingallinella et al., 2003). There are emergent technologies, such as adsorption on granular iron hydroxides or TiO_2, micro- and nanofiltration, bacteriogenic iron oxides, desalinization by evaporation (see, e.g., Bang et al., 2005; Castro de Esparza, 1999; Cumbal and SenGupta, 2005; Dávila-Jiménez et al., 2008; Driehaus et al., 1998; Guan et al., 2008; Haque et al., 2008; Liu et al., 2008; Sancha, 1996; Zhang et al., 2008).

Photocatalysis of arsenical systems were not described in our previous review (Litter, 1999). However, in the meantime, together with the increasing relevance of the subject, some papers appeared in the literature. Oxidative TiO_2 heterogeneous photocatalysis through pathway (c) (Figure 3) can be a convenient option for As(III) oxidation. The global reaction at alkaline pH is indicating a pH decrease with the reaction time:

$$HO^- + H_2AsO_3^- + \tfrac{1}{2} O_2 \rightarrow HAsO_4^{2-} + H_2O \qquad (46)$$

Monoelectronic steps with formation of As(IV) are suggested. The reduction potential of the couple As(IV)/As(III) is $E^0 \approx +2.4\,V$ (Kläning et al., 1989). Therefore, formation of As(IV) by attack of holes or $HO^{\bullet}$ (reaction (12)) is thermodynamically possible. However, reduction of As(III) or As(V) by TiO_2 conduction band electrons, which could lead to the less mobile elemental As, seems to be not feasible, as judged from the recent stopped-flow experiments developed by our research team (Meichtry et al., 2008a). This is in agreement with the highly negative reduction potential of the As(V)/As(IV) couple ($E^0 \approx -1.2\,V$, Kläning et al., 1989). The values for the subsequent monoelectronic couples until As(0) are not known.

Accordingly to the above conclusions, some examples of oxidative As(III) removal using TiO_2-heterogeneous photocatalysis under UV light are reported (Bissen et al., 2001; Dutta et al., 2005; Jayaweera et al., 2003; Lee and Choi, 2002; Ryu and Choi, 2004; Yang et al., 1999). In all cases, As(III) oxidation was very rapid, taking place in time scales of 10–100 min at various concentrations from the micromolar to the millimolar range. The reaction rate did not depend on pH, at least between 5 and 9. Addition of Fe(III), polyoxometalates, fluoride and humic acids, as well as platinization of TiO_2, increased the oxidation rate. The reaction was very fast in the presence of oxygen but not completely inhibited in its absence.

Mechanisms for As(III) oxidation were analyzed by several authors (Dutta et al., 2005; Ferguson et al., 2005; Ryu and Choi, 2004). In effect, the participation of As(IV) species is proposed, and the possible oxidants are $HO^{\bullet}$, trapped holes, or superoxide radicals. However, by the use of $HO^{\bullet}$ scavengers and superoxide dismutase, participation of $HO^{\bullet}$ is ruled out and superoxide radical is proposed to be the main oxidant, according to the following mechanism (Ferguson et al., 2005; Lee and Choi, 2002; Ryu and Choi, 2004):

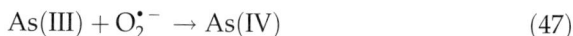

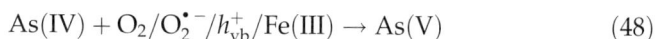

$$As(III) + O_2^{\bullet -} \rightarrow As(IV) \tag{47}$$

$$As(IV) + O_2/O_2^{\bullet -}/h_{vb}^{+}/Fe(III) \rightarrow As(V) \tag{48}$$

However, in the presence of Fe(III), the participation of $HO^{\bullet}$ in the oxidative pathway was proved (Lee and Choi, 2002).

Zhang and Itoh (2006) described a low-cost, environmentally friendly adsorbent for As(III) photocatalytic removal, formed by a mixture of TiO_2 and slag-iron oxide obtained from an incinerator of solid wastes. Arsenite is first oxidized to arsenate in a fast process, followed by a slow adsorption of arsenate, although the material shows an adsorbent capacity higher than that of pure anatase.

Ferguson and Hering (2006) reported a method to oxidize As(III) in a fixed-bed, flow-through reactor with TiO_2 immobilized on glass beads. The reactor residence time, the influent As(III) concentration, the number of TiO_2 coatings on the beads, the solution matrix, and the light source were varied to characterize the reaction and determine its feasibility for water treatment. A reactive transport model with rate constants proportional to the incident light at each bead layer fitted reasonably the experimental data. The reaction was also effective under natural sunlight.

Rapid and complete As(III) oxidation was obtained in our laboratory by TiO_2-photocatalysis and further addition of Fe(III) salts (Mateu, 2007). The influence of photocatalyst mass, pH, and As(III) concentration were analyzed. It was found that dissolved oxygen concentration was one of the most important factors for the reproducibility of the experiments. Addition of H_2O_2 increased the rate, while addition of Fe(III) at the beginning of irradiation caused complete As(III) removal from the suspension.

In a recent work (Morgada de Boggio et al., 2006, 2008, 2009), our research group studied the photocatalytic removal of As(III) and As(V) in PET plastic bottles impregnated with TiO_2 and exposed to artificial UV light for 6 h. Before or after irradiation, nongalvanized packing wire was added. After 24 h in the dark, an As removal higher than 80% was obtained. Experiments with well water samples of Las Hermanas (Tucumán Province,

Argentina) resulted in As removal higher than 94%. Fostier et al. (2008) obtained similar results.

As already said, As(V) or As(III) transformation by a direct reductive pathway driven by TiO_2 conduction band electrons is not thermodynamically possible. However, complete As(V) removal in the presence of methanol under N_2 at pH 3 was successful, indicating the participation of an indirect reductive mechanism (Yang et al., 1999). As an evidence, XPS measurements revealed the presence of elemental As deposited on the photocatalyst. Then, the authors proposed a viable scheme for TiO_2-photocatalytic As removal in two stages (As(III) $\rightarrow$ As(V) $\rightarrow$ As(0)) with only changing pH and methanol addition.

Summarizing, although the photocatalytic mechanisms for As removal have been analyzed, it is felt that application of photocatalysis is still in an early stage and more research studies for the possible application are necessary. The reductive mechanism is promissory in this sense.

8. CONCLUSIONS

Based on the information reported, the following can be concluded:

(a) Heterogeneous photocatalytic treatment of metals and metalloids can be a valuable option for removal of metals and metalloids from water in some especial cases, without use of expensive reagents and with the possibility of employing costless solar light. An overview of the literature on the subject for chromium, mercury, lead, uranium, and arsenic indicates that, in general, much important fundamental and applied research is still missing.

(b) On a thermodynamical basis, three types of mechanisms can take place, depending on the nature of the ion species, all of them involving one-electron charge-transfer processes to initiate the process: (a) a direct reduction driven by conduction band electrons, (b) an indirect reduction promoted by electron donors, or (c) an oxidative process produced by holes or hydroxyl radicals.

(c) For chromium (VI), a large quantity of papers has been published since 20 years ago involving photocatalytic transformation under UV light. The usual mechanism is direct reduction by conduction band electrons, which is very slow in the absence of electron donors but can be accelerated by organic electron donors. In many cases, the organic compounds are present simultaneously with Cr(VI) in wastewaters as a result of different industrial processes. Because Cr(VI) photocatalytic reduction is not inhibited by oxygen, this represents an additional advantage for the application. Recently, some photocatalytic reactions under visible light with good yields were observed, making possible the potential use of costless solar light. Mechanistically, several questions remain unanswered, and investigations merit to be continued.

(d) For mercury (II) photocatalysis, fewer examples are reported. The reaction occurs through direct reduction driven by conduction band electrons; electron donors cause an enhancement of the reaction rate. The features of the reaction are very dependent on the conditions of the medium and the nature of the starting mercuric compound; the final products can be metallic Hg, HgO, or calomel. Interesting and encouraging results have been found in the case of the extremely toxic organomercurial species.

(e) Lead (II) is a very motivating photocatalytic system. Although removal from water can take place by an oxidative pathway, over pure TiO_2 this route is poor and can be enhanced only by platinization or use of ozone, both expensive methodologies. On the other hand, the direct reductive route is unfeasible because reduction of Pb(II) to the monovalent state is highly energetic. The indirect route, driven by reducing species formed in the presence of alcohols or carboxylates is a viable route. These species can be present together with Pb(II) in wastewaters, rendering the technology appropriate for Pb(II) removal.

(f) Photocatalytic removal of uranium salts from water was scarcely studied. Photoreduction was possible only in the presence of hole scavengers in the absence of oxygen with the formation of uranium oxides; however, a rapid reoxidation and redissolution took place after exposure to air. Comprehensive studies should, however, be performed to fully demonstrate the possible application of the technology.

(g) The reaction of arsenic species in photocatalytic systems is also interesting because both oxidative and reductive mechanisms may lead to less toxic or solid phases. Although the oxidative system has been studied rather well, the reductive pathway must be also the goal of new research in order to find a best of application of this alternative route for As removal.

ACKNOWLEDGMENT

Work performed as part of Agencia Nacional de la Promoción de la Ciencia y la Tecnología PICT 00-512 Project and CYTED-IBEROARSEN 406RT0282. MIL is a member from CONICET.

REFERENCES

Aarthi, T., and Madras, G. *Catal. Comm.* **9**, 630 (2008).
Aguado, M.A., Cervera-March, S., and Giménez, J. *Chem. Eng. Sci.* **50**, 1561 (1995).
Aguado, M.A., Giménez, J., and Cervera-March, S. *Chem. Eng. Comm.* **104**, 71 (1991).
Aguado, M.A., Giménez, J., Simarro, R., and Cervera-March, S. *Sol. Energy* **49**, 47 (1992).

Amadelli, R., Maldotti, A., Sostero, S., and Carassiti, V. *J. Chem. Soc. Faraday Trans.* **87**, 3267 (1991).

Baba, R., Konda, R., Fujishima, A., and Honda, K. *Chem. Lett.* 1307 (1986).

Bahnemann, D.W., Cunningham, J., Fox, M.A., Pelizzetti, E., Pichat, P., and Serpone, N. Photocatalytic Treatment of Waters *in* G.R. Helz, R.G. Zepp, and D.G. Crosby (Eds.), "Aquatic and Surface Photochemistry". Lewis Publ., Boca Raton, FL (1994), p. 261.

Bahnemann, D.W., Mönig, J., and Chapman, R. *J. Phys. Chem.* **91**, 3782 (1987).

Bang, S., Patel, M., Lippincott, L., and Meng, X. *Chemosphere* **60**, 389 (2005).

Bard, A.J., Parsons, R., and Jordan, J. (Eds.), "Standard Potentials in Aqueous Solution", Marcel Dekker, Inc., New York (1985).

Baughman, G.L., Gordon, J.A., Lee Wolfe, N., and Zepp, R.G. "Chemistry of Organomercurials in Aquatic Systems", EPA-660/3-73-012 (1973) U.S. Environ. Protect. Agency, Office of Research and Development, Washington, D.C., 1973.

Behrends, T., and van Cappellen, P. *Chem. Geol.* **220**, 5642 (2005).

Bissen, M., Vieillard-Baron, M.-M., Schindelin, A.J., and Frimmel, F.H. *Chemosphere* **44**, 751 (2001).

Bockris, J.O.M., "Environmental Chemistry". Plenum Press, New York (1997).

Borrell-Damián, L., and Ollis, D.F. *J. Adv. Oxid. Technol.* **4**, 125 (1999).

Botta, S.G., Rodríguez, D.J., Leyva, A.G., and Litter, M.I. *Catal. Today* **76**, 247 (2002).

Botta, S., Navío, J.A., Hidalgo, M.,C., Restrepo, G.M., and Litter, M.I. *J. Photochem. Photobiol. A* **129**, 89 (1999).

Boxall, C., Le Gurun, G., Taylor, R.J., and Xiao, S. "The Handbook of Environmental Chemistry", Springer Berlin/Heidelberg, Chemistry and Materials Science, Volume 2M, Environmental Photochemistry Part II, ISBN: 3-540-00269-3, (2005), pp. 451–481.

Breitenkamp, M., Henglein, A., and Lilie, J. *Ber. Bunsen-Ges. Phys. Chem.* **80**, 973 (1976).

Bundschuh, J., Bonorino, G., Viero, A.P., Albouy, R., and Fuertes, A. (2000). Arsenic and other trace elements in sedimentary aquifers in the Chaco-Pampean Plain, Argentina: Origin, distribution, speciation, social and economic consequences. In: Bhattacharya, P., Welch, A.H. (Eds.), Arsenic in Groundwater of Sedimentary Aquifers, Pre-Congress Workshop, 31st Internat. Geol. Cong., Rio de Janeiro, Brazil, pp. 27–32.

Bundschuh, J., Farias, B., Martin, R., Storniolo, A., Bhattacharya, P., Cortes, J., Bonorino, G., and Albouy, R. *Appl. Geochem.* **19**, 231 (2004).

Bussi, J., Chanian, M., Vázquez, M., and Dalchiele, E.A. *J. Environ. Eng.* **128**, 733 (2002).

Cantrell, J.L., Kaplan, D.I., and Wietsma, T.W. *J. Hazard. Mater.* **42**, 201 (1995).

Cappelletti, G., Bianchi, C.L., and Ardizzone, S. *Appl. Catal. B* **78**, 193 (2008).

Castro de Esparza, M.L. "Remoción de arsénico a nivel domiciliario". CEPIS-OPS-OMS: Hojas de Divulgación Técnica No. **74**, 1 (1999).

Charlet, L., Silvester, E., and Liger, E. *Chem. Geol.* **151**, 85 (1998).

Chen, D., and Ray, A.K. *Chem. Eng. Sci.* **56**, 1561 (2001).

Chen, J., Ollis, D.F., Rulkens, W.H., and Bruning, H. *Colloids Surf A* **151**, 339 (1999).

Chen, L.X., Rajh, T., Wang, Z., and Thurnauer, M.C. *J. Phys. Chem. B* **101**, 10688 (1997).

Chenthamarakshan, C.R., and Rajeshwar, K. *Langmuir* **16**, 2715 (2000).

Chenthamarakshan, C.R., Yang, H., Savage, C.R. and Rajeshwar, K. *Res. Chem. Intermed.* **25**, 861 (1999).

Cho, Y., Kyung, H., and Choi, W. *Appl. Catal. B* **52**, 23 (2004).

Clechet, P., Martelet, C., Martin, J.-R., and Olier, R. *C. R. Acad. Sci., Paris, Ser. B* **287**, 405 (1978).

Colón, G., Hidalgo, M.C., and Navío, J.A. *J. Photochem. Photobiol. A* **138**, 79 (2001a).

Colón, G., Hidalgo, M.C., and Navío, J.A. *Langmuir* **17**, 7174 (2001b).

Cumbal, L., and SenGupta, A.K., *Environ. Sci. Technol.* **39**, 6508 (2005).

Custo, G., Litter, M.I., Rodriguez, D., and Vázquez, C. *Spectrochim. Acta Part B* **61**, 1119 (2006).

Das, D.P., Parida, K., and De, B.R. *J. Mol. Catal. A* **245**, 217 (2006).

Dávila-Jiménez, M.M., Elizalde-González, M.P., Mattusch, J., Morgenstern, P., Pérez-Cruz, M.A., Reyes-Ortega, Y., Wennrich, R., and Yee-Madeira, H. *J. Colloid Interface Sci.* **322**, 527 (2008).

de la Fournière, E.M., Leyva, A.G., and Litter, M.I. *Chemosphere* **69**, 682 (2007).

Di Iorio, Y., San Román, E., Litter, M.I., and Grela, M.A.*J. Phys. Chem. C.* 112 (42), 16532–16538, 2008. DOI: 10.1021/jp8040742.

Domènech, X. *in* D.F. Ollis, and H. Al-Ekabi (Eds.), "Photocatalytic Purification and Treatment of Water and Air", Photocatalysis for aqueous phase decontamination: Is TiO$_2$ the better choice?. Elsevier Sci. Publish. B.V., Amsterdam (1993), p. 337.

Domènech, J., and Muñoz, J. *Electrochim. Acta* **32**, 1383 (1987a).

Domènech, J., and Muñoz, J. *J. Chem. Res. (S)* 106 (1987b).

Domènech, X., and Muñoz, J. *J. Chem. Tech. Biotechnol.* **47**, 101 (1990).

Domènech, J., and Andrés, M. *Gazz. Chim. Ital.* **117**, 495 (1987a).

Domènech, J., and Andrés, M. *New. J. Chem.* **11**, 443 (1987b).

Domènech, J., Andrés, M., and Muñoz, J. *Tecnología del Agua* **29**, 35 (1986).

Driehaus, M., Jekel, M., and Hildebrandt, U. *J. Water SRT-Aqua* **47**, 30 (1998).

Dutta, P.K., Pehkonen, S.O., Sharma, V.K., and Ray, A.K. *Environ. Sci. Technol.* **39**, 1827 (2005).

Emeline, A.V., Ryabchuk, V.K., and Serpone, N. *J. Phys. Chem. B* **109**, 18515 (2005).

Emett, M.T., and Khoe, G.H. *Water Res.* **35**, 649 (2001).

Ferguson, M.A., and Hering, J.G. *Environ. Sci. Technol.* **40**, 4261 (2006).

Ferguson, M.A., Hoffmann, M.R., and Hering, J.G. *Environ. Sci. Technol.* **39**, 1880 (2005).

Fostier, A.H., Silva Pereira, M.S., Rath, S., Guimarães, J.R. *Chemospere* **72**, 319 (2008).

Fu, H., Lu, G., and Li, S. *J. Photochem. Photobiol. A* **114**, 81 (1998).

Fujishima, A., and Zhang, X. *C. R. Chimie* **9**, 750 (2006).

Giménez, J., Aguado, M.A., and Cervera-March, S. *J. Mol. Catal.* **105**, 67 (1996).

Goeringer, S., Chenthamarakshan, C.R., and Rajeshwar, K., *Electrochem. Commun.* **3**, 290 (2001).

Grela, M.A., and Colussi, A.J. *J. Phys. Chem.* **100**, 18214 (1996).

Grenthe, I., Fuger, J., Konings, R.J.M., Lemire, R.J., Muller, A.B., Trung, C.N., and Wanner, H. "Chemical Thermodynamics of Uranium". Elsevier, North-Holland imprint (2004).

Gu, B., Liang, L., Dickey, M.J., Yin, X., and Dai, S. *Environ. Sci. Technol.* **32**, 3366 (1998).

Guan, X.-H., Wang, J., and Chusuei, C.C. *J. Hazard. Mater.* **156**, 178 (2008).

Haque, N., Morrison, G., Cano-Aguilera, I., and Gardea-Torresdey, J.L. *Microchemical J.* **88**, 7 (2008).

Hidalgo, M.C., Colón, G., Navío, J.A., Macías, M., Kriventsov, V.V., Kochubey, D.I., and Tsodikov, M.V. *Catal. Today* **128**, 245 (2007).

Hodak, J., Quinteros, C., Litter, M.I., and San Román, E. *J. Chem. Soc., Faraday Trans.* **92**, 5081 (1996).

Hoffmann, M.R., Martin, S.T., Choi, W., and Bahnemann, D. *Chem. Rev.* **95**, 69 (1995).

Horváth, O., and Hegyi, J. *Progr. Colloid Polym. Sci.* **117**, 211 (2001).

Horváth, O., and Hegyi, J. *Progr. Colloid Polym. Sci.* **125**, 10 (2004).

Horváth, O., Bodnár, E., and Hegyi, J. *Colloids Surf A* **265**, 135 (2005).

Ingallinella, A.M., Fernández, R., and Stecca, L. *Rev. Ingeniería Sanitaria y Ambiental (AIDIS Argentina)*, Part 1: **66**, 36 (2003), and Part 2: **67**, 61 (2003).

Inoue, T., Fujishima, A., and Honda, K. *Chem. Lett.* 1197 (1978).

Inoue, T., Fujishima, A., and Honda, K. *J. Electrochem. Soc.* **127**, 324 (1980a).

Inoue, T., Fujishima, A., and Honda, K. *J. Electrochem. Soc.* **127**, 1582 (1980b).

Jayaweera, P.M., Godakumbra, P.I., and Pathiartne, K.A.S. *Curr. Sci. India* **84**, 541 (2003).

Jiang, F., Zheng, Z., Xu, Z., Zheng, S., Guo, Z., and Chen, L. *J. Hazard. Mater.* **134**, 94 (2006).

Kabra, K., Chaudhary, R., and Sawhney, R.L. *J. Hazard. Mater.* **149**, 680 (2007).

Kabra, K., Chaudhary, R., and Sawhney, R.L. *J. Hazard. Mater.* **155**, 424 (2008).

Kajitvichyanukul, P., Ananpattarachai, J., and Pongpom, S. *Sci. Technol. Adv. Mater.* **6**, 352 (2005).

Kaluza, U., and Boehm, H.P. *J. Catal.* **22**, 347 (1971).

Kanki, T., Yoneda, H., Sano, N., Toyoda, A., and Nagai, C. *Chem. Eng. J.* **97**, 77 (2004).

Katsoyiannis, I.A. *J. Hazard. Mater. B* **139**, 31 (2007).

Khalil, L.B., Mourad, W.E., and Rophael, M.W. *Appl. Catal. B* **17**, 267 (1998).

Khalil, L.B., Rophael, M.W., and Mourad, W.E. *Appl. Catal. B* **36**, 125 (2002).

Kläning, U.K., Bielski, B.H.J., and Sehesteds, K. *Inorg. Chem.* **28**, 2717 (1989).

Kobayashi, T., Taniguchi, Y., Yoneyama, H., and Tamura, H. *J. Phys. Chem.* **87**, 768 (1983).

Kryvoruchko, A.P., Yurlova, L.Y., Atamanenko, I.D., Kornilovich, B.Y. *Desalination* **162**, 229 (2004).

Ku, Y., and Jung, I.-L. *Water Res.* **35**, 135 (2001).

Kyung, H., Lee, J., and Choi, W. *Environ. Sci. Technol.* **39**, 2376 (2005).

Lau, L.D., Rodríguez, R., Henery, S., and Manuel, D. *Environ. Sci. Technol.* **32**, 670 (1998).

Lawless, D., Res, A., Harris, R., and Serpone, N. *Chem. Ind. (Milan)* **72**, 139 (1990).

Lee, H., and Choi, W. *Environ. Sci. Technol.* **36**, 3872 (2002).

Legrini, O., Oliveros, E., and Braun, A.M., *Chem. Rev.* **93**, 671 (1993).

Liger, E., Charlet, L., and van Cappellen, P. *Geochim. Cosmochim. Acta* **63**, 2939 (1999).

Lin, W.-Y., and Rajeshwar, K. *J. Electrochem. Soc.* **144**, 2751 (1997).

Lin, W.-Y., Wei, C., and Rajeshwar, K., *J. Electrochem. Soc.* **140**, 2477 (1993).

Linsebigler, A.L., Lu, G., and Yates Jr., J.T. *Chem. Rev.* **95**, 735 (1995).

Litter, M.I. *Appl. Catal. B* **23**, 89 (1999).

Litter, M.I. (Ed.), Prospect of Rural Latin American Communities for application of Low-cost Technologies for Water Potabilization, ASO Project AE 141/2001. Digital Grafic, La Plata, (2002), http://www.cnea.gov.ar/xxi/ambiental/agua-pura/default.htm.

Liu, G., Zhang, X., Talley, J.W., Neal, C.R., and Wang, H. *Water Res.* **42**, 2309 (2008).

Maillard-Dupuy, C., Guillard, C., and Pichat, P., *New J. Chem.* **18**, 941 (1994).

Manahan, S.E., "Environmental Chemistry", 4th ed. Lewis Publishers, Chelsea (1990).

Manohar, D.M., Krishnan, K.A., and Anirudhan, T.S. *Water Res.* **36**, 1609 (2002).

Martin, S.T., Herrmann, H., and Hoffmann, M.R. *J. Chem. Soc., Faraday Trans.* **90**, 3323 (1994).

Mateu, M., "Remoción de arsénico por fotocatálisis heterogénea", in English: "Arsenic removal by heterogeneous photocatalysis" Master Thesis, University of Buenos Aires, Argentina (2007).

Meichtry, J.M., Brusa, M., Mailhot, G., Grela, M.A., and Litter, M.I. *Appl. Catal. B* **71**, 101 (2007).

Meichtry, J.M., Dillert, R., Bahnemann, D., and Litter, M.I., unpublished results (2008a).

Meichtry, J.M., Rivera, V., Di Iorio, Y., Rodríguez, H.B., San Román, E., Grela, M.A., and Litter, M.I. submitted to *Photochem. Photobiol. Sci.* (2009, in press). DOI: 10.1039/b816441j.

Mellah, A., Chegrouche, S., and Barkat, M. *J. Colloid Interface Sci.* **296**, 434 (2006).

Mills, A., and Le Hunte, S. *J. Photochem. Photobiol. A* **108**, 1 (1997).

Mills, A., and Valenzuela, M.A. *Revista Mexicana de Física* **50**, 287 (2004).

Mishra, T., Hait, J., Noor Aman, R.K., Jana, S., and Chakravarty, *J. Colloid Interface Sci.* **316**, 80 (2007).

Miyake, M., Yoneyama, H., and Tamura, H. *Bull. Chem. Soc. Jpn.* **50**, 1492 (1977).

Mohapatra, P., Samantaray, S.K., and Parida, K. *J. Photochem. Photobiol. A* **170**, 189 (2005).

Morgada de Boggio, M.E., Levy, I.K., Mateu, M., and Litter, M.I. Low-cost technologies based on heterogeneous photocatalysis and zerovalent iron for arsenic removal in the Chaco-pampean Plain, Argentina in M.I. Litter (Ed.), "Final Results of OEA/AE141 Project: Research, Development, Validation and Application of Solar Technologies for Water Potabilization in Isolated Rural Zones of Latin America and the Caribbean", Cap.1. págs. 11–37. ISBN 978-987-22574-4-6. http://www.cnea.gov.ar/xxi/ambiental/agua-pura/default.htm. (2006).

Morgada de Boggio, M.F., Levy, I.K., Mateu, M., Bhattacharya, P., Bundschuh, J., and Litter, M.I. Low-cost technologies based on heterogeneous photocatalysis and zerovalent iron for arsenic removal in the Chacopampean plain, Argentina. *in* J. Bundschuh, M.A. Armienta,

P. Bhattacharya, J. Matschullat, P. Birkle, and A.B. Mukherjee (Eds.), "Natural Arsenic in Groundwater of Latin America – Occurrence, Health Impact and Remediation". Balkema Publisher, Lisse, The Netherlands, to be published as Vol. 1, interdisciplinary series: "Arsenic in the Environment" Chapter 73, Series Editors: Jochen Bundschuh and Prosun Bhattacharya, A.A. Balkema Publishers, Taylor and Francis Publishers, pp. 677–686. ISBN 978-0-415-40771-7 (hardback: alk. paper) – ISBN 978-0-203-88623-6 (ebook: alk. paper) (2009).

Morgada de Boggio, M.E., Mateu, M., Bundschuh, J., and Litter, M.I. *e-Terra*, http://e-terra@-geopor.pt, ISSN 1645-0388 **5** (5), (2008).

Muñoz, J. and Domènech, X. *J. Appl. Electrochem.* **20**, 518 (1990).

Murruni, L., Conde, F., Leyva, G., and Litter, M.I. *Appl. Catal. B* **84**, 563 (2008).

Murruni, L., Leyva, G., and Litter, M.I. *Catal. Today* **129**, 127 (2007).

Naftz, D., Morrison, S.J., Fuller, C.C., and Davis, J.A. (Eds.), "Handbook of Groundwater Remediation Using Permeable Reactive Barriers – Applications to Radionuclides, Trace Metals, and Nutrients". Academic Press, San Diego (2002).

Navío, J.A., Colón, G., Trillas, M., Peral, J., Domènech, X., Testa, J.J., Padrón, J., Rodríguez, D., and Litter, M.I. *Appl. Catal. B* **16**, 187 (1998).

Navío, J.A., Testa, J.J., Djedjeian, P., Padrón, J.R., Rodríguez, D., and Litter, M.I. *Appl. Catal. A* **178**, 191 (1999).

Noubactep, C., Schöner, A., and Meinrath, G. *J. Hazard. Mater. B* **132**, 202 (2006).

O'Loughlin, J.E., Kelly, D.S., Cook, E.R., Csencsits, R., and Kemmer, M.K. *Environ. Sci. Technol.* **37**, 721 (2003).

Papadam, T., Xekoukoulotakis, N.P., Poulios, I., and Mantzavinos, D., *J. Photochem. Photobiol. A* **186**, 308 (2007).

Prairie, M.R., and Stange, B.M. *AIChe Symp. Ser.* **89**, 460 (1993).

Prairie, M.R., Evans, L.R., and Martínez, S.L. *Chem. Oxid.* **2**, 428 (1992).

Prairie, M.R., Evans, L.R., Stange, B.M., and Martínez, S.L., *Environ. Sci. Technol.* **27**, 1776 (1993a).

Prairie, M.R., Stange, B.M., and Evans, L.R. *in* D.F. Ollis, and H. Al-Ekabi (Eds.), "Photocatalytic Purification and Treatment of Water and Air". TiO_2 Photocatalysis for the Destruction of Organics and the Reduction of Heavy Metals. Elsevier Sci. Publish. B.V., Amsterdam (1993b), p. 353.

Rader, W.S., Solujic, L., Milosavljevic, E.B., Hendrix, J.L., and Nelson, J.H. *J. Solar Energy Eng.* **116**, 125 (1994).

Rajeshwar, K. *J. Appl. Electrochem.* **25**, 1067 (1995).

Rajeshwar, K., Chenthamarakshan, C.R., Ming, Y., and Sun, W. *J. Electroanal. Chem.* **538/539**, 173 (2002).

Rajh, T., Tiede, D.M., and Thurnauer, M.C. *J. Non-Cryst. Solids* **205-207**, 815 (1996a).

Rajh, T., Ostafin, A.E., Micic, O.I., Tiede, D.M., and Thurnauer, M.C. *J. Phys. Chem.* **100**, 4538 (1996b).

Rengaraj, S., Venkataraj, S., Yeon, J.-W., Kim, Y., Li, X.Z., and Pang, G.K.H. *Appl. Catal. B* **77**, 157 (2007).

Ryu, J., and Choi, W. *Environ. Sci. Technol.* **38**, 2928 (2004).

Ryu, J., and Choi, W., *Environ. Sci. Technol.* **42**, 294 (2008).

Sabaté, J., Anderson, M.A., Aguado, M.A., Giménez, J., Cervera-March, S., and Hill Jr., C.G., *J. Molec. Catal.* **71**, 57 (1992).

San Román, E., Navío, J.A., and Litter, M.I. *J. Adv. Oxid. Technol.* **3**, 261 (1998).

Sancha, A.M. 9°. Cong. Arg. Saneam. Medio Amb., Córdoba, Argentina, 1996.

Schrank, S.G., José, H.J., and Moreira, R.F.P.M. *J. Photochem. Photobiol. A*. **147**, 71 (2002).

Selbin, J., and Ortego, J.D. *Chem. Rev.* **69**, 657 (1969).

Selli, E., De Giorgi, A., and Bidoglio, G. *Environ. Sci. Technol.* **30**, 598 (1996).

Serpone, N., Ah-You, Y.K., Tran, T.P., Harris, R., Pelizzetti, E., and Hidaka H. *Sol. Energy* **39**, 491 (1987).

Serpone, N., Borgarello, E., and Pelizzetti E. *in* M. Schiavello (Ed.), "Photocatalysis and Environment". Kluwer Academic Publishers, Dordrecht (1988), p. 527.

Serpone, N. *J. Photochem. Photobiol. A* **104**, 1 (1997).

Shim, E., Park, Y., Bae, S., Yoon, J., and Joo, H. *Int. J. Hydrogen Energy*, 33 (19), 5193–5198 (2008). doi:10.1016/j.ijhydene.2008.05.011.

Siemon, U., Bahnemann, D., Testa, J.J., Rodríguez, D., Bruno, N., and Litter, M.I. *J. Photochem. Photobiol. A* **148**, 247 (2002).

Skubal, L.R., and Meshkov, N.K., *J. Photochem. Photobiol. A* **148**, 211 (2002).

Smedley, P.L., Nicolli, H.B., Macdonald, D.M.J., Barros, A.J., and Tullio J.O. *Appl. Geochem.* **17**, 259 (2002).

Sun, B., Reddy, E.P., and Smirniotis, P.G. *Environ. Sci. Technol.* **39**, 6251 (2005).

Tanaka, K., Harada, K., and Murata, S. *Sol. Energy* **36**, 159 (1986).

Tennakone, K. *Sol. Energy Mater.* **10**, 235 (1984).

Tennakone, K., and Ketipearachchi, U.S. *Appl. Catal. B* **5**, 343 (1995).

Tennakone, K., and Wickramanayake, S. *J. Phys. Chem.* **90**, 1219 (1986).

Tennakone, K., and Wijayantha, K.G.U. *J. Photochem. Photobiol. A* **113**, 89 (1998).

Tennakone, K., Thaminimulle, C.T.K., Senadeera, S., and Kumarasinghe, A.R. *J. Photochem. Photobiol. A* **70**, 193 (1993).

Testa, J.J., Grela, M.A., and Litter, M.I. *Environ. Sci. Technol.* **38**, 1589 (2004).

Testa, J.J., Grela, M.A., and Litter, M.I. *Langmuir* **17**, 3515 (2001).

Thurnauer, M.C., Rajh, T., and Tiede, D.M. *Acta Chem. Scand.* **51**, 610 (1997).

Torres, J., and Cervera-March, S. *Chem. Eng. Sci.* **47**, 3857 (1992).

Tuprakay, S., and Liengcharernsit, W. *J. Hazard. Mater. B* **124**, 53 (2005).

Tzou, Y.M., Wang, S.L., and Wang, M.K. *Coll. Surf. A* **253**, 15 (2005).

Vaaramaa, K., and Lehto, J. *Desalination* **155**, 157 (2003).

Vohra, M.S., and Davis A.P. *Water Res.* **34**, 952 (2000).

Wang, L., Wang, N., Zhu, L., Yu, H., and Tang, H. *J. Hazard. Mater.* **152**, 93 (2008).

Wang, S., Wang, Z., and Zhuang, Q. *Appl. Catal. B* **1**, 257 (1992).

Wang, X., Pehkonen, S.O., and Ray, A.K. *Electrochim. Acta* **49**, 1435 (2004).

Wang, Z.-H., and Zhuang, Q.-X. *J. Photochem. Photobiol. A* **75**, 105 (1993).

WHO/SDE/WSH/03.04/118. "Uranium in Drinking-Water, Background Document for development of WHO Guidelines for Drinking-water Quality". Extract from Chapter 12 – Chemical fact sheets of WHO Guidelines for Drinking-Water Quality, 3rd ed. (2004).

World Health Organization. "Guidelines for Drinking-Water Quality [electronic resource]: Incorporating First Addendum. Vol. 1, Recommendations. – 3rd ed., electronic version for the Web. ISBN 92 4 154696 4, (2006), http://www.who.int/water_sanitation_health/dwq/gdwq0506_ann4.pdf.

Xu, X.,-R., Li, H.-B., and Gu, J.-D. *Chemosphere* **63**, 254 (2006).

Xu, Y., and Chen, X. *Chem. Ind.* 497 (1990).

Yang, H., Lin, W.-Y., and Rajeshwar, K. *J. Photochem. Photobiol. A* **123**, 137 (1999).

Yang, J.-K., and Lee, S.-M. *Chemosphere* **63**, 1677 (2006).

Yoneyama, H., Yamashita, Y., and Tamura, H., *Nature* **282**, 817 (1979).

Zhang, F.-S., and Itoh, I. *Chemosphere* **65**, 125 (2006).

Zhang, F.-S., Nriagu, J.O., and Itoh, H.J. *Photochem. Photobiol. A* **167**, 223 (2004).

Zhang, J.S., Stanforth, R., and Pehkonen, S.O. *J. Colloid Interface Sci.* **317**, 35 (2008).

Zheng, S., Xu, Z., Wang, Y., Wei, Z., and Wang B. *J. Photochem. Photobiol. A* **137**, 185 (2000).

CHAPTER **3**

Mineralization of Phenol in an Improved Photocatalytic Process Assisted with Ferric Ions: Reaction Network and Kinetic Modeling

Aaron Ortiz-Gomez[1,*], **Benito Serrano-Rosales**[2], **Jesus Moreira-del-Rio**[1], and **Hugo de-Lasa**[1]

Contents			
	1.	Introduction	70
		1.1 Photocatalysis: a promising low-cost alternative	70
		1.2 Moving toward an improved process: minimizing inefficiencies	71
		1.3 Organic oxidation and inorganic reduction	72
		1.4 Iron (Fe) ions in photocatalytic processes	74
		1.5 Kinetic modeling	78
		1.6 Recent advances in CREC	78
	2.	Experimental Methods Used in CREC	79
		2.1 Reaction setup	79
		2.2 Reactants	80
		2.3 Substrate analysis	80
		2.4 Catalyst elemental analysis: EDX and XPS	80
		2.5 Experiments in photo-CREC unit	81
	3.	Fe-Assisted Photocatalytic Mineralization of Phenol and its Intermediates	81
		3.1 Effect of Fe on the oxidation of phenol: optimum point and mechanism	82
		3.2 Fe-assisted mineralization of phenol and its intermediates	86

[1] The University of Western Ontario, London, Ontario, Canada

[2] Universidad Autonoma de Zacatecas, Zacatecas, Mexico

* Corresponding author.
E-mail address: aortizg2@uwo.ca

Advances in Chemical Engineering, Volume 36
ISSN 0065-2377, DOI: 10.1016/S0065-2377(09)00403-7

4. Kinetic Modeling: Unpromoted PC Oxidation and Fe-Assisted PC
   Oxidation of Phenol   92
   4.1 Overall kinetic model   92
   4.2 Parameter estimation   94
5. Conclusions   105
   Recommendations   106
   List of Symbols   106
   References   108

1. INTRODUCTION

1.1. Photocatalysis: a promising low-cost alternative

Every day, manufacturing industries produce water and gas effluent streams with significant amounts of organic and inorganic pollutants. Organic compounds in wastestreams include textile dyes, herbicides and pesticides, alkanes, haloalkanes, aliphatic alcohols, carboxylic acids, aromatics, surfactants, among many others (Guillard, 2003; Malato et al., 2004; Mukherjee and Ray, 1999). Inorganic compounds include complexes of metal ions such as mercury, cadmium, silver, nickel, lead, and other equally harmful species (Chen and Ray, 2001; Huang et al., 1996). Many of them are well known for their toxic effects on the environment and on human health. The elimination of these pollutants requires processes able to completely mineralize the organic pollutants and to convert the metallic contaminants into less harmful forms.

Heterogeneous photocatalysis has been proven to be a potential process to eliminate many of these hazardous organic pollutants present in air and water wastestreams and has therefore been the subject of extensive research over the last decades (Bahnemann, 2004; de Lasa et al., 2005). Photocatalytic (PC) processes, albeit advantageous for completely mineralizing complex harmful contaminants at relatively low cost (i.e., TiO_2 is inexpensive, 100–200 dollars per tonne; consumables are minimum, low lamp wattage or solar energy), take place at a rather slow rate. This has prompted the search of new means to improve their performance to tackle more efficiently the largely spread problem of polluted wastestreams. Owing to the particular characteristic of photocatalysis to produce nonselective hydroxyl radicals $HO^{\bullet}$, chemical species with high oxidative power, it was applied to environmental engineering for pollutant decontamination. Photocatalysis then emerged as a new process that could provide a solution to complete mineralization of organic contaminants and reduction of harmful inorganic metal ions.

Contaminants such as phenol, whose toxic effects on human health are well documented, have been widely used as a model pollutant to elucidate the complex PC reaction mechanisms and to evaluate the performance of many reactors designs and catalyst activities (Tryba et al., 2006a, b) Moreover, phenol and similar hydroxylated compounds are well-known contaminants present in wastewater effluents from many industrial processes (US Environmental Protection Agency, 2000 and references therein). Their toxic effects on human health are well documented, being related to severe illnesses such as leukemia (McDonald et al., 2001) and some serious human organ malfunctions (ATSDR, 1998). They are highly toxic and refractory pollutants not easily removed in biological wastewater treatment plants (Goi and Trapido, 2001). Hence, PC processes need to be improved to provide highly efficient solutions for heavily contaminated wastestreams and to minimize human exposure to species such as those mentioned above.

1.2. Moving toward an improved process: minimizing inefficiencies

An important research approach to improve the PC process performance is the use of metallic complexes as additives in PC reactions. Some metallic compounds, such as Hg(II) (Aguado et al., 1995), Cr(IV) (Chenthamarakshan and Rajeshwar, 2000), Zn(II) (Chenthamarakshan and Rajeshwar, 2002), Cd(II) (Chenthamarakshan and Rajeshwar, 2002), among others (Huang et al., 1996; Tan et al., 2003), have been shown to modify the oxidation rate of organic species depending on their oxidation states and concentrations. Most of those metallic compounds tested so far are equally harmful or even more so than the targeted organic pollutants, preventing them from being used as reaction enhancers.

Another important approach is the structural modification and doping of photocatalysts with metals or dyes to increase their PC activity (Araña et al., 2004; Bamwenda et al., 1995; Colmenares et al., 2006; Dai and Rabani, 2002; Hufschmidt et al., 2002; Karvinen and Lamminmaki, 2003; Kim et al., 2005; Leyva et al., 1998; Nagaveni et al., 2004; Zhou et al., 2006). These doping techniques reportedly improve catalyst activity in some cases, leading to higher mineralization rates than the untreated photocatalysts. The preparation techniques, however, often involve complex procedures that call for expensive reactants as metallic sources and high temperatures for calcination steps, therefore hindering their usage and production for large-scale applications. Additional equipment must also be added to the process for catalyst recovery if the catalyst cost is a factor or to produce catalyst-free water if it is intended for human consumption. These additional costs could eliminate one of the greatest advantages of PC processes: the low catalyst cost and low operating costs. Thus, the search for means to improve the rate of mineralization of hazardous pollutants has veered off to look for inexpensive techniques that can enhance the PC processes using metallic complexes and yet be environmental friendly.

Extensive research has shown that of all PC materials, TiO_2 is the most active for oxidation reactions. Its higher catalytic activity, along with its low chemical and biological activity, low cost, and high stability has made it the best option for PC reactions (Fox, 1990; Fujishima et al., 2000; Ray, 1997). UV light with wavelengths lower that 388 nm can be absorbed by TiO_2 generating a pair of charges e_{cb}^-/h_{vb}^+. These charges can either recombine dissipating the absorbed energy or promote different reduction–oxidation (redox) reactions (Herrmann, 1999). The redox reactions to take place depend on the chemical species present in the vicinity of the catalyst surface or in the bulk of the solution (Fujishima et al., 2000; Turch and Ollis, 1990). The recombination of charges is one of the main drawbacks of PC processes as it leads to UV radiation usage inefficiencies. Thus, reducing the e_{cb}^-/h_{vb}^+ recombination rates will result in higher production of hydroxyl radicals $HO^•$ and consequently to higher oxidation rates.

1.3. Organic oxidation and inorganic reduction

Since photocatalysis was discovered in the early 1970s, more than 6,200 papers related to this process have been published. Most of the work on this subject has focused on showing that organic molecules can be oxidized in PC reactors. So far, more than 800 organic molecules have been tested for oxidation in PC reactions (Blake, 2001). In most cases, the tested organic molecules were converted to CO_2, water, and mineral acids. Therefore, it can be definitely concluded that photocatalysis works for oxidation of organic molecules. The rate of oxidation depends on several factors that will be addressed in the upcoming section.

More recently, during the last decade, it was found that some metal cations in water could be reduced using photocatalysis. It was proposed that the photogenerated electron could be used for reducing inorganic metal ions. For instance, Hg(II) was reduced to Hg(0) over TiO_2 suspensions (Aguado et al., 1995), as well as Cr(IV) to Cr(III) (Chenthamarakshan and Rajeshwar, 2000; Colon et al., 2001), Zn(II) to Zn(0), and Cd(II) to Cd(0) (Chenthamarakshan and Rajeshwar, 2002). Additionally, Ag(I) was also reduced to Ag(0) (Huang et al., 1996) and Se(IV) and Se(VI) to elemental Se(0) (Tan et al., 2003). Therefore, it was concluded that photocatalysis could be applied for metal cations reduction.

For a metal cation in water to be reduced with a photogenerated electron e_{cb}^-, its reduction potential must be less negative than the reduction potential of the conduction band. In this category fall pairs such as Ni(II)/Ni(0), Pb(II)/Pb(0), Fe(III)/Fe(0), Cu(II)/Cu(I), Cu(II)/Cu(0), Cu(I)/Cu(0), Fe(III)/Fe(II), Ag(I)/Ag(0), and Hg(II)/Hg(0). Figure 1 reports the reduction potential of different metals compared with both the valence band potential and the conduction band potential. All metals located above the conduction band can theoretically be reduced (Chen and Ray, 2001).

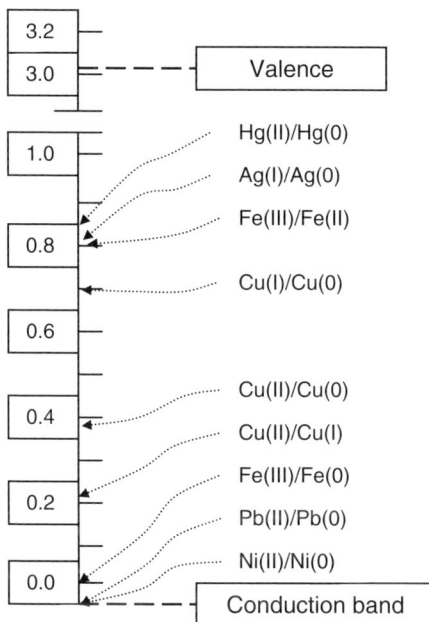

Figure 1 Reduction potential of different metal ions at pH 3 compared to TiO$_2$ conduction and valence bands (Chen and Ray, 2001).

In light of these findings, a new advantageous and characteristic of photocatalysis was discovered: the reduction of inorganic metal ions and the oxidation of organic molecules could be carried out simultaneously in a PC reactor. The photogenerated electron e_{cb}^- could help reduce the metallic ion while the photogenerated hole h_{vb}^+ could promote the oxidation of an organic molecule. This fact, as will be discussed later, would reduce the recombination of the generated charges, thus increasing the energy efficiency of the system (Colon et al., 2001; Hufschmidt et al., 2002; Vamathevan et al., 2002).

It was also found that there exists a synergic effect in the organic oxidation and inorganic reduction. Some studies show that the presence of some metal ions can affect the rate of oxidation of organic molecules. For instance, the rate of oxidation of phenol can be affected by the presence of silver (Huang et al., 1996) and silver can also affect the oxidation of textile dyes (Sokmen and Ozkan, 2002). The presence of Cr(VI) affects the rate of oxidation of salicylic acid (Colon et al., 2001).

Likewise, the presence of organic molecules can accclerate or decelerate the rate of reduction of metal ions. The presence of a dye in the photoreduction of Cr(IV) increases the rate of reduction compared to when there is no dye present (Li Puma and Lock Yue, 1999).

With regard to the oxidation reactions of organic compounds, although there is still controversy over the actual oxidation mechanism, there is general consensus that hydroxyl HO$^{\bullet}$ radicals are the primary oxidizing species in a PC reaction. However, the oxidation of an organic molecule can proceed via HO$^{\bullet}_{ads}$-attack or h^+_{vb}-attack (Zertal et al., 2004). The prevailing mechanism greatly depends on the substrate and on the catalyst surface characteristics (Carraway et al., 1994; Yang et al., 2006). The steps that proceed via HO$^{\bullet}_{ads}$-attack, for instance, are significantly affected by the presence of a hydroxyl radical HO$^{\bullet}_{ads}$ scavenger (e.g., isopropanol), while the steps that follow the h^+_{vb}-attack remain unaltered. In some cases, it was possible to switch from h^+_{vb}-attack to HO$^{\bullet}_{ads}$-attack in the presence of ions such as F$^-$ or SO$_4^{2-}$ (Yang et al., 2006). In spite of these observations, it is still difficult to distinguish between these two mechanisms as both, in many cases, lead to the same reaction products (Grela et al., 1996; Carraway et al., 1994). The presence of hydroxylated compounds as reaction intermediates and the detection of hydroxyl radicals through electron spin resonance (ESR) technique lead to the conclusion that in many cases the oxidations are via HO$^{\bullet}_{ads}$-attack (Turchi and Ollis, 1989; 1990). Other studies also report the detection of hydroxyl radicals with various techniques such as spin trapping with electron paramagnetic resonance (EPR) (Riegel and Bolton, 1995) that support the HO$^{\bullet}_{ads}$-attack mechanism.

1.4. Iron (Fe) ions in photocatalytic processes

As shown above, some metals can be reduced in their oxidation states in a PC process. Ferric ions Fe^{3+}, for instance, can be reduced to ferrous ions Fe^{2+} with the photogenerated electron e^-_{cb} given that the reduction potential is 0.77 eV, which lies within the reduction potential of e^-_{cb}. Of the metals shown in Figure 1, Fe is the most benign and is actually needed for a proper human metabolism. Fe is also one of the most abundant metals in the earth only after aluminum. Moreover, it is also the fourth most abundant element in the earth's crust. The major iron ores are hematite (Fe$_2$O$_3$), magnetite (Fe$_3$O$_4$), limonite [FeO(OH)], and siderite (FeCO$_3$) (Cotton et al., 1999). Thus, because it is naturally present in large amounts and it possesses some PC properties, it represents an excellent candidate for the enhancement of the PC process. There are some interesting contributions confirming that the PC properties of iron can be exploited in photoreactions.

In this regard, Fe ions have been commonly used in advanced oxidation processes in what is known as the Fenton's reaction. This involves the reaction between hydrogen peroxide H$_2$O$_2$ and ferric (Fe^{3+}) and ferrous (Fe^{2+}) ions to produce hydroxyl radicals HO$^{\bullet}$ and perhydroxyl radical HO$^{\bullet}_2$. Fe^{3+} ions form a complex with H$_2$O$_2$ (see reaction (17) below). This complex further decomposes to produce Fe^{2+} ions and perhydroxyl radicals HO$^{\bullet}_2$ (reaction (18)). Fe^{2+} ions are then reoxidized to Fe^{3+} by reacting with more H_2O_2 to produce a hydroxyl ion HO^- and a hydroxyl radical $HO^•$ (reaction (1)) (Pignatello et al., 1999; Kavitha and Palanivelu, 2004; Poulopoulos and Philippopoulous, 2004).

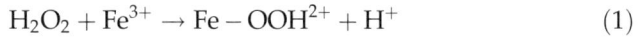

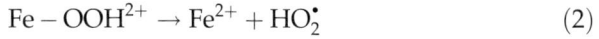

$$H_2O_2 + Fe^{3+} \rightarrow Fe - OOH^{2+} + H^+ \tag{1}$$

$$Fe - OOH^{2+} \rightarrow Fe^{2+} + HO_2^• \tag{2}$$

$$H_2O_2 + Fe^{2+} \rightarrow Fe^{3+} + HO^- + HO^• \tag{3}$$

The above reactions (1–3) can occur in the dark or under near UV or visible light ($\lambda = 436\,nm$) (Zepp et al., 1992). When the solution is irradiated, the rate of hydroxyl radical $HO^•$ formation is accelerated by the decomposition of H_2O_2 with radiation of less than $360\,nm$ (Pignatello et al., 1999). The decomposition of $Fe - OOH^{2+}$ is also accelerated with wavelengths lower than $313\,nm$ (Cermenati et al., 1997; Domenech et al., 2004).

The perhydroxyl radicals $HO_2^•$ can further react with either Fe^{2+} to produce more H_2O_2 (reaction (4)) or Fe^{3+} to produce molecular oxygen O_2 and a proton H^+ (reaction (5)).

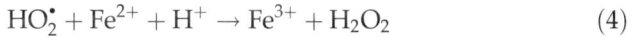

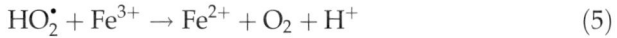

$$HO_2^• + Fe^{2+} + H^+ \rightarrow Fe^{3+} + H_2O_2 \tag{4}$$

$$HO_2^• + Fe^{3+} \rightarrow Fe^{2+} + O_2 + H^+ \tag{5}$$

$$H_2O_2 + HO^• \rightarrow HO_2^• + H_2O \tag{6}$$

$$HO^• + Fe^{2+} \rightarrow Fe^{3+} + HO^- \tag{7}$$

Additionally, reaction (6) can also take place in this system if H_2O_2 is in excess. Moreover, hydroxyl radicals can also be trapped by excess of ferrous ions (reaction (7)). Thus, despite the advantages such as commercial availability of the oxidant, no mass transfer problems, and formation of hydroxyl radicals from H_2O_2, this process presents several serious drawbacks. One of the most important one is that H_2O_2 has to be continuously added in controlled amounts as a source of hydroxyl radicals (Domenech et al., 2004).

Fe, on the other hand, has been used directly in PC processes as a dopant in semiconductors, in particular for TiO_2. The results seem to be somewhat contradictory nonetheless. These doped catalysts have been tested in the PC reactions of short-chain carboxylic acids such as maleic, formic, and oxalic acids, among others.

Some authors (Araña et al., 2001, 2002, 2003; Franch et al., 2005) have reported the activity of catalysts of TiO_2 doped with Fe in reactions with carboxylic acids such as maleic, formic, acetic, and acrylic acids. They found that catalysts with 0.5 wt% of Fe prepared by incipient wetness impregnation yielded higher mineralization rate for both formic and maleic acids than the undoped TiO_2. However, for acetic and acrylic acids, all Fe-doped TiO_2 catalysts showed lower activities than the Fe-free TiO_2 samples. For formic and maleic acids, they observed that as the Fe wt% increased, the mineralization rate decreased.

With the sol-gel preparation method (Navío et al., 1996, 1999), however, all doped catalysts yielded lower activities than the untreated TiO_2. These authors showed that only those doped catalysts prepared through incipient wetness impregnation and low Fe% produced better results than the untreated TiO_2.

The doping procedures for both methods (wet impregnation and sol-gel) require elaborate steps of impregnation and calcination (at 773 K). Moreover, in the case of samples yielding better results and prepared via impregnation method, it was detected that deposited iron leaches out of the lattice of TiO_2 by forming photoactive complexes of the type $[Fe - Acid]^{n+}$ with the model reactants. This suggests that one has to be very cautious while preparing doped photocatalysts to prevent the metal from leaching out of the catalyst structure.

Another study on the use of Fe showed that the oxidation rate of acetaldehyde was improved with TiO_2 catalysts doped with Fe and Si synthesized by thermal plasma (Oh et al., 2003). A Fe content lower than 15% rendered higher activities than the untreated catalyst. The catalyst preparation technique involved a complex procedure using a plasma torch, with all this likely leading to an expensive photocatalyst of mild prospects for large-scale applications.

From the studies above, one can observe that the use of Fe has been somewhat limited and that there are still many areas to explore for better utilization of Fe in PC reactions. More specifically, there is a need for the development of new inexpensive techniques or procedures to increase the photocatalyst activity. These new procedures should make the process more efficient without the economic burden that photocatalyst doping brings about. Also, one can notice that most studies have focused on carboxylic acid species containing less than four carbons. It is of utmost importance to explore this area with more refractory molecules such as phenol and other hydroxylated aromatics to determine whether Fe can truly be applied to enhance the PC mineralization.

In this regard, the PC oxidation of phenol produces similar hydroxylated aromatics as reaction intermediates since the oxidation occurs via hydroxyl radical attack. These intermediates might be equally harmful than the parent species. Several authors report that during the oxidation of phenol, they identified various hydroxylated intermediates such as 1,2,3-trihydroxybenzene (1,2,3-THB), *ortho-* and *para*-dihydroxybenzene (*o,p*-DHB) (Al-Ekabi and Serpone, 1988), 1,2,4-THB and 1,4-benzoquinone (1,4-BQ)

(Trillas et al., 1992; Winterbottom et al., 1997), p-DHB, and 1,4-BQ (Leyva et al., 1998; Tseng and Huang, 1990). These intermediate species were detected in experiments performed over a wide range of conditions and in different reaction setups. Therefore, the formation and concentration of reaction intermediates greatly depend on the conditions at which the reaction takes place. The pH of the solution plays a key role in the formation of oxidation intermediates. Salaices et al. (2004), for instance, reported the formation of significant amounts of p- and o-DHB at a pH of 4, while the latter was not formed at pH of 7. Moreover, o-DHB was not detected when the catalyst was changed from Degussa to Hombikat UV-100. Also, 1,2,4-THB and 1,4-BQ were identified in most experiments with their concentrations not varying significantly from run to run.

Additionally to phenolic intermediates, upon the aromatic ring opening, a series of carboxylic acids can be formed. Maleic acid, for instance, has been detected in the oxidation of byphenyls (Bouquet-Somrani et al., 1996), 1,2,4-THB (Li et al., 1991a), 1,2-dimethoxybenzene (Pichat, 1997), and acid orange 7 (Stylidi et al., 2003). Similarly, muconic acid has been reported as an intermediate of byphenyls oxidation (Bouquet-Sormani et al., 1996), while succinic and malonic acids were identified in polycarboxylic benzoic acid oxidation (Assabane et al., 2000). It is therefore expected that these or similar acids be formed during phenol oxidation.

The presence of Fe ions in PC reactions could have an important effect on the formation and distribution of all reaction intermediates. The use of Fe ions in the PC oxidation of phenol as reaction enhancer may not only affect the rate of oxidation, but also promote different reaction pathways leading to changes in the intermediate species distribution or to the formation of new ones.

A complete identification and quantification of aromatics and carboxylic acids, during the oxidation of phenol for both unpromoted PC reaction (no iron present) and PC reaction coupled with Fe ions, will allow the formulation of a comprehensive reaction network for both systems and thus, a systematic comparison between them. Also, this will permit the development of a more detailed kinetic model to incorporate most of the oxidation intermediates for both systems and will definitely help determine the role of Fe ions in PC reactions.

Regarding the kinetic modeling, few contributions propose kinetic models for the PC oxidation of phenol and other aromatics (Chen and Ray, 1998, 1999; Li et al., 1999b; Wei and Wan 1992;), with kinetic models being based mainly on the initial rates of reaction only. Such models fail to account for the formation of the different reaction intermediates, which may play an important role in the overall mineralization rate. More recently, Salaices et al. (2004) developed a *series–parallel* kinetic model based on observable aromatic intermediates. This model was applied to a wide range of pH, phenol concentration, and catalyst type. In this model, however, some steps were stated as hypothetical with no carboxylic acids being identified. To our knowledge, systematic studies dealing with all possible kinetic steps of PC oxidation of phenol and other hydroxylated aromatics with and without Fe ions have not been carried out.

1.5. Kinetic modeling

There is general consensus and evidence that PC reactions take place on the photocatalyst surface (Minero et al., 1992; Pellizetti, 1995). Several authors have studied the kinetics of pollutants degradation and found that the rate of reaction follows a Langmuir-Hinshelwood (LH)-type equation (Al-Ekabi et al., 1989; Okamoto et al., 1985; Turchi and Ollis, 1989, 1990). A LH equation type can be used as well for the degradation of phenol and its intermediates. In this regard, only a few studies have reported kinetic models for the oxidation of phenol and hydroxylated compounds (Li et al., 1999; Wei and Wan, 1992). Such kinetic models are based mainly on the initial reaction rates and fail to account for reaction intermediates that may play an important role in the kinetic expressions. In a more recent study, Salaices et al. (2004) developed a series–parallel kinetic reaction network for the photodegradation of phenol. This model involved LH equations for phenol and its major aromatic intermediates and was applied to various reaction pHs, different catalysts, and substrate concentrations with all conditions yielding good fitting of the model. Given the background provided by such study, phenol and other phenolic compounds are employed as model pollutants for establishing kinetics in the context of LH-type models.

1.6. Recent advances in CREC

CREC researchers have addressed recently the application of Fe cations as additives to enhance the performance of PC reactions employing inexpensive procedures. This was sought owing to the characteristic properties and behavior of Fe cations in a PC reaction. Phenol and similar aromatics were selected as model pollutants given their refractory nature in water treatment and health problems associated with their presence in drinking water.

In this respect, the effect of Fe cations on both the oxidation rate and complete mineralization rate of phenol and alike aromatic compounds was considered. Optimum conditions were reviewed to use Fe cations as reaction enhancers (henceforth PC reactions involving optimum Fe concentrations are called Fe-assisted PC reactions). This also involved the assessment of effect and mechanism of Fe ions on the PC reaction of phenol and other selected aromatic species. A systematic comparison between the kinetic reaction schemes for both unpromoted PC and Fe-assisted PC reactions for the selected model pollutants was also a primary emphasis. Last, the estimation of the enhancement through efficiency factor calculations was described.

Following this, efforts were devoted to the development of kinetic models to represent the PC oxidation of phenol. In this regard, models for the unpromoted PC reaction and a model for the Fe-assisted reaction are needed to account for the specific type of intermediates and their concentrations. This was necessary since Fe ions can alter the oxidation pathways yielding different amounts and various types of intermediates. For these studies, slurry-type PC reactors are the main experimental apparatus. While there is a considerable effort to find new PC reactors in which the catalyst is supported instead of being suspended to avoid catalyst loss or to ease its separation from the slurry, it is well established that slurry reactors provide more certainty of the photocatalyst loading used in a given experiment and the extent of its irradiation (Aguado et al., 2000; Guillard, 2003; Preis et al., 1997) and in general yield higher reaction rates (Mathews and McEvoy, 1992; Wyness et al., 1994). Hence, the slurry Photo-CREC and Photo-CREC Solar Simulator are considered in this chapter as the primary systems to study and understand the PC oxidation pathways.

2. EXPERIMENTAL METHODS USED IN CREC

2.1. Reaction setup

The main experimental setup used in CREC studies is shown in Figure 2. It is comprised of a slurry-annular PC reactor (1), a mixing tank (2), and a pump (3). The entire system operates in batch mode. The reacting media is

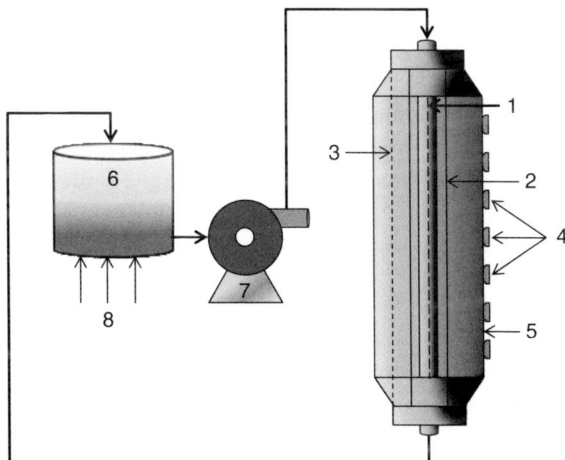

Figure 2 Schematic representation of the Photo-CREC Water-II Reactor: (1) MR or BL lamp, (2) replaceable 3.2-cm-diameter glass inner tube, (3) replaceable 5.6-cm-diameter glass inner tube, (4) fused-silica windows, (5) UV-opaque polyethylene outer cylinder, (6) stirred tank, (7) centrifugal pump, and (8) air injector.

first fed into the mixing tank and then pumped to the upper entrance of the reactor. After passing through the reactor annular section, the reacting media is pumped back to the mixing tank. A sampling port and an air supply are placed in the mixing tank. The Photo-CREC handles 6 L and has one lamp positioned in the center of the reactor. The lamp used in this reactor is a low-energy UV lamp. This reactor is equipped with fused-silica windows for absorbed irradiation measurements (de Lasa et al., 2005).

2.2. Reactants

The following reactants were used as received from suppliers without any further treatment: $FeSO_4$ $7H_2O$, phenol, TiO_2 P25 (Degussa), catechol, hydroquinone, 1,4-BQ, 1,2,4-benzentriol, fumaric acid, maleic acid, oxalic acid, and formic acid. Methyl viologen dichloride hydrate 98%, 4-chlorophenol 99%+, and Fe_2SO_3 H_2O were heated up for 2 h to 200°C and placed in a desiccator until they reached room temperature before preparing the solutions. H_2SO_4 was used in all experiments to control the pH of the reacting media.

2.3. Substrate analysis

2.3.1. Fe analysis
Fe content analyses were performed using a colorimetric technique as described in Karamanev et al. (2002).

2.3.2. Model pollutant analysis
The analyses of aromatic components were performed on a 1525 Binary Waters HPLC with a dual absorbance detector using a Symmetry C18 column and a mobile phase of methanol and water. Carboxylic acids analyses were performed using the same HPLC system with an Atlantis dC18 column and mobile phase. pH was monitored with a Corning 430 pH meter. For most experiments, the total organic carbon was also analyzed using a Shimadzu 5050 TOC analyzer equipped with a NDIR detector coupled with an autosampler ASI 5000.

The identification of both aromatic and carboxylic intermediates was performed comparing the retention times of model reactants with those of the reaction intermediates detected in the samples. Additionally, comparisons with spectra provided in catalogs from column manufacturers were made to corroborate the identification of reaction intermediates.

2.4. Catalyst elemental analysis: EDX and XPS

Samples for catalyst elemental analysis were prepared as follows: After taken from the continuous stirred tank (CST), they were let settle down overnight.

The supernatant was removed after 12 h of settling. The catalyst-rich solution was used for analysis. A few drops of this concentrate were placed over silicon films and the water was let to evaporate. Dried samples were then analyzed.

Selected samples from various experiments were analyzed in energy dispersive X-ray spectroscopy (EDX). These analyses were carried out on a Hitachi S-4500 field emission scanning electron microscopy equipped with an EDAX^{TM} Phoenix model EDX spectrometer. An electron beam of 15 kV was used. These samples were also analyzed in a Kratos Axis Ultra X-ray photoelectron spectroscopy (XPS).

2.5. Experiments in photo-CREC unit

The detailed experimental procedure for this study is explained elsewhere (Ortiz-Gomez et al., 2008). In general, a predefined amount of reactant was weighed. It was then added to a known volume of water, whose pH was adjusted with a H_2SO_4 solution. In the experiments with Fe ions, the Fe solution containing the desired amount of Fe (either as Fe^{3+} or Fe^{2+}) was premixed with TiO_2 in 100 mL for 30 min, then it was added to the previous mixture. H_2SO_4 solution was added to adjust the pH. The reactants were allowed to be in contact with the catalyst for 30 min or more before the UV lamp was turned on. During this period of time (dark period), the reacting media was pumped around the system. After this period, the lamp was turned on. All other operating conditions (air flowrate, reacting media flowrate, catalyst weight $-0.14 \, g \, L^{-1}$, room temperature) were kept constant, except for the pH, which was not adjusted after the reaction started. Samples were taken at different time intervals to track the concentration of the reactants and intermediates.

3. Fe-ASSISTED PHOTOCATALYTIC MINERALIZATION OF PHENOL AND ITS INTERMEDIATES

In order to study the Fe-assisted PC mineralization of phenol, experiments were typically performed for the PC oxidation of phenol to determine the optimum pH at which the rate of oxidation was the highest possible in the reactor setup employed for this study. Then a series of tests were carried out to identify the aromatic and carboxylic intermediates of phenol oxidation at the optimum pH value. Subsequently, more tests with those aromatics identified as phenol oxidation intermediates were performed to unveil the intermediates produced during their corresponding PC oxidations and also to compare with the Fe-assisted PC reactions.

Once the kinetic reaction scheme for the unpromoted PC reaction of phenol and other aromatic species was determined, the next step was to evaluate the influence of Fe cations on both the rate of oxidation and the rate of mineralization of phenol and similar hydroxylated aromatics.

A group of experiments was also performed to evaluate the extent of influence of Fe ions on the rate of oxidation of phenol. This new set was carried out at the optimum pH value determined in the previous runs. Once an optimum value of Fe ions was determined and the extent of influence assessed, more runs were performed using Fe ions from a different metallic complex to evaluate the effect of the oxidation state of the metal on the rate of oxidation (Ortiz-Gomez et al., 2008). Henceforth, the experiments with the optimum amount of Fe ions are referred to as Fe-assisted PC reactions.

Series of tests were also performed at optimum pH and optimum Fe ion concentrations to identify the oxidation intermediates for the Fe-assisted PC oxidation of phenol. Likewise, another series of experiments were performed to evaluate the effect of Fe ions on the rate of oxidation of those species identified as intermediates in the Fe-assisted PC oxidation of phenol.

All these data allowed the systematic comparison with the species detected and photoconversion rates observed in the unpromoted PC reactions.

3.1. Effect of Fe on the oxidation of phenol: optimum point and mechanism

A first set of experiments was aimed at evaluating the effect of Fe on the phenol oxidation rate at a concentration of 20 ppm C and at a pH of 3.2. The initial concentration of Fe cations was varied using different amounts of a solution containing Fe^{3+} ions.

It was found that the rate of phenol oxidation was significantly affected by the presence of ferric ions depending on their concentrations. This effect was a strong function of their initial concentrations, as shown in Figure 3. High Fe^{3+} ion concentrations considerably reduce the photo-oxidation rate revealing a negative effect of high ferric ion concentrations. To the contrary, Fe^{3+} ion concentrations below 10 ppm promote higher oxidation rates than those obtained in the unpromoted PC reaction. Moreover, as the concentration decreases below 10 ppm, the rate of phenol oxidation reaches a maximum value at 5 ppm of Fe^{3+} ions. Below this maximum, the rate decreases as well. Hence, it was determined that for the reaction conditions in Photo-CREC system, 5 ppm of ferric ions is the optimum Fe concentration. For additional information about the experiments with different iron concentrations, we advice the reader to refer to Ortiz-Gomez (2006).

The effect of 5 ppm of ferric ions is demonstrated for the oxidation of 50 ppm C in phenol, as shown in Figure 4. It can be observed that the addition of 5 ppm of Fe^{3+} ions promotes a higher oxidation rate than unpromoted PC reaction performed at pH 3.2.

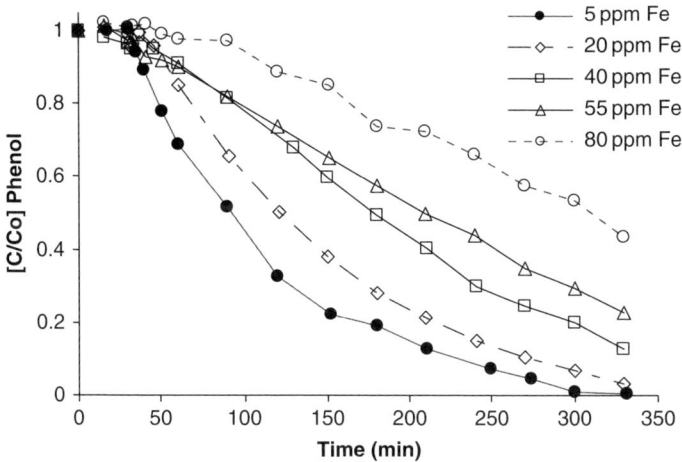

Figure 3 Influence of ferric ions on the rate of disappearance of phenol. 20 ppm C in phenol (Ortiz-Gomez et al., 2008).

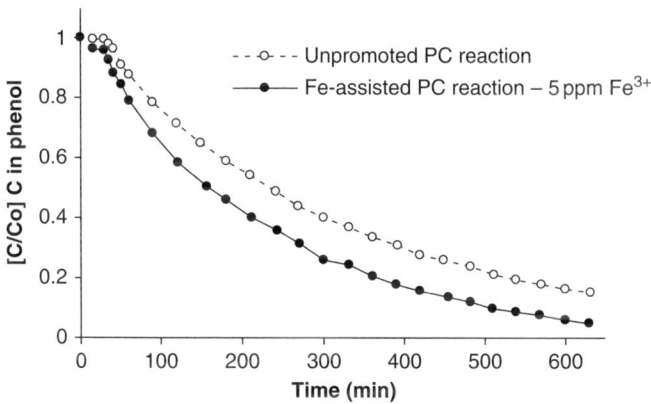

Figure 4 Influence of 5 ppm of ferric ions on the rate of oxidation of 50 ppm C in phenol (Ortiz-Gomez et al., 2008).

To determine the influence of the oxidation state of Fe ions on the rate of degradation of phenol, a new set of experiments using ferrous ions Fe^{2+} was carried out. It was found that regardless of the Fe oxidation state, be this state 3+ or 2+, adsorbed Fe onto TiO_2 promotes a higher oxidation rate than the unpromoted PC reaction. The concentration changes for the oxidation of 50 ppm C in phenol using Fe^{3+} and Fe^{2+} ions are shown in Figure 5a (5 ppm of Fe ions) and b (2 ppm of Fe ions). One can observe that photoconversion profiles are essentially identical and that the effect of the oxidation state of Fe, ferric or ferrous, yields the same results. Similar experiments were

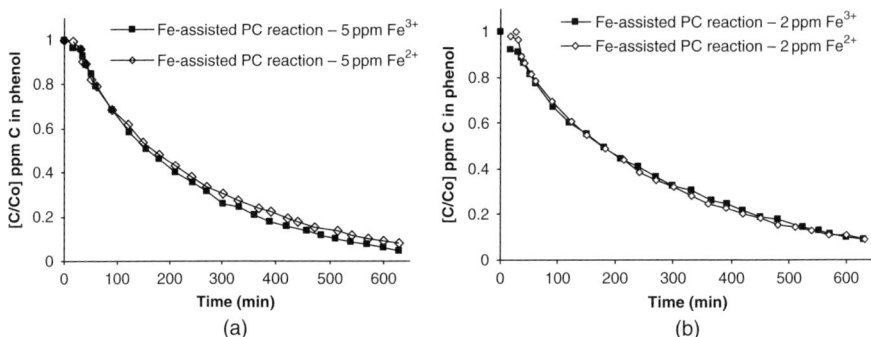

Figure 5   (a) Influence of ferric (Fe^{3+}) and ferrous (Fe^{2+}) on the oxidation rate of 50 ppm C in phenol. (b) Influence of 2 ppm of ferric and ferrous ions on the oxidation rate of 50 ppm C in phenol (Ortiz-Gomez et al., 2008).

carried out for different phenol concentrations with all cases showing and confirming the same behavior.

One can notice as well that the effect of Fe ions lasted throughout the reaction time. That is, its effect does not fade away with time. To the contrary, it has a major influence on the last stages of the PC conversion.

A number of studies using several metallic complexes have suggested that the enhancement in the oxidation rate by metals, such as Ag, Hg, and Cr, is due to the fact metal cations efficiently trap the photogenerated electrons e_{cb}^-, thus reducing the e_{cb}^-/h_{vb}^+ recombination rate and leading to higher rates of hydroxyl radical formation by the hole h_{vb}^+ (Colon et al., 2001; Oh et al., 2003). This mechanism seems appropriate for some metal cations; since once they are completely reduced to their lower oxidation state, their effect on the oxidation rate is diminished. Furthermore, in cases such as that of Ag and Cr they may bring the oxidation reaction to a halt by depositing the reduced cations on the catalyst surface and preventing the UV irradiation from reaching the catalyst surface (Chenthamarakshan and Rajeshwar, 2002; Colon et al., 2001; Navío et al., 1999).

The results of this study, however, reveal that Fe ions promote an enhancement through a different mechanism, as its change in the oxidation state does not have an influence on the oxidation reaction.

In the experiments with high concentrations of Fe^{3+} ions (shown in Figure 3), the change in the oxidation state of Fe^{3+} ions in the bulk of the liquid was also measured. In this regard, it was found that ferric ions are reduced to ferrous ions once the PC reaction was initiated. This cation reduction occurs in a relatively short time period compared with the phenol mineralization times. For instance, 20 ppm of ferric ions is completely reduced to ferrous ions in about 80 min while complete degradation of phenol is achieved in 330 min. This phenomenon is observed in all experiments and for all tested Fe^{3+} concentrations. In every case, ferric ions are reduced to ferrous ions and remain in this state throughout the rest of the photoconversion reaction. Moreover, it is observed that the total Fe concentration is lower than the initial concentration. This confirms that Fe cations adsorb onto TiO_2 during the initial adsorption period in which the photocatalyst is first in contact with the Fe solution.

These results suggest that it is actually the Fe ions as Fe^{2+} ions that promote higher photoconversion reaction rates, as Fe^{3+} is quickly reduced to Fe^{2+}. Ferrous ions may be, however, quickly reoxidized to Fe^{3+} on the catalyst surface when an electron acceptor molecule, such as O_2, scavenges the e_{cb}^- forming a superoxide radical ($O_2^{-\bullet}$). This process appears to be similar to that reported for the platinization of TiO_2, where Pt reduces the recombination rate by transferring the electron to the O_2 molecule to form a superoxide radical (Hufschmidt et al., 2002).

It is important to point out that the optimum Fe concentration was determined from the total amount of iron added to the water system. Larger Fe concentrations yielded lower oxidation rates. This can be explained considering that higher Fe ion concentrations than 5 ppm might also increase the amount of iron remaining on the catalyst surface during the irradiation period. Those higher Fe concentrations on the surface might become "recombination centers" short-circuiting the photogenerated charges and thus, leading to higher recombination rates and higher inefficiencies. Araña et al. (2001, 2002) reported a similar effect on the Fe-doped catalysts, where those doped with high Fe content (2 wt%) yielded lower reaction rates than the ones with low Fe content (0.5 wt%). This was attributed to a similar phenomenon.

In a study with Fe-doped TiO_2, Araña et al. (2003) concluded that a catalyst containing 0.5 wt% of Fe showed an improvement in the oxidation of carboxylic acids. To prepare their $Fe-TiO_2$-doped catalysts, they followed a lengthy procedure that included a 48-h period of mixing of the solution containing the catalyst and Fe ions. After mixing, the catalyst was dried at 393 K for 24 h and then calcined at 773 K. The results obtained with those catalysts are similar to the results presented in this study, which were obtained with a much simpler and considerably less costly technique. This demonstrates then that a simple preimpregnation of the catalyst will render the same or better results than complex doping techniques.

Regarding iron addition as described above, one important consideration is the removal of the iron ions (Fe^{2+} or Fe^{3+}) in treated streams. These cation ions can easily be removed from treated water increasing the pH. This pH adjustment causes iron cation precipitation via the formation of iron hydroxide floccules. These floccules may entrap TiO_2-suspended particles facilitating both iron cations and TiO_2 removal from the treated water.

3.2. Fe-assisted mineralization of phenol and its intermediates

3.2.1. Oxidation of phenol: Fe-assisted PC reaction and its comparison with the unpromoted PC reaction

Given the findings reported in Ortiz-Gomez et al. (2008), all remaining experiments were developed with Fe^{3+} ions considering that the presence of iron as Fe^{2+} or Fe^{3+} is not significant and that both species lead to the same results.

Once the optimum amount of Fe ions was determined for the oxidation of 20 ppm C in phenol, new experiments were carried out to evaluate the effect of adding 5 ppm of ferric ions on different phenol concentrations (20, 30, 40, and 50 ppm C in phenol). This allows one to evaluate how Fe ions change the formation of reaction intermediates and, most importantly, how they influence the overall mineralization rate of phenol. These results are compared with those from unpromoted PC reaction.

In the experiments for the Fe-assisted PC reaction of phenol as a model reactant using 5 ppm of Fe^{3+}, various intermediates that included aromatics and carboxylic acids are identified. The major observed aromatic intermediates are o-DHB, p-DHB, and 1,4-BQ. The carboxylic acid intermediates are maleic acid (MeAc), fumaric acid (FuAc), oxalic acid (OxAc), and formic acid (FoAc). These are the same intermediaries detected in the unpromoted PC reaction of phenol (Ortiz-Gomez et al., 2007). The concentration profiles for the unpromoted PC oxidation of 30 ppm of phenol and its aromatic intermediates are shown in Figure 6a, while the concentration profiles for the Fe-assisted PC oxidation of 30 ppm C of phenol using 5 ppm of Fe^{3+} ions are shown in Figure 6b. It can be observed that in the presence of ferric ions, the formation of o-DHB is favored whereas that of p-DHB is suppressed. The increase in concentration of o-DHB is around the same magnitude of the decrease in p-DHB concentration. More detailed comparisons between the intermediates for both systems are presented in Figures 7 and 8.

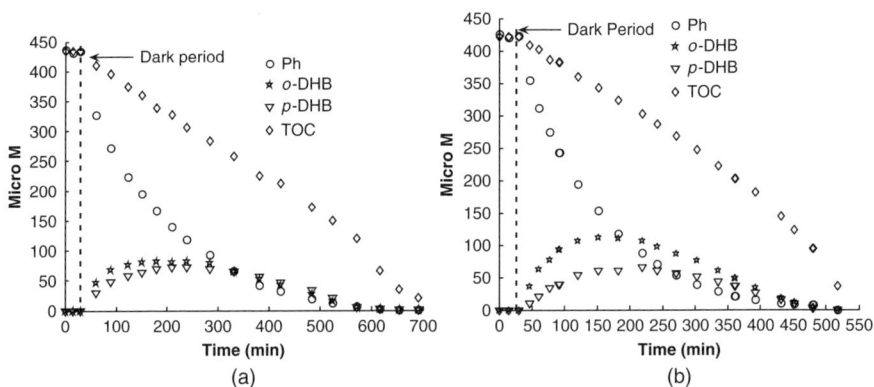

Figure 6 Concentration profiles for the (a) unpromoted PC oxidation of 30 ppm C in phenol and (b) Fe-assisted PC oxidation of 30 ppm C in phenol using 5 ppm Fe^{3+} ions (Ortiz-Gomez et al., 2008).

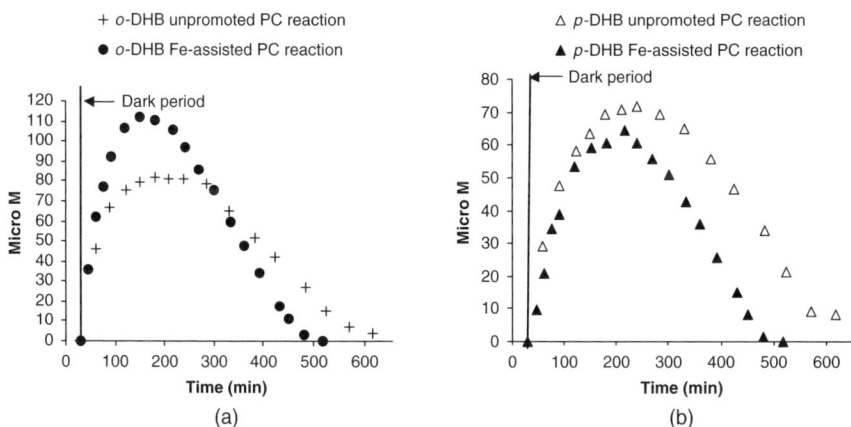

Figure 7 (a) o-DHB concentration profiles for both unpromoted PC reaction (30 ppm C in phenol) and Fe-assisted PC (30 ppm C in phenol + 5 ppm Fe^{3+} ions). (b) p-DHB concentration profiles for both PC reaction (30 ppm C in phenol) and Fe-assisted PC (30 ppm C in phenol + 5 ppm Fe^{3+} ions) (Ortiz-Gomez et al., 2008).

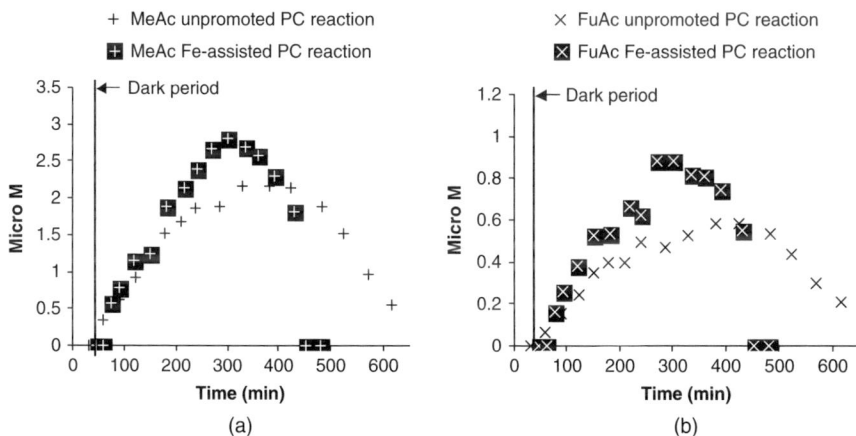

Figure 8 (a) MeAc concentration profiles for both unpromoted PC reaction (30 ppm C in phenol) and Fe-assisted PC (30 ppm C in phenol + 5 ppm Fe^{3+} ions). (b) FuAc concentration profiles for both PC reaction (30 ppm C in phenol) and Fe-assisted PC (30 ppm C in phenol + 5 ppm Fe^{3+} ions) (Ortiz-Gomez et al., 2008).

As shown in Figure 8a and b, the concentrations of carboxylic acids are also influenced by the presence of Fe^{3+} ions. For the Fe-assisted PC reaction, MeAc and FuAc concentrations slightly supersede the concentration for the unpromoted PC runs in the middle of the experiment and decrease more rapidly later. Similar trends were observed for OxAc and FoAc.

An additional effect, which is the most important of all the phenomena observed in these experiments, is the effect the Fe^{3+} ions observe on the overall mineralization rate of phenol. The Fe^{3+} not only enhances the rate of oxidation of phenol and changes the formation of its oxidation intermediates but also accelerates the rate of its overall mineralization. Figure 9a and b compare the TOC profiles for the oxidation of 20 and 30 ppm C of phenol in both unpromoted PC and Fe-assisted PC systems. It can be seen that during the Fe-assisted PC reaction, the overall mineralization of phenol is faster than that for the unpromoted PC reaction. One can also observe that for both cases, during the first part of the reaction time, the two profiles follow a similar trend. The TOC profile in the Fe-assisted PC systems, however, drops off faster than that in the PC system. More importantly, one can notice that in the last part of the reaction period, there is a considerable change on the slope of the TOC profile in the Fe-assisted PC reactions.

In this respect, the overall mineralization rate of phenol has often been approximated with a zero-order reaction rate (Salaices et al., 2004) as it follows a fairly straight line. For the Fe-assisted PC reaction, however, this approximation cannot be applied given the sharp change of slope in the last part of the photoconversion reaction. Thus, a more complex kinetic rate equation needs to be developed to account for this behavior.

From the results above, it can be observed that as in the unpromoted PC oxidation, in the Fe-assisted PC oxidation of phenol all aromatic intermediates are detected in small amounts from the time when the photoreaction is initiated. One can notice as well that there is a rapid reduction in the TOC

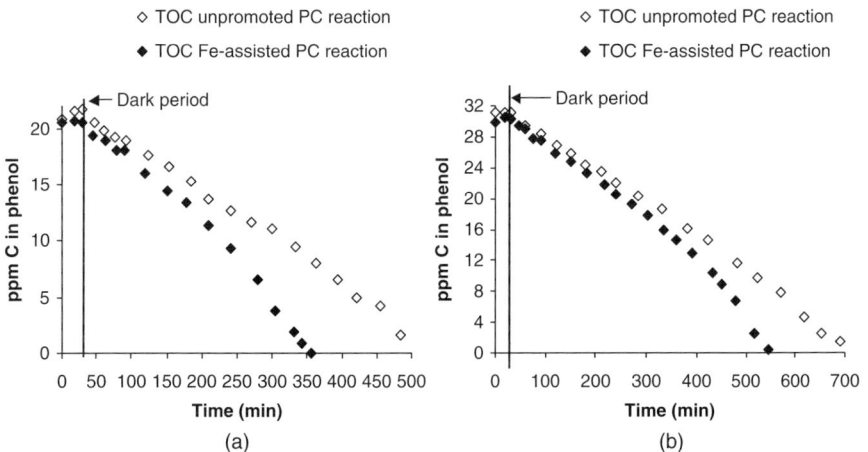

Figure 9   (a) TOC profiles for both unpromoted PC reaction (20 ppm C in phenol) and Fe-assisted PC reaction (20 ppm C in phenol + 5 ppm Fe^{3+} ions). (b) TOC profiles for both PC reaction (30 ppm C in phenol) and Fe-assisted PC reaction (30 ppm C in phenol + 5 ppm Fe^{3+} ions) (Ortiz-Gomez et al., 2008).

content from the early stages of the reaction, a phenomenon also observed in the unpromoted PC reaction. This can be equally attributed to a complete mineralization. As observed from the carboxylic acid profiles, they are also detected in the system once the lamp is turned on. Therefore, it can be concluded that in Fe-assisted PC reaction hydroxyl radicals oxidize phenol in a similar fashion but at a different rate than in the unpromoted PC oxidation. Phenol is simultaneously hydroxylated into o-DHB, p-DHB, and carboxylic acids and mineralized into CO_2. The aromatic intermediates can be further oxidized into carboxylic acids and CO_2. The carboxylic acids can be formed from the oxidation of all aromatics. Hence, the same concept of nonuniform distribution of hydroxyl radicals over TiO_2 surface leading to different degrees of oxidation can help explain the observed phenomena, with phenol having aromatic ring cleavage and formation of carboxylic acids and CO_2 as soon as the photoreaction is commenced.

Thus, considering the Fe-assisted PC conversion of phenol, the observable effect is that phenol produces many intermediate species from the start of the reaction regardless of the pathways involved in the production of such intermediates. It can thus be concluded that the Fe-assisted PC oxidation of phenol can be equally represented with a series–parallel reaction scheme, as it was for the unpromoted PC reaction. All the steps described above are summarized in Figure 10, a reaction scheme based on observable species. It must be emphasized that although the reaction network describes both unpromoted PC and Fe-assisted PC reactions, the values of the kinetic constants will be different for both systems.

3.2.2. Oxidation of *ortho*-dihydroxybenzene: Fe-assisted PC reaction and its comparison with the unpromoted PC reaction

Separate series of experiments were performed using o-DHB as model reactant to understand its behavior in the Fe-assisted reaction. In this case, in the Fe-assisted PC oxidation of o-DBH using 5 ppm Fe^{3+}, the only

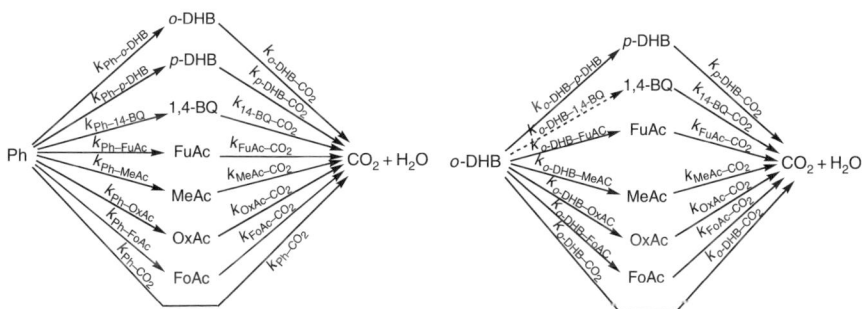

Figure 10 Series–parallel reaction scheme for oxidation of (a) phenol and (b) o-DHB involving all detected species. This applies to unpromoted PC and Fe-assisted PC reactions (Ortiz-Gomez et al., 2008).

aromatic observed was p-DHB. The same carboxylic acids in the unpromoted PC reaction were identified: FuAc, MeAc, FoAc, and OxAc.

It is shown that for o-DHB PC oxidation, the addition of 5 ppm Fe^{3+} ions also promotes an important improvement on the PC oxidation rate with respect to the unpromoted PC reaction, 1,4-BQ.

All observations lead to the conclusion that the oxidation of o-DHB follows a similar pattern than phenol oxidation in both unpromoted PC and Fe-assisted PC with all intermediates, aromatic species, and carboxylic acids being detected as soon as the photoreaction is initiated. This shows that some of the o-DHB molecules are quickly oxidized to CO_2, as suggested by the early decrease in TOC. Likewise, some o-DHB molecules are partially oxidized to carboxylic acids. These results demonstrate that the oxidation of o-DHB can also be represented with a series–parallel reaction scheme.

3.2.3. Oxidation of *para*-dihydroxybenzene and 1,4-benzoquinone: Fe-assisted PC reaction and its comparison with the unpromoted PC reaction

The next set of experiments involved the Fe-assisted PC oxidation of p-DHB and 1,4-BQ as the model pollutants. In these cases, similar patterns were observed leading to the conclusion that these two compounds behave in a similar way to phenol and o-DHB with both exhibiting a series–parallel reaction scheme (Ortiz-Gomez et al., 2008)

3.2.4. Overall refined series–parallel reaction mechanism for PC and Fe-assisted PC reaction

Each of the series–parallel reaction schemes presented in the previous sections for the Fe-assisted oxidations of phenol, o-DHB, p-DHB, and 1,4-BQ, were established considering the chemical species detected in the bulk of the water solution. Given that o-DHB, p-DHB, and 1,4-BQ are intermediates in the Fe-assisted PC oxidation of phenol, an overall reaction scheme for the Fe-assisted PC oxidation of phenol can be formulated. This overall reaction network incorporates all the detected intermediaries.

The underlying assumption in this reaction scheme is that all chemical species behave the same as a model pollutant or as an intermediate. To exemplify, o-DHB is an intermediate in the Fe-assisted oxidation of phenol and when used as a model reactant, o-DHB forms p-DHB. It is hypothesized therefore that o-DHB as an intermediate in phenol Fe-assisted PC conversion forms p-DHB. Hence, the same assumptions for the development of the overall reaction scheme in the unpromoted PC reaction are considered for the Fe-assisted PC system.

The overall reaction network for the oxidation of phenol for the Fe-assisted PC reaction is depicted in Figure 11. The main differences with the unpromoted PC reaction scheme are highlighted with dashed arrows.

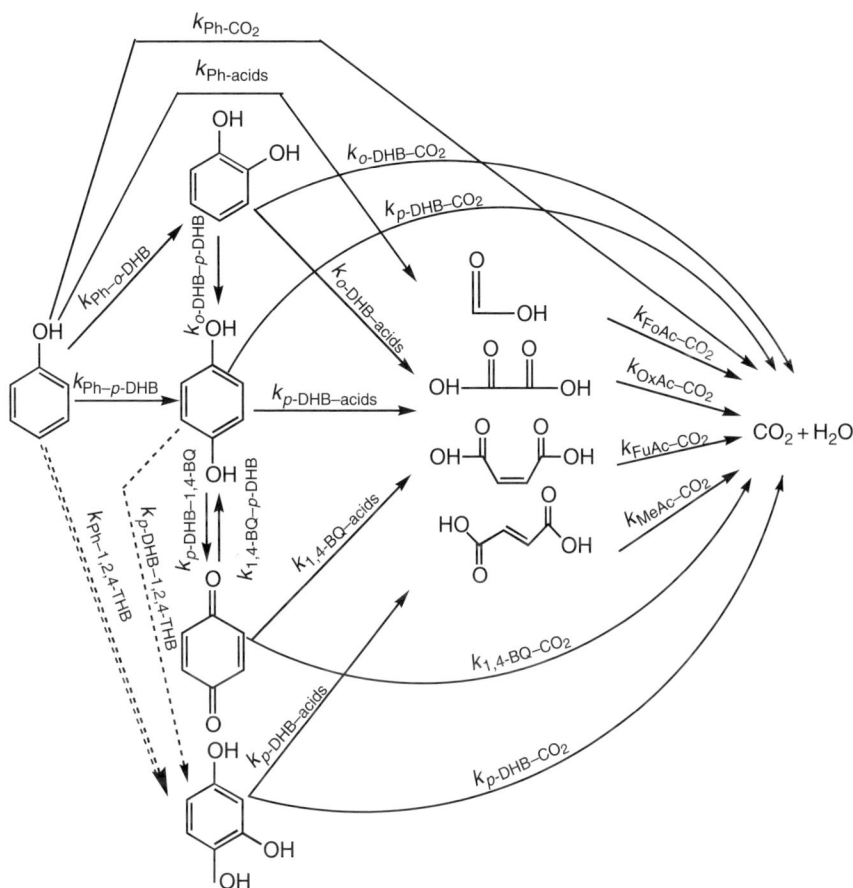

Figure 11 Detailed series–parallel reaction scheme for the unpromoted PC and Fe-assisted PC oxidation of phenol. Broken arrow applies only to PC reaction. Double broken arrow is established as a possible kinetic step not observed at experimental conditions for the reaction (Ortiz-Gomez et al., 2008).

A similar diagram was presented in a previous study for the unpromoted PC oxidation of phenol (Salaices et al., 2004). The reaction scheme introduced in this chapter incorporates all carboxylic acids detected in the oxidations of the various aromatic species, as well as the existing relationships among the intermediate species. A very important fact is that this newly developed reaction network describes the Fe assisted PC oxidation of phenol as well as the unpromoted PC reaction. One important difference between the reaction scheme for the unpromoted PC reaction and that of the Fe-assisted PC reaction is the step relating the formation of 1,2,4-THB from

p-DHB. The 1,2,4-THB was not detected in any of the Fe-assisted runs, and this step is represented in Figure 11 with a broken arrow. All other steps apply to both unpromoted PC and Fe-assisted reactions alike.

The proposed reaction scheme encompasses a series of new contributions over those presented in previous studies. These include the following:

(a) the consideration that both unpromoted PC and Fe-assisted PC oxidation of phenol lead to the formation of the same aromatics o-DHB, p-DHB, and 1,4-BQ, and the same carboxylic acids FuAc, MeAc, OxAc, and FoAc, as reaction intermediates;

(b) the observation that both the unpromoted PC and the Fe-assisted PC oxidation of the three aromatic intermediates (o-DHB, p-DHB, and 1,4-BQ) produce the same aromatic and carboxylic intermediates, except for p-DH that in the Fe-assisted PC reaction no 1,2,4-THB is detected;

(c) the finding that in the oxidation of o-DHB, the p-DHB is formed as an intermediate in small amounts, being the only detected aromatic intermediary;

(d) the observation that the 1,4-BQ is reduced at a high rate to form p-DHB as soon as the catalyst is irradiated. Although the reaction p-DHB $\Leftrightarrow$ 1,4-BQ is a reversible reaction, the formation of p-DHB is more favorable during the PC reaction. The same behavior is observed in both unpromoted PC and Fe-assisted PC reactions;

(e) the finding that 1,2,4-THB formation from phenol can be presented as a possible step, in agreement with a series–parallel scheme for the unpromoted PC oxidation of p-DHB;

(f) the postulate that the reaction step relating the 1,4-BQ formation directly from phenol can be traced to the conversion of p-DHB.

(g) regarding the formation of 1,2,4-THB in the unpromoted PC oxidation of p-DHB, it was observed that during the unpromoted PC oxidation of 1,2,4-THB as a model pollutant (results not shown here), its reaction rate was very fast compared with the oxidation of the other aromatics. Moreover, the Fe-assisted PC reaction of 1,2,4-THB was even faster than the unpromoted PC reaction. Thus, it can be expected that if 1,2,4-THB is formed in the Fe-assisted oxidation of p-DHB, it will disappear very quickly as well, keeping its concentration below the detectable limit.

4. KINETIC MODELING: UNPROMOTED PC OXIDATION AND Fe-ASSISTED PC OXIDATION OF PHENOL

4.1. Overall kinetic model

For the overall series–parallel reaction scheme, a set of differential equations can be developed to describe the rates of formation and disappearance of phenol and all its aromatic and carboxylic intermediates. It is well known that PC reactions occur on the catalyst surface, therefore the rates of formation and disappearance of all components can be modeled using a LH-type rate equation, which takes into account the adsorption of the reactants on the catalyst surface as well as the reaction kinetics. The general form of a LH equation for this system is given by (Ollis et al., 1989)

$$r_i = \frac{k_i^k K_i^A C_i}{1 + \sum_{j=1}^{n} K_j^A C_j} \tag{8}$$

where r_i is the rate of reaction of component i in mol/(g_{cat} min), k_i^k is the kinetic constant for component i in mol/(g_{cat} min), K_i^A is the adsorption constant for component i in LM^{-1}. j is a subscript to denote each component in the denominator term while n is the number of chemical species. In addition, considering that the system in which the experiments were carried out operates in batch mode, a balance equation for each component i can be expressed as follows:

$$\frac{V dC_i}{W dt} = r_i \tag{9}$$

where V is the volume of the reactor in L, W is the weight of the catalyst in g, and t is the time in minutes. By combining Equations (8) and (9), the general form for the rates of reaction for each chemical species is obtained:

$$\frac{dC_i}{dt} = \frac{\frac{W}{V} k_i^k K_i^A C_i}{1 + \sum_{j=1}^{n} K_j^A C_j} \tag{10}$$

Let $k_i = \frac{W}{V} k_i^k K_i^A$, then Equation (10) can be simplified to

$$\frac{dC_i}{dt} = \frac{k_i C_i}{1 + \sum_{j=1}^{n} K_j^A C_j} \tag{11}$$

All rate constants in Equation (11) represent apparent constants. The intrinsic kinetic constant can be calculated using the following relationship (Salaices et al., 2004; Wolfrum and Turchi, 1992):

$$k_i^I = \frac{V_{CST} + V_{PFR}}{V_{PFR}} \tag{12}$$

Thus, by developing one equation with the form of Equation (11) for each component, one can obtain a set of differential equations to represent the PC oxidation of phenol. One should notice, however, that in the following kinetic modeling, the 1,2,4-THB intermediate considered in Figure 11 was omitted in the analysis given that it was not detected directly in the oxidation of phenol. As a result, the 1,2,4-THB formation and consumption steps are not accounted for in the following rate equations.

First, for phenol the rate of reaction is given by

$$\frac{dC_{Ph}}{dt} = \frac{-(k_{Ph\rightarrow Ac} + k_{Ph\rightarrow o\text{-}DHB} + k_{Ph\rightarrow p\text{-}DHB} + k_{Ph\rightarrow CO_2})C_{Ph}}{1 + K_{Ph}^A C_{Ph} + K_{o\text{-}DHB}^A C_{o\text{-}DHB} + K_{p\text{-}DHB}^A C_{p\text{-}DHB} + K_{14-BQ}^A C_{14-BQ} + K_{Ac}^A C_{Ac}}$$

(13)

where $k_{Ph\rightarrow Ac}$ is a lumped kinetic constant that includes all the kinetic constants for the production of acids from phenol and is given by

$$k_{Ph\rightarrow Ac} = k_{Ph\rightarrow FoAc} + k_{Ph\rightarrow OxAc} + k_{Ph\rightarrow MeAc} + k_{Ph\rightarrow FuAc}$$

(14)

and the last term in the denominator involving all the adsorption terms for all carboxylic acids, which is defined as

$$K_{Ac}^A C_{Ac} = K_{FoAc}^A C_{FoAc} + K_{OxAc}^A C_{OxAc} + K_{MeAc}^A C_{MeAc} + K_{FuAc}^A C_{FuAc}$$

(15)

Similar equations can be written for each intermediate as described in Ortiz-Gomez et al. (2008).

4.2. Parameter estimation

The validation of the proposed kinetic model can be done through the estimation of parameters in the equations by fitting the experimental data. The mathematical model with the best parameter estimates can be used to predict the behavior of a system where that model is assumed to describe the process. Since the Ordinary Differential Equation (ODE) system describing the process cannot be solved analytically, the problem is to determine the parameters using a different algorithm that calls for the iterative integration of the ODEs set and the minimization of an objective function (Englezos and Kalogerakis, 2001). For additional details, please refer to Ortiz-Gomez et al. (2007).

4.2.1. Constrained relationships for the estimation of parameters
The analysis of the set of equations reveals that there are two important relationships that must be established before all parameters can be estimated simultaneously. One of them is between $k_{o\text{-}DHB\rightarrow p\text{-}DHB}$ and $(k_{o\text{-}DHB\rightarrow Ac} + k_{o\text{-}DHB\rightarrow CO_2})$ and the second one between $k_{14\text{-}BQ\rightarrow p\text{-}DHB}$ and $k_{p\text{-}DHB\rightarrow 14\text{-}BQ}$, given that these relationships might affect the parameter estimation if they are not clearly defined beforehand. Bearing in mind that the estimation of parameters is based on the minimization of an objective function, the minimization can lead to unfeasible solutions or solutions that

do not fully represent the real phenomena if these relationships are not constrained. For instance, the solver may converge to a solution where $k_{o\text{-DHB}\rightarrow p\text{-DHB}}$ is much larger than $(k_{o\text{-DHB}\rightarrow \text{Ac}} + k_{o\text{-DHB}\rightarrow CO_2})$ and still be a minimum of the objective function. However, from the unpromoted o-DHB reaction, it is quite apparent that o-DHB forms p-DHB in low amounts with most of o-DHB being converted into acids and CO_2. Thus, a very large $k_{o\text{-DHB}\rightarrow p\text{-DHB}}$ would not be an acceptable solution despite the fact that it might lead to a lower value of the objective function. The estimation of these relationships is shown elsewhere (Ortiz-Gomez et al., 2008).

The values of the estimated parameters and their confidence intervals (CI) are presented in Table 1. From these results, a ratio R of $k_{o\text{-DHB}\rightarrow p\text{-DHB}}$ to $k_{o\text{-DHB}\rightarrow CO_2}$ can be established and used to constrain the estimation of parameters in the complete reaction system.

Likewise, the values of the estimated parameters for the second constraint are reported in Table 2. These results show that the reverse reaction kinetic constant $k_{14\text{-BQ}\rightarrow p\text{-DHB}}$ is extremely large compared to forward reaction kinetic constant, $k_{p\text{-DHB}\rightarrow 14\text{-BQ}}$. Similar results were obtained for different 1,4-BQ and p-DHB concentrations. Even more, the estimated value is much lower than its CI. Thus, this kinetic constant is statistically insignificant and should be dropped from the kinetic model altogether. Hence, it can be assumed that in all cases the production of 1,4-BQ from p-DHB will be extremely small. Only in cases where there is some 1,4-BQ at the beginning of the reaction, it will immediately convert to p-DHB once the PC reaction is initiated, and the step considering the formation of p-DHB from 1,4-BQ has to be included. However, if the concentration of 1,4-BQ at the beginning of the reaction is negligible, all terms involving 1,4-BQ concentrations and their related constants can be safely omitted from the kinetic models since 1,4-BQ is not formed in significant amounts.

Table 1 Estimated parameters for the unpromoted PC and Fe-assisted PC oxidation of *ortho*-dihydroxybenzene (20 ppm C in *o*-DHB)

	Unpromoted PC oxidation		Fe-assisted PC oxidation	
	Estimate	CI	Estimate	CI
$k_{o\text{-DHB}\rightarrow CO_2}$ (1/min)	6.19e-04	5.0e-04	1.03e-03	6.5488e-04
$k_{o\text{-DHB}\rightarrow p\text{-DHB}}$ (1/min)	1.49e-04	1.0e-04	1.2270e-04	8.5995e-05
$K^A_{o\text{-DHB}}, K^A_{p\text{-DHB}}$ (1/μM)	5.22e-02	4.28e-2	5.57e-02	3.93e-2
$R = k_{o\text{-DHB}\rightarrow CO_2} / k_{o\text{-DHB}\rightarrow p\text{-DHB}}$	4.15		8.39	

Table 2 Estimated parameters for both unpromoted PC and Fe-assisted PC oxidations of 1,4-benzoquinone

	Unpromoted PC oxidation		Fe-assisted PC oxidation	
	Estimate	CI	Estimate	CI
$k_{p\text{-DHB}\rightarrow 1,4\text{-BQ}}$ (1/min)	2.34e-14	1.01e-04	2.33e-14	1.11e-04
$k_{1,4\text{-BQ}\rightarrow p\text{-DHB}}$ (1/min)	2.70e-03	1.20e-03	2.70e-03	1.30e-03
$k_{p\text{-DHB}\rightarrow CO_2}$ (1/min)	1.41e-04	1.00e-05	2.51e-04	1.00e-04
$k_{1,4\text{-BQ}\rightarrow CO_2}$ (1/min)	1.52e-04	2.00e-04	1.32e-04	2.01e-04
$K^A_{1,4\text{-BQ}}, K^A_{p\text{-DHB}}$ (1/μM)	3.20e-03	3.1e-03	4.9e-03	4.1e-03

Kinetic model #1 (KM#1): aromatics only   The first proposed model considers that those aromatic intermediates produced in small amounts can be neglected and that all remaining aromatics are converted directly into CO_2 and water (e.g., formation of carboxylic acids is neglected). A schematic representation of this reaction network is given in Figure 12.

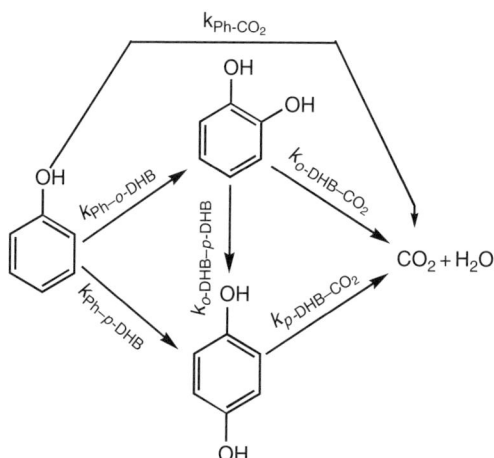

Figure 12   Schematic representation of reaction network for kinetic model #1 (KM#1) (Ortiz-Gomez et al., 2008).

This proposed model assumes the following:

(a) The concentration and rate constants for 1,4-BQ, 1,2,4-THB, and all carboxylic acids are neglected.
(b) There is an immediate conversion of phenol to CO_2, as demonstrated from the TOC profile, thus the constant $k_{Ph \to CO_2}$ is retained.
(c) The ratio R is included in the model to constrain the ratio of o-DHB producing p-DHB and CO_2.

The resulting differential equations are as follows

$$\frac{dC_{Ph}}{dt} = \frac{-(k_{Ph \to CO_2} + k_{Ph \to o\text{-}DHB} + k_{Ph \to p\text{-}DHB})C_{Ph}}{1 + K_{Ph}^A C_{Ph} + K_{o\text{-}DHB}^A C_{o\text{-}DHB} + K_{p\text{-}DHB}^A C_{p\text{-}DHB}} \qquad (16)$$

$$\frac{dC_{o\text{-}DHB}}{dt} = \frac{k_{Ph \to o\text{-}DHB}C_{Ph} - \left[\left(\frac{k_{o\text{-}DHB \to p\text{-}DHB}}{R} + k_{o\text{-}DHB \to CO_2}\right)C_{o\text{-}DHB}\right]}{1 + K_{Ph}^A C_{Ph} + K_{o\text{-}DHB}^A C_{o\text{-}DHB} + K_{p\text{-}DHB}^A C_{p\text{-}DHB}} \qquad (17)$$

$$\frac{dC_{p\text{-}DHB}}{dt} = \frac{k_{Ph \to p\text{-}DHB}C_{Ph} + \frac{k_{o\text{-}DHB \to p\text{-}DHB}}{R}C_{o\text{-}DHB} - k_{p\text{-}DHB \to CO_2}C_{p\text{-}DHB}}{1 + K_{Ph}^A C_{Ph} + K_{o\text{-}DHB}^A C_{o\text{-}DHB} + K_{p\text{-}DHB}^A C_{p\text{-}DHB}} \qquad (18)$$

The results for the estimation of parameters using the data for three different concentrations (20, 30, and 40 ppm C in phenol) are shown in Figure 13. Note that this model provides a good fit of the experimental data for both systems. The estimated rate parameters and their

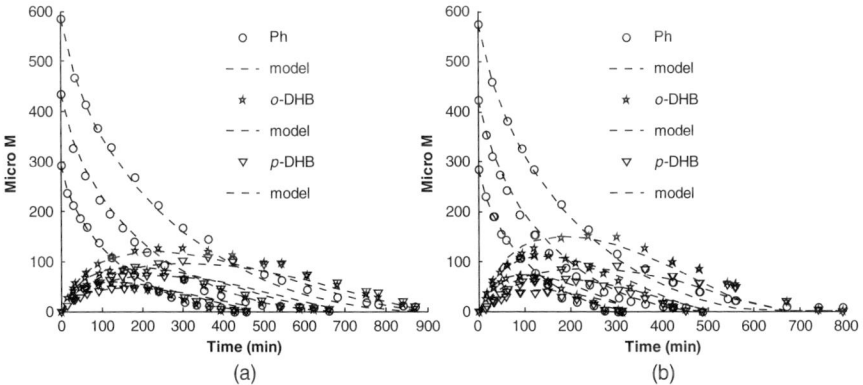

Figure 13 Experimental and predicted profiles of phenol, *ortho*-dihydroxybenzene, and *para*-dihydroxybenzene using KM#1 for (a) unpromoted PC oxidation and (b) Fe-assisted PC oxidation. Simultaneous parameter evaluation of 20, 30, and 40 ppm C in phenol (Ortiz-Gomez et al., 2008).

Table 3 Estimated parameters with KM#1 tcqazIn spite of these observations, it ation of phenol

	Unpromoted PC oxidation		Fe-assisted PC oxidation	
	Estimate	CI	Estimate	CI
$k_{Ph \rightarrow CO_2}$ (1/min)	1.14e-04	1.01e-04	1.15e-04	1.05e-04
$k_{Ph \rightarrow o\text{-}DHB}$ (1/min)	3.90e-04	2.46e-04	6.75e-04	5.66e-04
$k_{Ph \rightarrow p\text{-}DHB}$ (1/min)	2.49e-04	1.58e-04	2.98e-04	2.54e-04
$k_{o\text{-}DHB \rightarrow CO_2}$ (1/min)	5.03e-04	3.16e-04	6.97e-04	5.78e-04
$k_{p\text{-}DHB \rightarrow CO_2}$ (1/min)	5.39e-04	3.31e-04	8.02e-04	6.62e-04
K_{Ph}^A (1/μM)	1.03e-02	8.00e-03	1.26e-02	1.02e-02
$K_{o\text{-}DHB}^A, K_{p\text{-}DHB}^A$ (1/μM)	3.78e-02	2.82e-02	3.78e-02	3.370e-02

Simultaneous parameter evaluation for 20, 30, and 40 ppm C in phenol.

corresponding CI for this case are given in Table 3. This was done with the additional assumption that the adsorption constants for o-DHB and p-DHB are equal.

A comparison between the kinetic parameters obtained for both unpromoted PC and Fe-assisted PC is shown in Figure 14 (they are displayed in the same order as shown in the tables). One can observe that for the simultaneous concentrations of 20, 30, and 40 ppm C in phenol, the estimates obtained for the Fe-assisted PC oxidation are also higher than those estimated for the unpromoted PC reaction. This comparison allows to corroborate the model adequacy by producing estimates consistent with the experimental observations. That is, kinetic parameters in the Fe-assisted PC oxidation must be larger than in the unpromoted PC reaction.

Kinetic Model #2 (KM#2): lumped acids and CO_2 production   The kinetic model #1 considers only the oxidation of the major aromatic intermediates. As shown in the previous section, when most of the major intermediates have been depleted, there is still a substantial concentration of other remaining organic intermediates, as the TOC profile indicates. Therefore, it is of particular interest to calculate and predict the total mineralization times. Also, with TOC measurements, it is possible to approximate the amount of CO_2 produced in the course of the reaction. In this new series–parallel model, the formation and disappearance of carboxylic acids as well as the production of CO_2 has been incorporated.

The representation of the experimental data with this new approach allows one to infer important information that can be applied for the estimation of parameters in this new model. When the summation of the

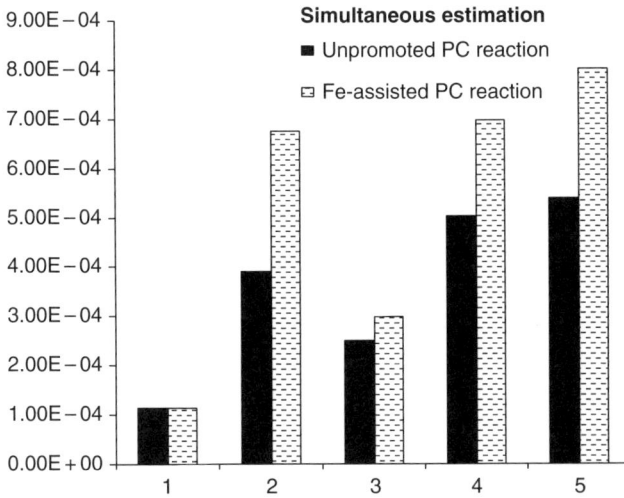

Figure 14 Estimates for simultaneous concentrations (20, 30, and 40 ppm C in phenol) (Ortiz-Gomez et al., 2008).

organic carbon due to the detected aromatic components (OC_{AR}) is compared with the TOC, it is observed that as the reaction proceeds both profiles divert from one another, as reported in Figure 15. The difference between these lines is employed to represent the amount of organic carbon contained in the carboxylic acids (OC_{AC}). Therefore, OC_{AC} can be calculated by subtracting OC_{AR} from TOC. If this amount is considered as a *lumped* concentration, a new kinetic model can be developed using this information.

Therefore, a number of relationships can be applied. The sum of all organic carbon (OC) in aromatics is given by

$$OC_{AR} = \sum_{i=1}^{n_A} C_{aromatics} \qquad (19)$$

where n_A is the number of aromatics. Thus, the difference between TOC and OC_{AR} is the organic carbon due to carboxylic acids

$$OC_{Ac} = TOC - OC_{AR} \qquad (20)$$

The amount of CO_2 produced during the course of the reaction is approximated by the difference between the initial concentration of total organic carbon and the concentration of TOC at any given time, as expressed in the following equation:

$$CO_2^{Prod} = [TOC_0] - [TOC] \qquad (21)$$

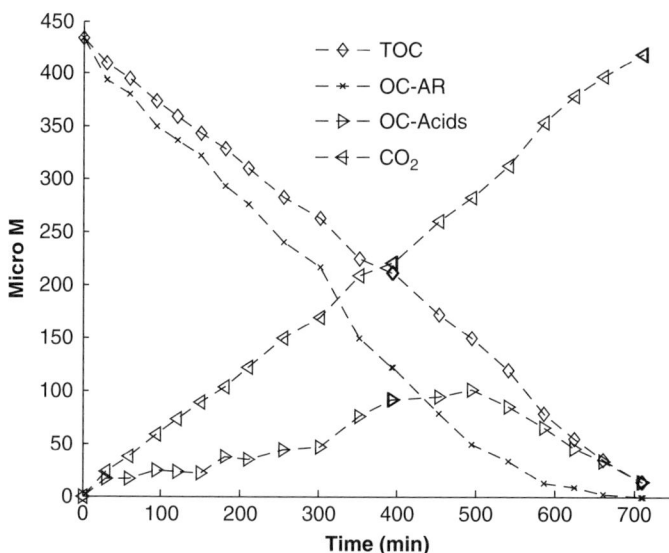

Figure 15 Comparison between OC_{AR} and TOC for the unpromoted PC oxidation of 30 ppm C in phenol. TOC and CO_2 profiles are in a different scale (six times higher than that shown on Y-axis) (Ortiz-Gomez et al., 2008).

where CO_2^{Prod} is the amount of CO_2 produced and $[TOC_0]$ is a vector containing the initial amount of TOC in all its elements. The number of data points for each experiment dictates its length. The CO_2^{Prod} profile is also shown in Figure 15. In Figure 15, both TOC and CO_2 profiles are reported in $\mu mol\,L^{-1}$ (μM) and are shown in a different scale for ease of comparison. The scale is six times higher than that shown on the Y-axis.

A summary of main assumptions for this model is as follows:

(a) 1,4-BQ and 1,2,4-THB terms are neglected.
(b) All carboxylic acids are *lumped* in one term
(c) All aromatic intermediates produce both carboxylic acids and CO_2 during the reaction.
(d) The amount of CO_2 produced is incorporated in the model. CO_2 is not adsorbed onto the catalyst surface and its concentration is considered to be cumulative throughout the course of the reaction.

Figure 16 shows a schematic representation of this new reaction network.

The estimation of parameters in this model was done testing different combinations to determine those parameters that were statistically significant and to obtain a kinetic model that could describe the experimental data over a wide range of concentrations with narrow CI. The model that provided the best estimates with the narrowest CI was obtained from the

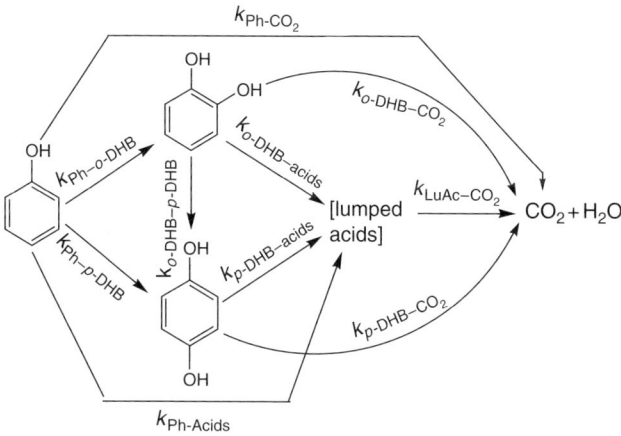

Figure 16   Schematic representation of reaction network for kinetic model #2 (KM#2) (Ortiz-Gomez et al., 2008).

reaction scheme presented in Figure 17. One can observe that some reaction steps have been neglected.

The results obtained with the simplified KM#2 for the unpromoted PC oxidation of phenol are shown in Figure 18a and b.

The concentration profiles for the same components are given when the estimation of parameters is performed for all 30, 40, and 50 ppm C in phenol. CO_2 profiles are shown in a different scale for ease of comparison. One can notice that the simplified KM#2 provides very good fitting for both single and simultaneous evaluations, which proves the validity of the model for the unpromoted PC oxidation of phenol.

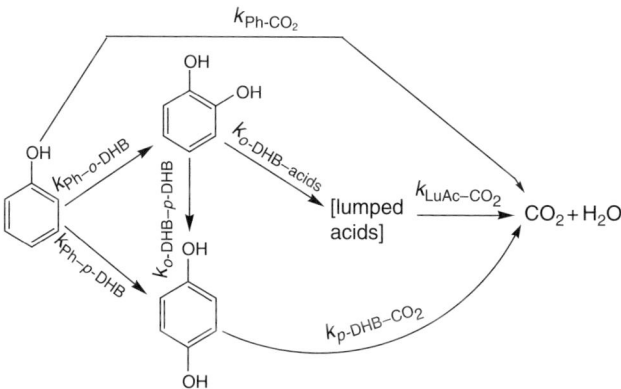

Figure 17   Reaction scheme obtained after testing various scenarios based on the original scheme used to develop kinetic model #2 (Ortiz-Gomez et al., 2008).

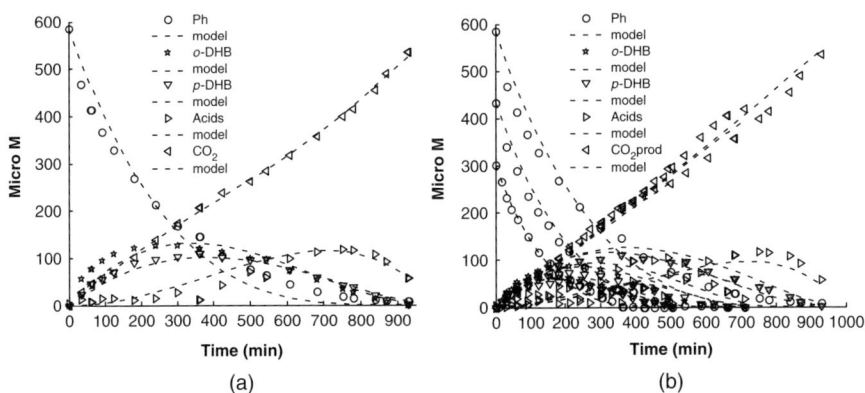

Figure 18 (a) Experimental and predicted profiles for the unpromoted PC oxidation of 40 ppm C in phenol using KM#2. (b) Experimental and predicted profiles for the unpromoted PC oxidation of 30, 40, and 50 ppm C in phenol using KM#2 (Ortiz-Gomez et al., 2008).

The statistically meaningful estimated parameters for this example are presented in Table 4. In this case, it was assumed that the adsorption constants for phenol, o-DHB, and p-HB were equal ($K^A_{Ph} = K^A_{p\text{-DHB}} = K^A_{o\text{-DHB}}$). All other combinations led to similar solutions with very wide CIs.

Likewise, the results using the simplified KM#2 for the estimation of parameters in the Fe-assisted PC reaction are reported in Figure 19. This figure shows the experimental and predicted concentration profiles obtained for the simultaneous evaluation of 30, 40, and 50 ppm C in phenol. It can be seen that the proposed model provides a good fit to the experimental data for this system as well. The estimated parameters and their corresponding CI for this case are given in Table 5.

Thus, it can be concluded that both KM#1 and KM#2 are adequate for a single concentration level or for wide range of concentrations, showing the usefulness of such models in the prediction of formation and disappearance of the model reactant and its oxidation intermediates. A comparison between the kinetic parameter estimates obtained for both unpromoted PC and Fe-assisted PC is given in Figure 20a and b. One can observe that the estimated parameters are close for most of the kinetic constants as predicted by the KM#2 model. This corroborates that this kinetic approach is suitable for the prediction of concentration profiles of phenol and its oxidation intermediates.

This section shows that both models such as KM#1 and KM#2 are suitable for the prediction of concentration profiles. Although many reaction intermediates are experimentally identified, not all of them can be included in the kinetic modeling due to their low concentrations and thus, their low

Table 4 Estimated parameters using reduced KM#2 for the simultaneous estimation of 30, 40, and 50 ppm C in phenol in the unpromoted PC oxidation

	Unpromoted PC oxidation	
	Simultaneous 30, 40, 50 ppm C in phenol	
	Estimate	CI
$k_{Ph \to CO_2}$ (1/min)	7.86e-04	6.66e-04
$k_{Ph \to o\text{-DHB}}$ (1/min)	9.58e-04	8.58e-04
$k_{Ph \to p\text{-DHB}}$ (1/min)	8.42e-04	7.92e-04
$k_{o\text{-DHB} \to LuAc}$ (1/min)	5.31e-04	4.13e-04
$k_{p\text{-DHB} \to CO_2}$ (1/min)	8.85e-04	7.52e-04
$k_{LuAc \to CO2}$ (1/min)	2.05e-04	1.75e-04
$K_{Ph}^A, K_{o\text{-DHB}}^A, K_{p\text{-DHB}}^A$ (1/µM)	7.23e-02	6.27e-02

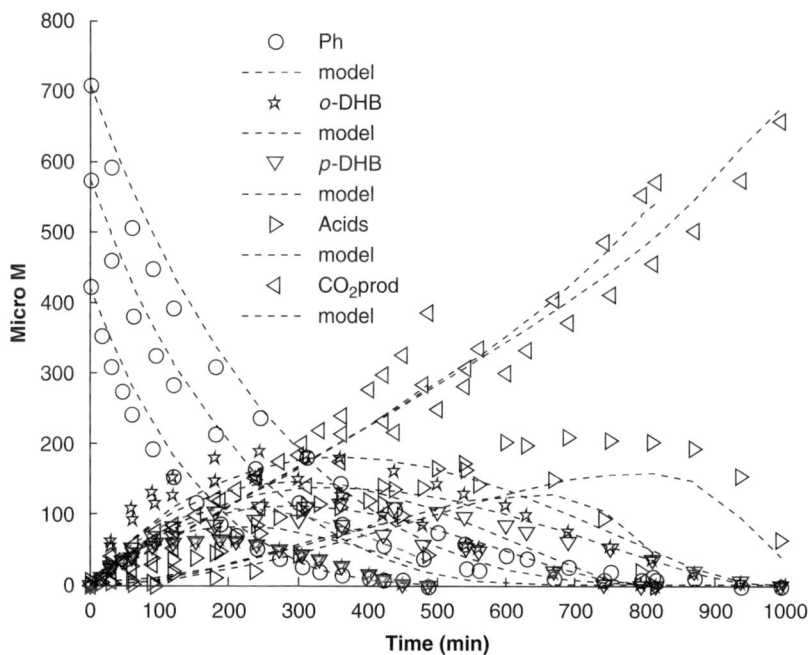

Figure 19 Experimental and predicted profiles for the Fe-assisted PC oxidation of 30, 40, and 50 ppm C in phenol using reduced KM#2 (Ortiz-Gomez et al., 2008).

significance in the parameter estimation. This shows as well that in cases like this, complex kinetic networks may be developed including all experimentally identified intermediates, but the kinetic models need to be

Table 5 Estimated parameters using reduced KM#2 for the Fe-assisted PC oxidation of 40 ppm C in phenol and for the simultaneous estimation of 30, 40, and 50 ppm C in phenol

	Fe-assisted PC oxidation	
	Simultaneous 30, 40, 50 ppm C in phenol	
	Estimate	CI
$k_{Ph \to CO_2}$ (1/min)	6.12e-04	5.03e-04
$k_{Ph \to o\text{-DHB}}$ (1/min)	1.30e-03	1.25e-04
$k_{Ph \to p\text{-DHB}}$ (1/min)	9.87e-04	7.92e-04
$k_{o\text{-DHB} \to LuAc}$ (1/min)	7.10e-04	4.56e-04
$k_{p\text{-DHB} \to CO_2}$ (1/min)	1.10e-03	8.58e-04
$k_{LuAc \to CO_2}$ (1/min)	2.60e-04	2.06e-04
$K_{Ph}^A, K_{o\text{-DHB}}^A, K_{p\text{-DHB}}^A$ (1/μM)	6.33e-03	5.47e-02

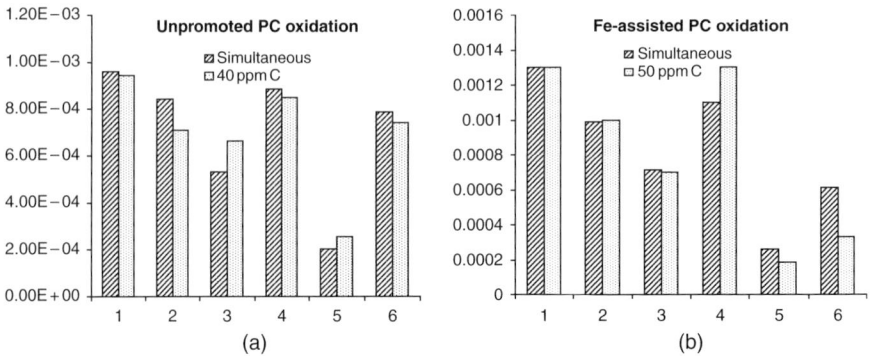

Figure 20   (a) Comparison between estimates obtained using KM#2 for unpromoted PC oxidation for 40 ppm C in phenol and simultaneous estimation. (b) Comparison between estimates obtained using KM#2 for Fe-assisted PC oxidation for 50 ppm C in phenol and simultaneous estimation (Ortiz-Gomez et al., 2008).

simplified due to the large number of parameters and their low statistical significance. It is clear that identifying as many reaction intermediates as possible helps understand the kinetic pathways but such information may not be readily applied in the development of kinetic models. Lastly, although these kinetic models are more comprehensive than those shown in other studies, they still need to be revisited to incorporate the effect of the radiation field in the kinetic constants so that they could be extensively used in reactor design.

5. CONCLUSIONS

The following are the main conclusions drawn from this study.

(a) It is shown that Fe cations have a strong influence on the phenol PC reaction. High concentrations lead to a decrease in the mineralization rates while low concentration promotes a significant increase. It is found that 5 ppm of Fe^{3+} ions renders the highest phenol oxidation rate. Moreover, it is demonstrated that ferric ions Fe^{3+} and ferrous ions Fe^{2+} promote the same enhancement. It is proposed that Fe cations increase the mineralization rate by facilitating the transfer of photogenerated electrons e_{cb}^- to electron scavengers. This occurs through a continuous oxidation–reduction cycle of Fe cations adsorbed onto the catalyst surface.

(b) Additionally, it is demonstrated that the Fe-assisted PC oxidation of phenol can be represented with a series–parallel kinetic reaction scheme. Likewise, the Fe-assisted PC oxidations of reaction intermediates o-DHB, p-DHB, and 1,4-BQ can be represented with a series–parallel reaction network. An overall kinetic reaction scheme for the Fe-assisted PC reaction of phenol is proposed. Its similarities and differences with that derived for the unpromoted PC reaction, as well as the main contributions, are discussed in detail.

(c) Lastly, a kinetic model based on the overall kinetic reaction network proposed for both unpromoted PC reaction and Fe-assisted PC reaction is developed using LH-type kinetics for each chemical species. Given that the concentrations of some of the intermediates in both systems are very small, this allows the simplification of the overall kinetic model. Subject to simplifying assumptions, two simplified models (KM#1 and KM#2) are obtained.

(d) The KM#1 is developed to predict the rate of reaction of phenol and its major aromatic intermediates. A parameter representing the ratio of o-DHB converted into p-DHB to o-DHB converted into acids and CO_2 was incorporated in the kinetic model originally presented by Salaices et al. (2004). This parameter was represented as R. The resulting model provided very good and statistically meaningful fitting for both unpromoted PC and Fe-assisted PC oxidations. In this case, the estimated parameters are consistent with the experimental data, with the parameters from the Fe-assisted PC reaction being larger than those of the unpromoted PC system.

(e) The KM#2 is developed considering a lumped acid concentration and the formation of CO_2. This model predicts the formation and disappearance of aromatic intermediates, carboxylic acids, and CO_2 in the course of the PC reaction. It provides very good fit of the experimental data and works very well for a wide range of phenol concentrations (20–50 ppm C in phenol). The estimated parameters in this case are also consistent with the experimental data.

(f) In this study, it was demonstrated that an environmentally friendly and highly abundant metal such as Fe could be readily used for the enhancement of the mineralization rates of a wide variety of recalcitrant organic pollutants.

RECOMMENDATIONS

Based on the assertions above, various recommendations could be outlined for future work in this research area. The method employed for the enhancement of PC oxidation and the type of the reactors used, albeit low cost, present two significant limitations: loss of catalyst and use of UV radiation for catalyst activation. Therefore, two major areas must continue to receive special attention: catalyst recovery and use of visible radiation to make use of an inexhaustible radiation source, the solar radiation.

To increase catalyst recovery, efforts have been directed toward immobilization of the photocatalyst on a wide variety of materials. However, there is an inherent decrease of catalytic activity due to a reduced surface area. As such, Fe could be used to minimize or compensate for the catalyst deactivation due to the support preparation. New methods must be developed to ensure Fe ions stay adsorbed onto the photocatalyst surface to avoid further reactivation posttreatments.

Regarding the use of solar radiation, new materials including doped TiO_2, capable of being activated with visible irradiation, must be developed. Development of photocatalysts with such features will undoubtedly help take advantage of the large regions with high solar radiation intensities. Also careful attention should be given to the feasibility of using sophisticated catalyst-doping procedures in the preparation of these improved semiconductor materials.

LIST OF SYMBOLS

Variables

C	Pollutant concentration, μM
r_i	Rate of reaction, $\mu\ mol/(g_{cat}\ min)$
R	Ratio of $k_{o\text{-DHB}\rightarrow CO_2}$ to $k_{o\text{-DHB}\rightarrow p\text{-DHB}}$
t	Time, min
V	Volume, L
V_{CST}	Volume of CST, L
V_{PFR}	Volume of PFR, L
W	Weight of catalyst, g

Parameters

k Apparent kinetic constant
k^I Intrinsic kinetic constant
k^k Kinetic constant, $M/(g_{cat}\,min)$
K^A Adsorption constant $1/\mu M$

Subscripts

Ac	Carboxylic acids
CO_2	Carbon dioxide
FoAc	Formic acid
FuAc	Fumaric acid
1,2,4-THB	1,2,4-Trihydroxybenzene
1,4-BQ	1,4-Benzoquinone
i	Chemical species
j	Chemical species
MeAc	Maleic acid
Ph	Phenol
OxAc	Oxalic acid
o-DHB	*ortho*-dihydroxybenzene
p-DHB	*para*-dihydroxybenzene
CST	Continuous stirred tank
PFR	Plug flow reactor

Acronyms

1,4-BQ	1,4-Benzoquinone
1,2,3-THB	1,2,3-Trihydroxybenzene
1,2,4-THB	1,2,4-Trihydroxybenzene
CI	Confidence intervals
Fe-assisted PC	Photocatalytic reaction assisted with Fe ions
HPLC	High-performance liquid chromatography
KM#1	Kinetic model #1
KM#2	Kinetic model #2
Unpromoted PC	Photocatalytic reaction in the absence of Fe ions.
OQY	Overall quantum yield
o-DHB	*ortho*-dihydroxybenzene
p-DHB	*para*-dihydroxybenzene
ppm C	Parts per million of carbon
Photo-CREC	Chemical reactor designed at the Chemical Reaction Engineering Centre
PC	Photocatalytic
TOC	Total organic carbon
TiO_2	Titanium dioxide
UV	Ultraviolet

REFERENCES

Agency for Toxic Substances and Disease Registry (ATSDR). Public Health Service, US Department of Health and Human Services, Atlanta, GA (1998).

Aguado, M.A., Cervera-March, S., and Gimenez, J. *Chem. Eng. Sci.* **50**(10), 1561 (1995).

Aguado, J., Grieken, R.V., López-Muñoz, M.J., and Marugán, J. *Catal. Today* **75**, 95 (2000).

Al-Ekabi, H., and Serpone, N. *J. Phys. Chem.* **92**, 5726 (1988)

Al-Ekabi, H., Serpone, N., Pelizzetti, E., Minero, C., Fox, M.A., and Draper, B.R. *Langmuir* **5**, 250 (1989).

Araña, J., González, O., Doña, J.M., Herrera, J.A., Garriaga, C., Pérez, J., Carmen, M., and Navío-Santos, J. *J. Mol. Catal. A: Chem.* **197**, 157 (2003).

Araña, J., González, O., Doña, J.M., Herrera, J.A., Tello Rendon, E., Navío-Santos, J., and Perez Peña, J. *J. Mol. Catal. A: Chem.* **215**, 153 (2004).

Araña, J., González, O., Miranda, M., Doña, J.M., and Herrera, J.A. *Appl. Catal. B: Environ.* **36**, 113 (2002).

Araña, J., González, O., Miranda, M., Doña, J.M., Herrera, J.A., and Pérez, J. *Appl. Catal B: Environ.* **36**, 49 (2001).

Assabane, A., Ichou, Y.A., Tahiri, H., Guillard, C., and Herrmann, J.M. *Appl. Catal. B: Environ.* **24**, 71 (2000).

Bahnemann, D. *Solar Energy* **77**, 445 (2004).

Bamwenda, G.R., Tsubota, S., Nakamura, T., and Haruta, M. *J. Photochem. Photobiol. A: Chem.* **89** (1995).

Blake, Daniel. M., Bibliography of work on the Heterogeneous Photocatalytic removal of Hazardous Compounds from Water and Air, National Renewable Energy Laboratory (NREL). Update No. 4 to October (2001).

Bouquet-Somrani, C., Finiels, A., Graffin, P., and Olivé, J.L. *Appl. Catal. B: Environ. B* **8**, 101 (1996).

Carraway, E.R., Hoffman, A.J., and Hoffman, M.R., *Environ. Sci. Tech.* **28**, 766 (1994).

Cermenati, L., Pichat, P., Guillard, C., and Albini, A. *J. Phys. Chem. B* **101**, 2650 (1997).

Chen, D., and Ray, A.K. *Water Res.* **32**, 11, 3223 (1998).

Chen, D., and Ray, A.K. *Appl. Catal. B* **23**, 143 (1999).

Chen, D., and Ray, A.K. *Chem. Eng. Sci.* **56**, 1561 (2001).

Chenthamarakshan, C.R., and Rajeshwar, K. *Langmuir* **16**, 2715 (2000).

Chenthamarakshan, C.R., and Rajeshwar, K. *Electrochem. Comm.* **2**, 527 (2002).

Colmenares, J.C., Aramendia, M.A., Marinas, A., Marinas, J.M., and Urbano, F.J. *Appl. Catal. A: Gen.* **306**, 120 (2006).

Colon, G., Hidalgo, M.C., and Navío, J.A. *J. Photochem. Photobiol. A: Chem.* **138**, 78 (2001).

Cotton, F.A., Wilkinson, G., Murillo, C., and Bochmann, M. "Advanced Inorganic Chemistry", Chapter 17, Wiley, New York (1999), p. 692.

Dai, Q., and Rabani, J. *J. Photochem. Photobiol. A: Chem.* **148**, 17 (2002).

de Lasa, H., Serrano, B., and Salaices, M. "Photocatalytic Reaction Engineering". Springer, New York (2005).

Domenech, X., Jardim, W.F., and Litter, M.I. Advanced oxidation Processes for Contaminant Elimination. *in* M.A. Blesa, and B. Sánchez (Eds.), "Elimination of Contaminants Through Heterogeneous Photocatalysis", CIEMAT (2004).

Englezos, P., and Kalogerakis., N. "Applied Parameter Estimation for Chemical Engineers", Chapter 2, 7 Marcel Dekker, Inc. New York (2001).

Fox, M.A. *in* N. Serpone, and E. Pelizzeti (Eds.), "Photocatalysis – Fundamentals and Applications". Wiley-Interscience, New York (1990), p. 421.

Franch, M.I. Ayllón, A.J., Peral, J., and Domenech, X. *Catal. Today* **101**, 245 (2005).

Fujishima, A., Rao, T.N., and Tryk, D.A. *J. Photochem. Photobiol. C: Photochem. Rev.* **1**, 1 (2000).

Goi, A., and Trapido, M. *Proc. Estonian Acad. Sci. Chem.* **50**(1), 5 (2001).

Grela, M.A., Coronel, M.E.J., and Colussi, A.J., *J. Phys. Chem.* **100**, 16940 (1996).

Guillard, C. *J. Photochem. Photobiol. A.* **158**, 27 (2003).

Herrmann, J.M. *Catal. Today* **53**, 115 (1999).

Huang, M., Tso, E., Datye, A.K., Prairie, M.R., and Stange, B.M. *Environ. Sci. Tech.* **30**, 3084 (1996).

Hufschmidt, D., Bahnemann, D., Testa, J.J., Emilio, C.A., and Litter, M.I. *J. Photochem. Photobiol. A: Chem.* **148**, 223 (2002).

Jakob, L., Hashem, T.M., Burki, S., Guindy, N., and Braun, A.M. (1993a).

Jakob, L., Oliveros, E., Legrini, O., and Braun, A.M. *in* Ollis, D., and Al-Ekabi, H. (Eds.), "Photocatalytic Purification and Treatment of Water and Air". Elsevier, New York (1993b), p. 511.

Karamanev, D.G., Nikolov, L.N., and Mamatarkova, V. *Min. Eng.* **15**, 341 (2002).

Karvinen, S., and Lamminmaki, R.-J. *Solid State Sci.* **5**, 1159 (2003).

Kavitha, V., and Palanivelu, K. *Chemosphere* **55**, 1235 (2004).

Kim, T.K., Lee, M.N., Lee, S.H., Park, Y.C., Jung, C.K., and Boo, J.H. *Thin Solid Films* **475**, 171 (2005).

Leyva, E., Moctezuma, E., Ruíz, M.G., and Torres-Martinez, L. *Catal. Today* **40**, 367 (1998) .

Li, X., Cubbage, J.W., Jenks, and W.S. *J. Org. Chem.* **64**, 8525 (1999b).

Li, X., Cubbage, J.W., Tetzlaff, T.A., and Jenks, W.S. *J. Org. Chem.* **64**, 8509 (1999a).

Li Puma, G, and Lock Yue, P., *Ind. Eng. Chem. Res.* **38**, 3246 (1999).

Malato, S., Blanco-Galvez, J., and Estrada-Gasca, C. *Photocatal. Guest Editorial Solar Energy* **77**, 443 (2004).

Mathews, R.W., and McEvoy, S.R. *Solar Energy* **49**(6), 507 (1992).

McDonald, T.A., Holland, N.T., Skibola, C., Duramad, P., and Smith, M.T. *Leukemia* **15**, 10 (2001).

Minero, C., Catozzo, F., Pelizzetti, E. *Langmuir*, **8**, 481 (1992).

Mukherjee, P.S., and Ray, A.K. *Chem. Eng. Tech.* **22**(3), 253 (1999).

Nagaveni, K., Hegde, M.S., and Madras, G. *J. Phys. Chem. B* **108**, 20204 (2004).

Navío, J.A., Colón, G., Litter, M.I. and Bianco, G.N. *J. Mol. Catal. A: Chem.* **106**, 267 (1996).

Navío, J.A., Testa, J.J., Djedjeian, P., Padrón, J.R., Rodríguez, D., and Litter, M. *Appl. Catal. A: Gen.* **178**, 191 (1999).

Oh, S.M., Kim, S.S., Lee, J.E., Ishigaki, T., and Park, D.W. *Thin Solid Films* **435**, 252 (2003).

Okamoto, K., Yamamoto, Y., Tanaka, H., and Tanaka, M. *Bull. Chem. Soc. Jpn.* **58**, 2015 (1985).

Ollis, D.F., Pelizzeti, E., and Serpone, N. *in* N. Serpone, and E. Pelizzeti (Eds.) "Photocatalysis Fundamentals and Applications". Wiley Interscience, New York (1989), p. 603.

Ortiz-Gomez, A. Ph.D., "Enhanced Mineralization of Phenol and other Hydroxylated Compounds in a Photocatalytic Process assisted with Ferric Ions". Thesis Dissertation, The University of Western Ontario, Ontario (2006).

Ortiz-Gomez, A., Serrarro-Rosales, B., and de Lasa, H. *Chem. Eng. Sci.* **63**, 520 (2008).

Ortiz-Gomez, A., Serrano-Rosales, B., Salaices, M., and de Lasa, H. *Ind. Eng. Chem. Res.* **46**, 23, 7394 (2007).

Pellizetti, E. *Solar Energy Mater. Solar Cells* **38**, 453 (1995).

Pichat, P. *Water Sci. Tech.* **4**, 73 (1997).

Pignatello, J.J., Liu, D., and Huston, P. *Environ. Sci. Tech.* **33**, 1832 (1999).

Poulopoulos, S.G., and Philippopoulous, C.J. *J. Environ. Sci. Health A* **39**(6), 1385 (2004).

Preis, S., Terentyeva, Y., and Rozkov, A. *Water. Sci. Tech.* **35**(4), 165 (1997).

Ray, A.K. *Dev. Chem. Eng. Mineral. Proc.* **5**(1/2), 115 (1997).

Riegel, G., and Bolton, J.R. *J. Phys. Chem.* **99**(12), 4215 (1995).

Rothenberger, G., Monser, J., Gratzel, M., Serpone, N., and Sharma, D.K. *J. Am. Chem. Soc.* **107**, 8054 (1985).

Salaices, M., Serrano, B., and de Lasa, H. *Chem. Eng. Sci.* 59, 3 (2004).

Sokmen, M., and Ozkan, A. *J. Photochem. Photobiol. A.* **147**, 77 (2002).

Stylidi, M., Kondaridis, D.I., and Verykios, X.E. *Appl. Catal. B: Environ.* **40**, 271 (2003).

Tan, T., Beydoun, D., and Amal, R. *J. Photochem. Photobiol. A.* **159**, 273 (2003).

Trillas, M., Pujol, M., and Doménech, X. *J. Chem. Tech. Biotech.* **55**, 85 (1992).

Tryba, B., Morawski, A.W., Inagaki, M., and Toyoda, M. *Appl. Catal. B: Environ.* **63**, 215 (2006a).

Tryba, B., Morawski, A.W., Inagaki, M., and Toyoda, M. *J. Photochem. Photobiol. A: Chem.* **179**, 224 (2006b).

Tseng, J., and Huang, C.P. Mechanistic aspects of the photocatalytic oxidation of phenol in aqueous solutions. *in* D.W. Tedder, and F.G. Pohland (Eds.), "Emerging Technologies in Hazardous Waste Management", Chapter 2. Amer. Chem. Soci., Washington, DC, (1990), pp. 12–39.

Turchi, C.S., and Ollis, D.F. *J. Catal.* **119**, 483 (1989).

Turchi, C.S., and Ollis, D.F. *J. Catal.* **122**, 178 (1990).

US Environmental Protection Agency. "Phenol-Hazard Summary". Reviewed on 2000. US Environmental Protection Agency website (www.epa.gov) (2000).

Vamathevan, V., Amal, R., Beydoun, D., Low, G., and McEvoy, S. *J. Photochem. Photobiol. A: Chem.* **148**, 233 (2002).

Wei, T.Y., and Wan, C., *J. Photochem. Photobiol. A: Chem.* **69**, 241 (1992).

Winterbottom, J.M., Khan, Z., Boyes, A.P., and Raymahasay, S. *Environ. Prog.* **62**, 125 (1997).

Wolfrum, E.J., and Turchi, G.S. *J. Catal.* **136**, 626 (1992).

Wyness, P., Klausner, J.F., and Goswami, D.Y. *J. Solar Energy Eng.* **116**, 8 (1994).

Yang, S., Lou, L., Wang, K., and Chen, Y. *App. Catal. A: Gen.* **301**, 152 (2006).

Zepp, R.G., Faust, B.C., and Holgne, J. *Environ. Sci. Tech.* **26**, 313, (1992).

Zertal, A., Molnár-Gábor, D., Malouki, M.A., Shili, T., and Boule, P. *Appl. Catal. B: Environ.* **49**, 83 (2004).

Zhou, J., Zhang, Y., Zhao, X.S., and Ray, A.J. *Ind. Eng. Chem. Res.* **45**, 3503, (2006).

CHAPTER 4

Photocatalytic Water Splitting Under Visible Light: Concept and Catalysts Development

R.M. Navarro[*]**, F. del Valle, J.A. Villoria de la Mano, M.C. Álvarez-Galván,** and **J.L.G. Fierro**

Contents
1. Introduction 111
2. Photoelectrochemistry of Water Splitting 113
   2.1 Concept 113
   2.2 Configurations 114
   2.3 Energy requirements 117
   2.4 Solar spectrum and water-splitting efficiency 119
3. Photocatalysts for Water Splitting Under Visible Light 124
   3.1 Material requirements 124
   3.2 Strategies for developing efficient photocatalysts under visible light 125
   3.3 Photocatalyst development 130
4. Concluding Remarks and Future Directions 140
Acknowledgments 141
References 141

1. INTRODUCTION

Sunlight in the near-infrared, visible, and ultraviolet regions radiates a large amount of energy and intensity that would contribute significantly to our electrical and chemical needs. At a power level of $1,000\,\mathrm{Wm}^{-2}$, the incidence of solar energy on the earth's surface by far exceeds all human energy needs (Lewis and Nocera, 2006; Lewis et al., 2005). Against the backdrop of the

Instituto de Catálisis y Petroleoquímica (CSIC), C/Marie Curie 2, 28049, Madrid, Spain

[*] Corresponding author.
E-mail address: r.navarro@icp.csic.es

Advances in Chemical Engineering, Volume 36
ISSN 0065-2377, DOI: 10.1016/S0065-2377(09)00404-9

daunting carbon-neutral energy needs for sustainable development in the future, the large gap between our present use of solar energy and its enormous undeveloped potential defines a compelling imperative for science and technology in the twenty-first century. To make a material contribution to the energy supply, solar energy must be captured, converted, and stored to overcome the daily cycle and the intermittency of solar radiation. Undoubtedly, the most attractive method for this solar conversion and storage is in the form of an energy carrier such as hydrogen. The conversion of solar radiation into hydrogen offers the advantages of being transportable as well as storable for extended periods of time. This point is important because energy demand is rarely synchronous with incident solar radiation. Hydrogen is a clean energy carrier because the chemical energy stored in the H—H bond is easily released when it combines with oxygen, yielding only water as a reaction product. Accordingly, a future energy infrastructure based on hydrogen has been perceived as an ideal long-term solution to energy-related environmental problems (Bockris, 2002; Lewis and Nocera, 2006; Ogden, 2003).

The conversion of solar radiation into a chemical carrier such as hydrogen is one of the most important scientific challenges for scientists in the twenty-first century (Lewis and Nocera, 2006; Service, 2005). Solar energy can be used to produce hydrogen in the form of heat (thermochemical, Funk, 2001; Kogan et al., 2000), light (photoelectrochemical, Fujishima and Honda, 1972; photosynthetic, Miyake et al., 1999 or photocatalytic, Bard, 1980), or electricity (electrolysis, Levene and Ramsden, 2007). Solar energy used as light is the most efficient solar path to hydrogen since it does not have the inefficiencies associated with thermal transformation or with the conversion of solar energy to electricity followed by electrolysis. Therefore, the most promising method of hydrogen generation using a source of renewable energy is that based on water decomposition by means of photoelectrochemical or photocatalytic technologies using solar energy. This process mimics photosynthesis by converting water into H_2 and O_2 using inorganic semiconductors that catalyze the water-splitting reaction [Equation (1)]:

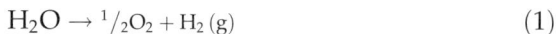

$$H_2O \rightarrow {}^1/_2 O_2 + H_2 (g) \tag{1}$$

The reaction involves the standard Gibbs free energy change (ΔG^0) greater than 237 KJ mol^{-1}, equivalent to 2.46 eV per molecule (1 eV = 96.1 KJ mol^{-1}). This energy is equivalent to the energy of photons with wavelengths shorter than 500 nm. However, direct water splitting cannot be achieved by sunlight because the water molecule cannot be electronically excited by sunlight photons (Balziani et al., 1975). Hence, appropriate systems have to be developed to efficiently absorb solar energy in order to split water in an indirect way. Since the electrochemical decomposition of water to hydrogen and oxygen is a two-electron stepwise process, it is possible to use photocatalytic surfaces capable of absorbing solar energy to generate electrons and holes that can respectively reduce and oxidize the water molecules adsorbed on photocatalysts. The groundbreaking work of Fujishima and Honda in 1972 showed that hydrogen generation via water splitting was possible using photocatalysts based on semiconductors capable of adsorbing light energy. Since this pioneering work, there have been many papers published on the impact of different semiconductor materials on the performance in photocatalytic water splitting. These studies clearly prove that the energy conversion efficiency of water splitting is determined principally by the properties of the semiconductors used as photocatalysts. There has been significant progress in recent years in the research conducted into the development of efficient photocatalysts under visible light, but the maximum efficiency achieved up to now (2.5%) is still far from the required efficiency for practical applications. Consequently, the commercial application for hydrogen generation from solar energy and water will be determined by future progress in material science and engineering applied to the development of efficient semiconductors used as photocatalysts.

2. PHOTOELECTROCHEMISTRY OF WATER SPLITTING

2.1. Concept

Photochemical water splitting, like other photocatalytic processes, is initiated when a photosemiconductor absorbs light photons with energies greater than its band-gap energy (E_g). This absorption creates excited photoelectrons in the conduction band (CB) and holes in the valence band (VB) of the semiconductor, as schematically depicted in Figure 1a. As indicated in Figure 1b, once the photogenerated electron–hole pairs have been created in the semiconductor bulk, they must separate and migrate to the surface (paths a and b in Figure 1b) competing effectively with the electron–hole recombination process (path c in Figure 1b) that consumes the photocharges generating heat. At the surface of the semiconductor, the photoinduced electrons and holes reduce and oxidize adsorbed water to produce gaseous oxygen and hydrogen by the following reactions (described for water splitting in acid media):

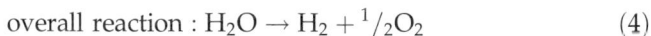

$$\text{oxidation}: H_2O + 2h^+ \rightarrow 2H^+ + {}^1\!/_2 O_2 \tag{2}$$

$$\text{reduction}: 2H^+ + 2e \rightarrow H_2 \tag{3}$$

$$\text{overall reaction}: H_2O \rightarrow H_2 + {}^1\!/_2 O_2 \tag{4}$$

Water splitting into H_2 and O_2 is classified as an "up-hill" photocatalytic reaction because it is accompanied by a large positive change in the Gibbs free energy ($\Delta G^0 = 237\,k\,J\,mol^{-1}$, 2.46 eV per molecule). In this reaction,

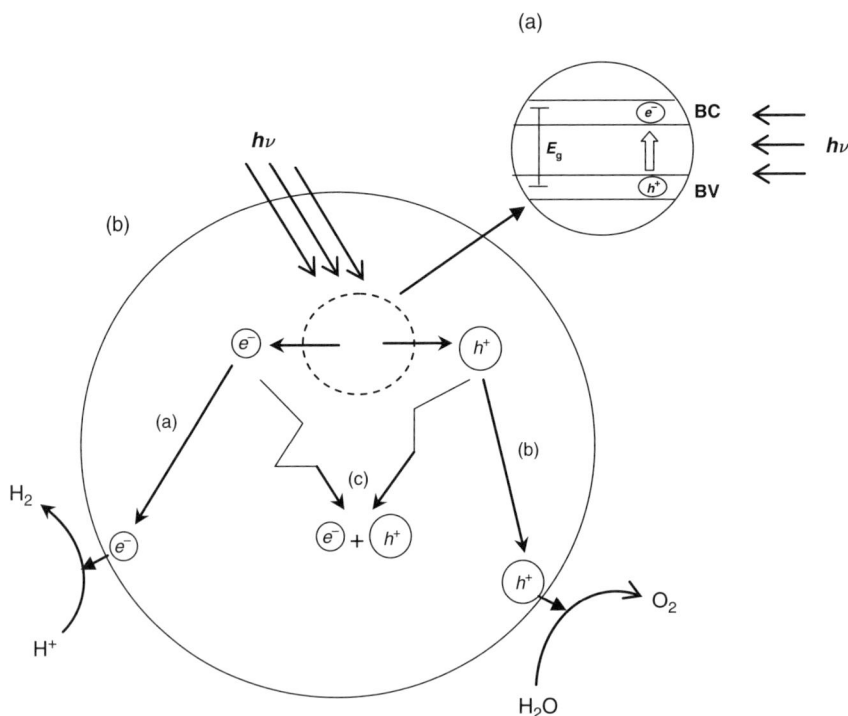

Figure 1 The principle of photocatalytic water splitting: (a) photoelectronic excitation in the phototcatalyst-generating electron–hole pairs and (b) processes occurring on photocatalyst particle following photoelectronic excitation (Mills and Le Hunte, 1997).

photon energy is converted into chemical energy (hydrogen, Figure 2), as seen in photosynthesis by green plants. This reaction is therefore sometimes referred to as artificial photosynthesis.

2.2. Configurations

Photocatalysts for photochemical water splitting can be used for this purpose according to two types of configurations: (i) photoelectrochemical cells and (ii) photocatalytic systems.

The photoelectrochemical cell for water decomposition (Figure 3) involves two electrodes immersed in an aqueous electrolyte, of which one is a photocatalyst exposed to light (photoanode in Figure 3).

The photogenerated electron–hole pairs, produced as a result of light absorption on the photoanode, are separated by an electric field within the semiconductor. On the one hand, the photogenerated holes migrate to the surface of the semiconductor where they oxidize water molecules to oxygen

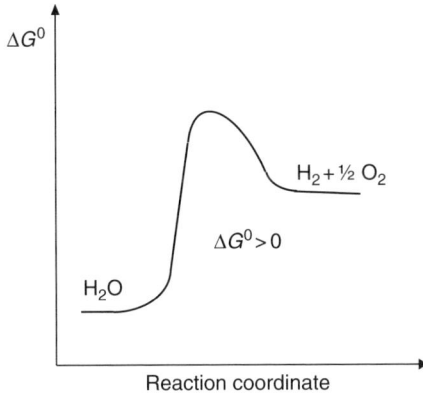

Figure 2   Change in the Gibbs free energy for water splitting (uphill reaction) (Kudo, 2003).

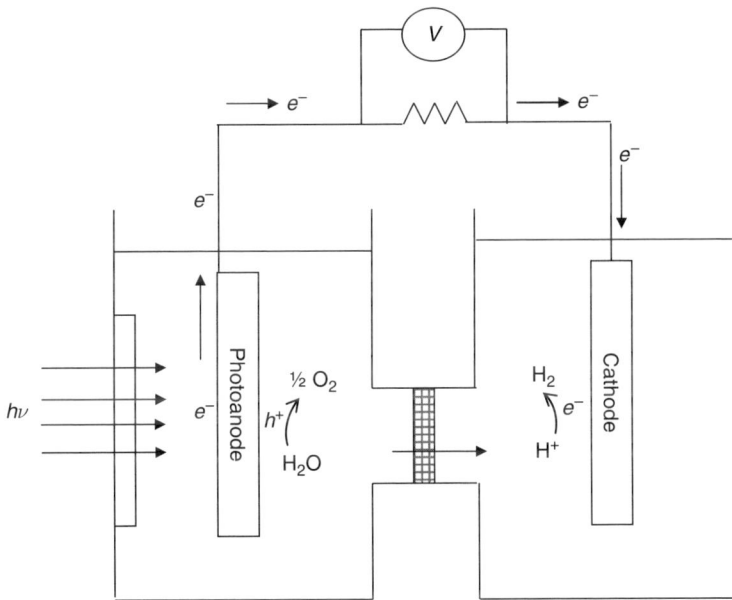

Figure 3   Diagram of photoelectrochemical cell for water splitting (Matsuoka et al., 2007).

[Equation (2)]. On the other hand, the photogenerated electrons move through an electrical circuit to the counter electrode and there reduce H^+ to hydrogen [Equation (3)]. In order to achieve practical current densities, an additional potential driving-force is usually required through the imposition of external bias voltage or the imposition of an internal bias voltage by using different concentrations of hydrogen ions in the anode/cathode compartments. Water splitting using a photoelectrochemical cell was first reported by Fujishima and Honda (Fujishima et al., 1972, 1975) using an electrochemical system consisting of a TiO_2 semiconductor electrode connected through an electrical load to a platinum black counter electrode. Photoirradiation of the TiO_2 electrode under a small electric bias leads to the evolution of H_2 and O_2 at the surface of the Pt counter electrode and TiO_2 photoelectrode, respectively.

An alternative photoelectrochemical system is based on the use of semiconducting materials as both photoelectrodes (Nozik, 1976). In this case, n- and p-type materials are used as the photoanode and photocathode, respectively. The advantage of such a system is that photovoltages are generated on both electrodes, consequently resulting in a substantial increase in solar energy conversion and in the formation of an overall photovoltage that is sufficient for water decomposition without the application of external bias. However, the corresponding price to be paid is the concomitant increase in the device complexity derived from the fact that the photocurrents through the two electrodes must be carefully matched since the overall current flowing in the cell must obviously be the same.

The principles of photochemical water splitting can be extended to the design of systems using photocatalytic semiconductors in the form of particles or powders suspended in aqueous solutions (Bard, 1979, 1980). In this system, each photocatalyst particle functions as a microphotoelectrode performing both oxidation and reduction of water on its surface (Figure 4).

Such particulate systems have the advantage of being much simpler and less expensive to develop and use than photoelectrochemical cells. Furthermore, a wide variety of materials can be used as photocatalysts in

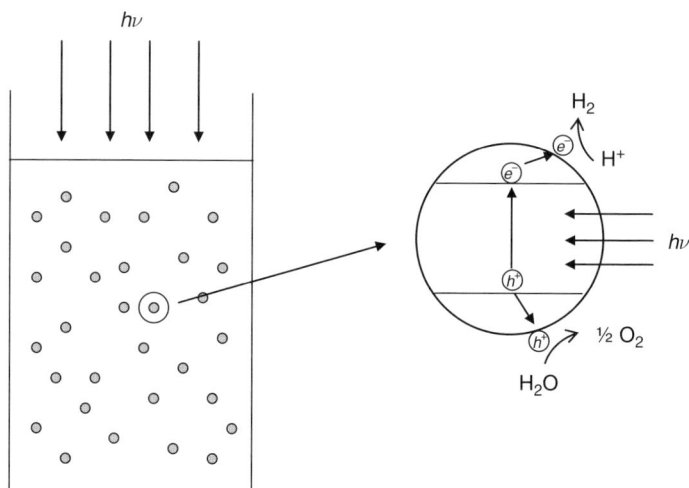

Figure 4   Scheme of the photocatalyst particulate suspension system for water splitting.

particulate systems. These particulates may be rather difficult to prepare in the form of single crystals or in form of high-quality polycrystalline phases which are necessary in the case of photoelectrodes. Another advantage is that electrical conductivity does not need to be as high as that required by photoelectrodes. Moreover, the efficiency of light absorption in suspensions or slurries of powders can be very high because of the high semiconductor surface exposed to light. For instance, 100 mg of the photocatalyst powder of particle diameter $\sim$0.1 μm consists of more than 10^{11} particles that are mobile and independent of each other. However, particulate photocatalytic systems also have disadvantages given the separation of charge carriers is not as efficient as with a photoelectrode system, and there are difficulties associated with the effective separation of the stoichiometric mixture of oxygen and hydrogen to avoid the backward reaction.

2.3. Energy requirements

For photochemical water reduction to occur, the flat band potential of the photocatalytic semiconductor must exceed the proton reduction potential (0.0 V against the normal hydrogen electrode, NHE, at pH $=$ 0, Figure 5). Furthermore, to facilitate water oxidation [Equation (2)], the VB edge must exceed the oxidation potential of water ($+$1.23 V against the NHE, at pH $=$ 0, Figure 5). Therefore, according to these values, a theoretical semiconductor band-gap energy of 1.23 eV is required to drive the overall water-splitting reaction according to Equation (4).

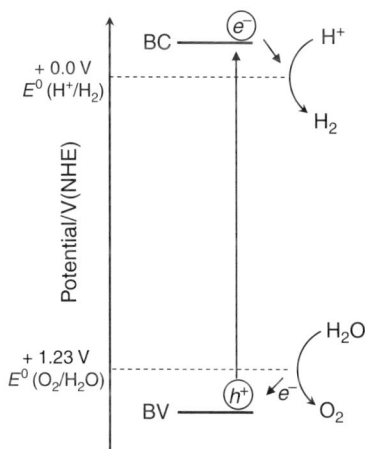

Figure 5 Potential energy diagram for photocatalytic water splitting using a single semiconductor system (Kudo, 2003).

As overall water splitting is generally difficult to achieve due to the uphill nature of the reaction, the photocatalytic activities of photocatalysts have sometimes been examined in the presence of reducing reagents (alcohols, sulfides, sulfites, EDTA) or oxidizing ones (persulfate, Ag^+, Fe^{3+}...) to facilitate either water reduction or oxidation. The basic principle of photocatalytic reactions using redox (or sacrificial) reagents is depicted schematically in Figure 6. When the photocatalytic reaction is carried out in aqueous solutions including easily reducing reagents such as alcohols and sulfides, photogenerated holes irreversibly oxidize the reducing reagent instead of water, thus facilitating water reduction by CB electrons as shown in Figure 6a. On the other hand, electron acceptors such as Ag^+ or Fe^{3+} consume the photogenerated electrons in the CB, and the O_2 evolution reaction is enhanced as shown in Figure 6b. The reactions using redox reagents are not "overall" water-splitting reactions but are often carried out as test reactions for photocatalytic H_2 or O_2 evolution (Kato et al., 2004; Kim et al., 2004; Kudo et al., 1999; Yoshimura et al., 1993). However, one should realize that the results obtained using redox reagents do not guarantee activity for overall water splitting using pure water.

Semiconductors with band gaps smaller than 1.23 eV can be combined to drive water-oxidation/reduction processes separately via multiphoton processes. An example of this is the two-step water-splitting system using a reversible redox couple (A/R) shown in Figure 7 (Fujihara et al., 1988; Kudo et al., 1999; Sayama et al., 1987; Tennakone and Wickramanayake, 1986) with two photocatalysts with different band-gap and band

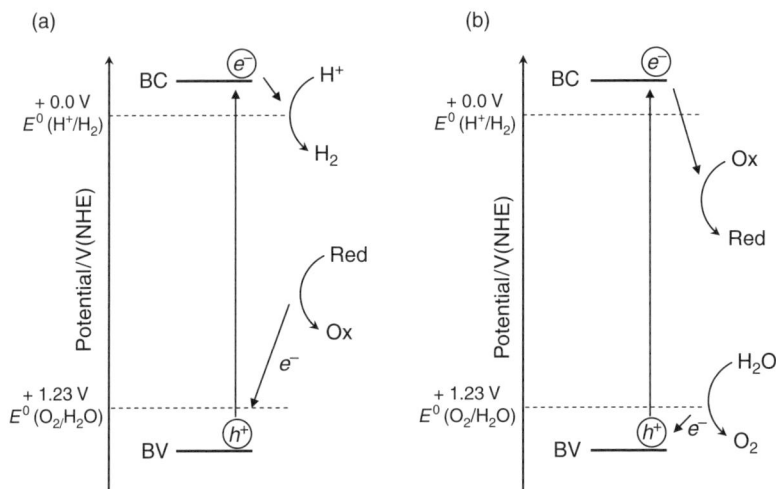

Figure 6 Schematic diagram of photocatalytic water splitting in the presence of redox (sacrificial) reagents: (a) reducing reagent (Red) for H_2 evolution and (b) oxidizing reagent (Ox) for O_2 evolution (Maeda and Domen, 2007).

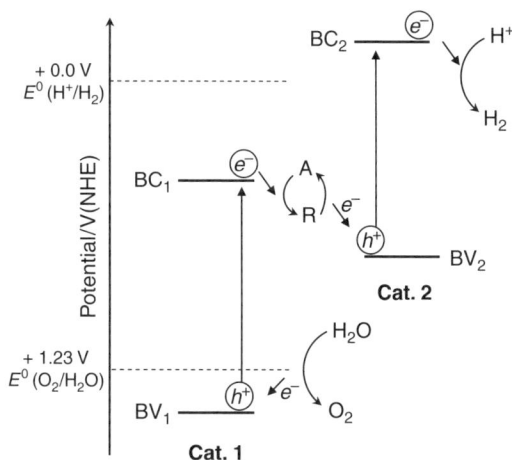

Figure 7   Diagram of a dual photocatalysts system (z scheme) employing a redox shuttle (Abe et al., 2005).

positions: one for O_2 evolution (cat. 1 in Figure 7) and another one for H_2 evolution (cat. 2 in Figure 7). In this two-step water-splitting configuration, water is reduced to H_2 by photoexcited electrons over the photocatalyst for H_2 evolution (cat. 2, Figure 7), and the electron donor (R in Figure 7) is oxidized by holes to its electron acceptor form (A in Figure 7). Simultaneously, this electron acceptor (A) is reduced to its electron donor form (R) by the photoexcited electrons on the photocatalyst for O_2 evolution (cat. 1 in Figure 7), while the holes on this photocatalyst oxidize water to O_2. However, the photoproduction efficiency achieved with this water-splitting configuration is low because of the use of more than one photonic processes that increase the overall energy losses associated with the light energy conversion on each photocatalysts (Tennakone and Wickramanayake, 1986).

2.4. Solar spectrum and water-splitting efficiency

The mean normal incident solar irradiance just outside the earth's atmosphere is $1353\,W\,m^{-2}$ (Bird et al., 1985). The spectral distribution of sunlight that reaches the earth's surface is modified by scattering by aerosol and by absorption by ozone (in the UV region) and by water vapor and CO_2 (in the IR region). The extent of these effects depends mainly on the air mass (AM) ratio, which is the ratio between the optical path length of sunlight through the atmosphere and the path length when the sun is at the zenith.

Therefore, AM 0 sunlight means extraterrestrial sunlight, AM 1 means that the sun is overhead, and AM 25 refers to a setting or rising sun. Figure 8 shows a representative solar global spectral irradiance under the global AM

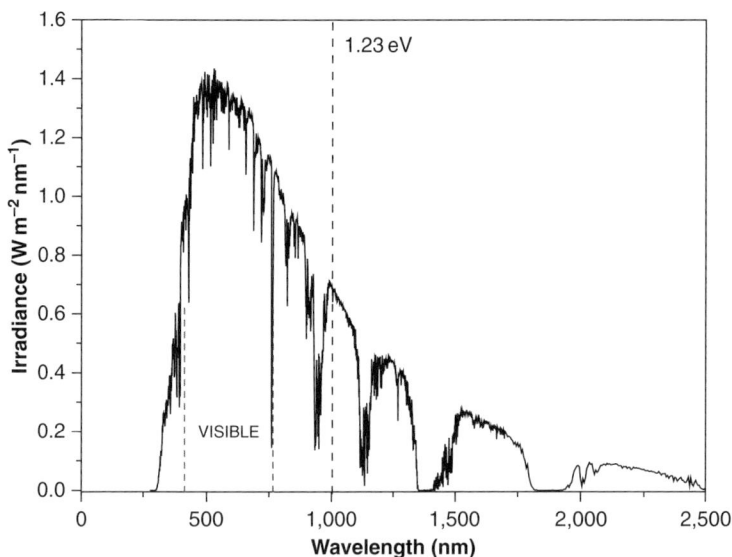

Figure 8   Solar spectral irradiance (AM 1.5) in terms of radiation energy versus photon wavelength (Bird et al., 1985).

condition 1.5 (AM 1.5). AM 1.5 corresponds to a situation when the sun is at a zenith angle of 48.19°, and it is representative of temperate latitudes in the northern hemisphere. The energy distribution in the spectrum of AM 1.5 sunlight is given in Table 1.

The efficiency to convert solar energy into chemical energy (H_2) will be the main determining factor of hydrogen production costs using photocatalyst technology. As mentioned in the previous section, the thermodynamic

Table 1   Energy distribution in the terrestrial solar spectrum (AM 1.5)

Spectral region	Interval boundaries		Solar irradiance	
	Wavelength (nm)	Energy (eV)	W m^{-2}	Percentage of total
Near UV	315–400	3.93–3.09	26	2.9
Blue	400–510	3.09–2.42	140	14.6
Green/ yellow	510–610	2.42–2.03	153	16.0
Red	610–700	2.03–1.77	132	13.8
Near IR	700–920	1.77–1.34	208	23.5
Infrared	920–> 1400	1.34 – <0.88	283	29.4

potential for water splitting requires a minimum energy of 1.23 eV per photon. This energy is equivalent to the energy of a photon with a wavelength of around 1,010 nm, and hence about 70% of all solar photons are theoretically available for water splitting (see fraction of energy above 1.23 eV in Table 1). However, all solar photonic processes involve unavoidable energy losses that in practice involve values of energy per photon higher than the theoretical limit of 1.23 eV. The energy losses associated with solar energy conversion using photocatalysts include several factors associated with the following effects: (i) transport of electrons/holes from the position of their generation at the near-surface outward to the photocatalyst–water interface; (ii) there are irreversible processes of energy loss associated with the recombination of photogenerated electron–hole pairs; (iii) all solar photons with energies lower than E_g cannot be absorbed and thus are lost to the conversion process; (iv) solar photons with energy higher than E_g can be absorbed, but the excess energy ($E_{photon} - E_g$) is lost as heat and, consequently, only a fraction of photon energy is efficiently used for conversion, and (v) the energy of the excited state in photocatalysts is thermodynamically an internal energy and not Gibbs energy [only a fraction (up to about 75%, Fonash, 1981) of the excited state energy can be used to split water].

The ideal limiting efficiency, η_{limit}, of any photonic process is given by (Archer and Bolton, 1990)

$$\eta_{limit} = \frac{J_g \Delta \mu_x \phi_c}{E_s} \tag{5}$$

where:

-J_g is the absorbed photon flux (photons s^{-1} m^{-2}) with energy higher than the band gap of the photocatalyst
-$\Delta \mu_x$ is the chemical potential of the excited state relative to the ground state of photocatalysts and represents the maximum energy available for water splitting
ϕ_c is the quantum yield (QY) of the conversion process and represents the fraction of excited states contributing to water splitting
-E_s is the total incident solar irradiance (W m^{-2}).

From Equation (5), it is clear that the basic parameter that decides the light harvesting ability of the photocatalyst is its band gap. The ideal limiting efficiencies for conversion of solar radiation calculated by Equation (5) as a function of the band-gap wavelength for standard AM 1.5 solar irradiation in a single band-gap device are represented in Figure 9.

As it can be seen in the Figure 9, the maximum ideal efficiency corresponds to semiconductors with a band-gap wavelength of $800 < \lambda_g < 950$ nm ($1.3 < E_g < 1.4$ eV). Obviously, the optimal band gap of semiconductors in terms of efficiency changes when the overall energy losses associated with

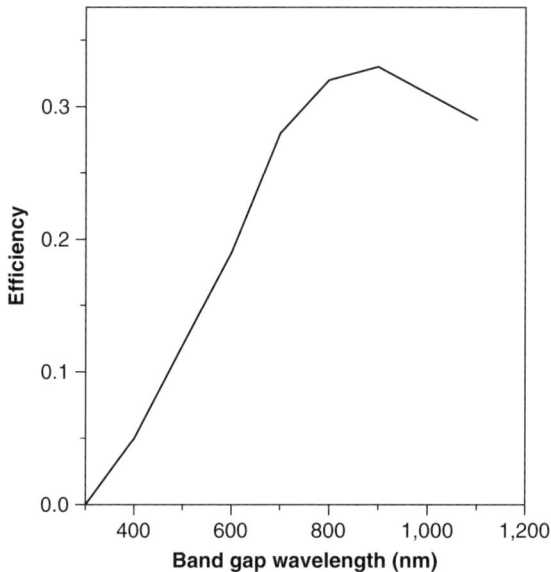

Figure 9 The ideal limiting solar conversion efficiency for a single-band gap photocatalyst (Archer and Bolton, 1990).

solar energy conversion on photocatalysts are taken into account. In this case, the efficiency, η_r, is defined according to Equation (6) (Bolton et al., 1985) to include overall energy losses:

$$\eta_r = \eta_g \phi_c \eta_{chem} \qquad (6)$$

where:

-η_g is the fraction of the incident solar irradiance that has a photon energy above the photocatalyst band-gap energy. η_g is given by

$$\eta_g = \frac{J_g E_g}{E_s} \qquad (7)$$

where:

-J_g is the absorbed photon flux (photons s^{-1} m^{-2}) with energy higher than the photocatalyst band-gap energy (E_g)
-E_s is the total incident solar irradiance (W m^{-2})
-η_{chem} is the chemical efficiency, that is, the fraction of the excited state energy used to split water. η_{chem} is given by

$$\eta_{chem} = \frac{E_g - E_{loss}}{E_g} \qquad (8)$$

where:

-E_{loss} is the energy loss per water molecule in the photowater-splitting process. E_{loss} involves fundamental losses imposed by thermodynamics (entropy change associated with the creation of excited states) as well as losses due to nonidealities in the conversion process (transport of electron/holes, electron/holes recombination, kinetic losses, etc.). E_{loss} takes a minimum value of around 0.3–0.4 eV in ideal photocatalysts due to thermodynamic losses, and this value rises to ~0.8 eV in practical photocatalysts (Bolton, 1996).

The calculation of photoconversion efficiency by Equations (6)–(8) using a single band-gap photocatalyst and assuming a global energy loss (E_{loss}) equal to 0.8 eV gives a maximum efficiency of about 17% that is achievable with photocatalysts with optimum band-gap wavelength (λ_g) of around 600 nm (2.0–2.2 eV) (Bolton, 1996). It should be noted from Equations (6)–(8) that lowering the global energy losses in the photocatalyst (E_{loss}) augments conversion efficiency, whereby semiconductors can be used with a higher band gap (up to the limit value of $E_g \sim 1.6$ eV that corresponds to the minimum achievable thermodynamic value of $E_{loss} \sim 0.3$–0.4 eV). The above analysis has not taken into account further losses arising from incomplete absorption, QYs less than unity, reflection losses or losses in collecting gases. These losses may be estimated at around 5% according to the evaluations made by Bolton et al. (1985).

From Equations (6)–(8), it might appear that there are many factors to be determined in the measurement of solar hydrogen photoproduction efficiencies; however, in practice, efficiencies are measured using another form of Equation (6) (Serrano and de Lasa, 1997):

$$\eta_r = \frac{\Delta H^0 R_{H_2}}{E_s A} \tag{9}$$

where ΔH^0 is the standard enthalpy for water splitting, R_{H_2} is the rate of generation of H_2 (mol s^{-1}), E_s is the incident solar irradiance (W m^{-2}), and A is the irradiated area (m^2). In the case of photoelectrochemical cells, the electrical power input must be subtracted from the rate of production of the evolved hydrogen. Equation (9) must then be modified to

$$\eta_r = \frac{\Delta H^0 R_{H_2} - IV}{E_s A} \tag{10}$$

where I is the cell current (A) and V is the bias voltage applied to the cell.

3. PHOTOCATALYSTS FOR WATER SPLITTING UNDER VISIBLE LIGHT

3.1. Material requirements

Taking into account the photoelectrochemistry of water dissociation analyzed in the previous section, the photocatalysts used for the photodissociation of water must satisfy several functional requirements with respect to semiconducting and electrochemical properties: (i) suitable solar visible-light absorption capacity with a band gap of around 2.0–2.2 eV and band edge potentials suitable for overall water splitting, (ii) capacity for separating photoexcited electrons from reactive holes, (iii) minimization of energy losses related to charge transport and recombination of photoexcited charges, (iv) chemical stability against corrosion and photocorrosion in aqueous environments, (v) kinetically suitable electron transfer properties from photocatalyst surface to water, and (vi) being easy to synthesize and cheap to produce.

As stated in the previous section, the basic parameter deciding the light-harvesting ability of the photocatalyst is its electronic structure that determines the band-gap energy. Figure 10 illustrates the band positions of various semiconductors regarding the potentials (NHE) for water-oxidation/reduction processes (Xu and Schoonen, 2000).

From the perspective of band positions, among the semiconductors represented in Figure 10, those that fulfil the thermodynamic requirements for overall water splitting are $KTaO_3$, $SrTiO_3$, TiO_2, ZnS, CdS, and SiC. However, it is important to stress that the potential of the band structure

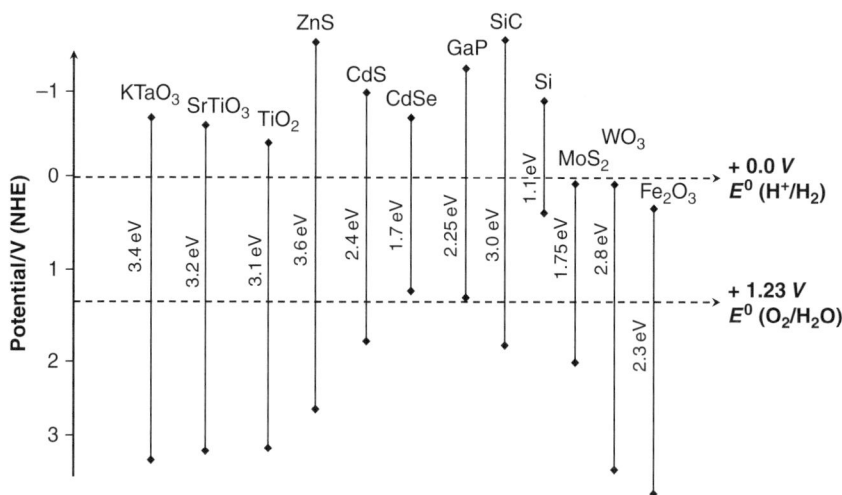

Figure 10 Band-gap energy and relative band position of different semiconductors with respect to the potentials (NHE) for water oxidation/reduction processes.

of the semiconductor is just the thermodynamic requirement. As commented in Section 1.4, there is an activation barrier in the charge transfer process between photocatalyst and water molecules derived from the energy losses associated with solar energy conversion on photocatalysts: thermodynamic losses, transport of electron/holes, electron/holes recombination, and kinetic losses. The existence of these energy losses increases the optimal band gap for high-performance photocatalysts from the theoretical value of 1.23 to 2.0–2.2 eV (Bolton, 1996).

Another essential requirement for the photocatalyst is its resistance to reactions at the solid/liquid interface that may result in a degradation of its properties. These reactions include electrochemical corrosion, photocorrosion, and dissolution (Morrison, 1980). A large group of photocatalysts with suitable semiconducting properties for solar energy conversion (CdS, GaP, etc.) are not stable in the water-oxidation reaction because the anions of these materials are more susceptible to oxidation than water, causing their degradation by oxidation of the material (Ellis et al., 1977; Williams, 1960).

Water-splitting reactions at the photocatalyst interface can occur if charge carriers generated by absorbed light can reach the solid–liquid interface during their lifetime and are capable of finding suitable reaction partners – protons for electrons and water molecules for holes. For that reason, the generation and separation of photoexcited carriers with a low recombination rate is also an essential condition to be fulfilled by the photocatalysts. The transport of photoexcited carriers strongly depends on both the microstructural and surface properties of photocatalysts. In general, the high crystalline level of the photocatalyst has a positive effect on photoactivity, as the density of defects caused by grain boundaries, which act as recombination centers of electrons and holes, decreases when particle crystallinity increases (Ikeda et al., 1997; Kominami et al., 1995; Reber and Meier, 1984). Surface properties such as surface area and active reaction sites are also important. The surface area, determined by the size of the photocatalyst particles, also influences the efficiency of charge carrier transport. To have an efficient charge transport, the diffusion length of charge carriers must be long compared to the size of the particles. Therefore, the possibility of the charge carrier reaching the surface increases as the size of the photocatalysts decreases (Ashokkumar, 1998). Nevertheless, the improvement in the efficiency associated with the high crystalline level of the photocatalyst prevails over the improvement associated with small-sized particles (Kudo et al., 2004).

3.2. Strategies for developing efficient photocatalysts under visible light

Although the properties required by photocatalysts for water splitting have been identified (band edge potentials suitable for water splitting, band-gap energy around 2.0–2.2 eV, and stability under reaction conditions), it is difficult to obtain materials that meet all these requirements. In addition and in the development of efficient photocatalysts for water splitting, it is important to control the interinfluence between electronic, microstructural, and surface properties of photocatalysts by means of a careful design of both bulk and surface properties.

Several research approaches are pursued in the quest for more efficient and active photocatalysts for water splitting: (i) to find new single-phase materials, (ii) to tune the band-gap energy of UV-active photocatalysts (band-gap engineering), and (iii) to modify the surface of photocatalysts by deposition of cocatalysts to reduce the activation energy for gas evolution. Obviously, the previous strategies must be combined with the control of the synthesis of materials to customize the crystallinity, electronic structure, and morphology of materials at nanometric scale, as these properties have a major impact on photoactivity.

3.2.1. Band-gap engineering

In the development of active photocatalysts under visible light, it is essential to control their electronic energy structure. The strategies for controlling the energy structure of photocatalysts for water splitting may be classified in three ways: (i) cation or anion doping, (ii) use of mixed semiconductor composites, and (iii) use of semiconductor alloys.

Cation or anion doping Ion doping has been extensively investigated for enhancing the visible-light response of wide band-gap photocatalysts (UV-active). Examples include Sb- or Ta- and Cr-doped TiO_2 and $SrTiO_3$ (Kato et al., 2002; Ishii et al., 2004), ZnS doped with Cu or Ni (Kudo and Sekizawa, 1999; Kudo and Sekizawa 2000), or C-doped TiO_2 (Khan et al., 2002).

The replacement of cations in the crystal lattice of a wide band-gap semiconductor may create impurity energy levels within the band gap of the photocatalyst that facilitates absorption in the visible range, as depicted in Figure 11. Although cation-doped photocatalysts can induce visible-light response, most of these photocatalysts do not have photoactivity because dopants in the photocatalysts act not only as visible-light absorption centers, with an absorption coefficient dependent on the density of dopants, but also as recombination sites between photogenerated electrons and holes (Choi et al., 1994). Furthermore, the impurity levels created by dopants in the photocatalysts are usually discrete, which would appear disadvantageous for the migration of the photogenerated holes (Blasse et al., 1981). It is therefore important to carefully control both the content and the depth of the cation substitution in the structure of the host photocatalysts to develop visible-light-active photocatalysts. The metal ions may be incorporated into the photocatalysts by chemical methods (impregnation or precipitation) or using the advanced ion-implantation technique [based on the incorporation

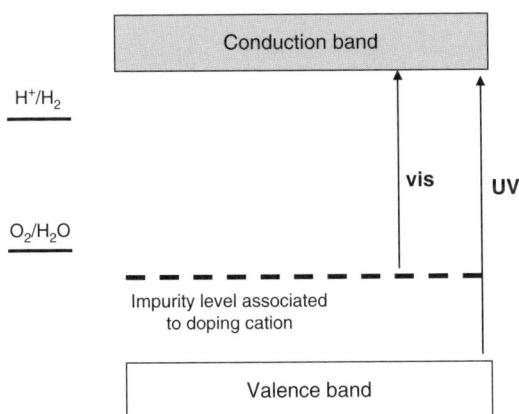

Figure 11 Band structure of cation-doped photocatalyst with visible light response from a semiconductor with wide band gap (UV response) (Kudo, 2003).

of cations by the impact of high-energy ions accelerated by high voltage (50–200 keV)] (Anpo, 2000; Anpo and Takeuchi, 2003; Yamashita et al., 2002)). The advanced ion-implantation technique is more effective than chemical methods for controlling the insertion of dopants into the photocatalyst structure (Anpo, 2000).

Anion doping is another way of enhancing the visible-light response of wide band-gap (UV-active) photocatalysts based on oxide semiconductors. In wide band-gap oxide photocatalysts, the top of the VBs consists of O 2p atomic orbitals. In recent years, interesting papers have been published on the development of visible-light photocatalysts by doping with anions such as N (Asahi et al., 2001; Hitoki et al., 2002), S (Umebayashi et al., 2002), or C (Khan et al., 2002) as substitutes for oxygen in the oxide lattice. In these anion-doped photocatalysts, the mixing of the p states of the doped anion (N, S, or C) with the O 2p states shifts the VB edge upward and narrows the band-gap energy of the photocatalyst as depicted in Figure 12. In contrast to the cationic dopant technique, the anionic replacement usually forms fewer recombination centers, and therefore, it is more effective for enhancing photocatalyst activity. Nevertheless, in the anion-doped materials, it is necessary to control the number of oxygen defects due to the difference in the formal oxidation numbers of oxygen and the dopant anions, as these defects will act as recombination centers that may reduce the efficiency of the anion-doped photocatalyst.

Composite semiconductors Semiconductor mixing (composite) is another strategy for developing photocatalysts with visible-light response from photocatalysts with a wide band gap. This strategy is based on the coupling of a wide band-gap semiconductor with a narrow band semiconductor with

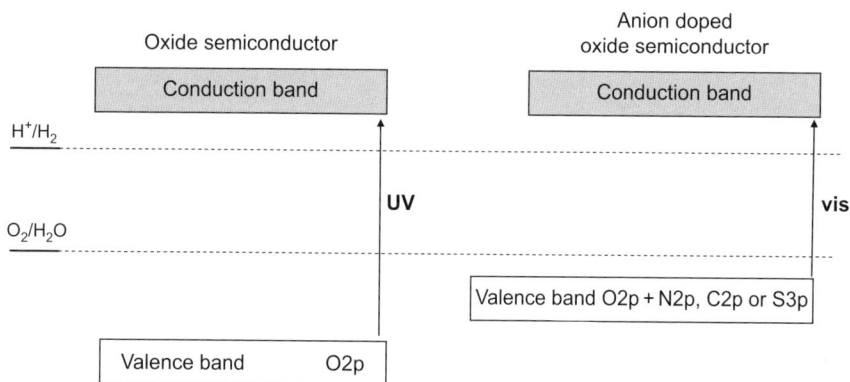

Figure 12   Band structure of anion-doped semiconductor with visible light response from a semiconductor with wide band gap (UV response) (Kudo et al., 2004).

a more negative CB level. In this way, CB electrons can be injected from the small band-gap semiconductor into the large band semiconductor, thereby extending the absorption capacity of the mixed photocatalyst (Figure 13).

An additional incentive for using composite semiconductor photocatalysts derives from the possibility of mitigating the carrier recombination by means of the interparticle electron transfer. In the photocatalyst composites, the semiconductor particles are in electronic contact with no mixing at molecular level. Illustrative examples of this approach include the following composites: $CdS-TiO_2$ (Serpone et al., 1984), $CdS-K_4Nb_6O_{17}$ (Yoshimura et al., 1988), or $Ca_2Fe_2O_4-PbBi_2Nb_{1.9}W_{0.1}O_9$ (Sung Lee, 2005).

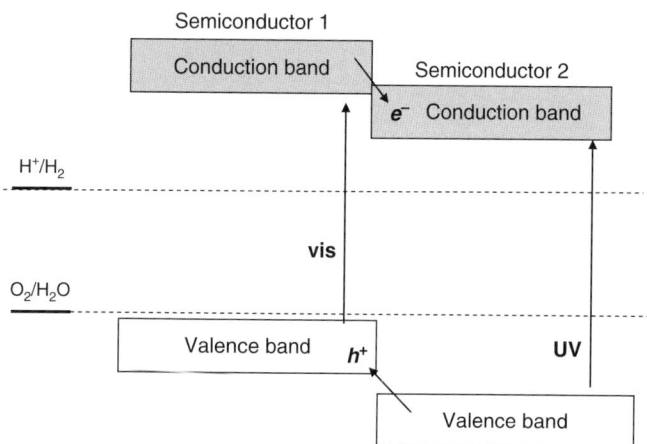

Figure 13   Band structure of the photocatalyst composite with enhanced visible-light response made from the mixture between wide and narrow band-gap photocatalysts.

The following conditions need to be met for the successful coupling of semiconductors: (i) the CB level of the semiconductor of the narrow band gap should be more negative than that of the semiconductor with a wide band gap, (ii) the CB level of the semiconductor with a wide band gap should be more negative than the water reduction potential, and (iii) electron injection should be fast and efficient.

Semiconductor alloys The third approach to the extension of the visible-light response of wide band-gap photocatalysts involves making solid solutions between wide and narrow band-gap semiconductors with a similar lattice structure as depicted in Figure 14.

Solid solutions of two or more semiconductors are formed where the lattice sites are interdispersed with the solid solution components. In these systems, the band gap can be customized by means of changes in the solid solution composition. Examples of semiconductor alloys include GaN—ZnO (Maeda et al., 2005) ZnO-GeO (Domen and Yashima, 2007), ZnS—CdS (Kakuta et al., 1985), ZnS—AgInS$_2$ (Tsuji et al., 2004), and CdS—CdSe (Kambe et al., 1984).

3.2.2. Surface modification by deposition of cocatalysts

The deposition of noble metals (e.g., Pt, Rh) or metal oxides (e.g., NiO, RuO$_2$) onto photocatalyst surfaces is an effective way of enhancing photocatalyst activity (Sato and White, 1980; Subramanian et al., 2001). The cocatalyst improves the efficiency of photocatalysts, as shown in Figure 15, as a result of (i) the capture of CB electrons or VB holes from the photocatalysts (Maruthamuthu and Ashokkumar, 1988), thereby reducing the possibility of electron–hole recombination and (ii) the transference

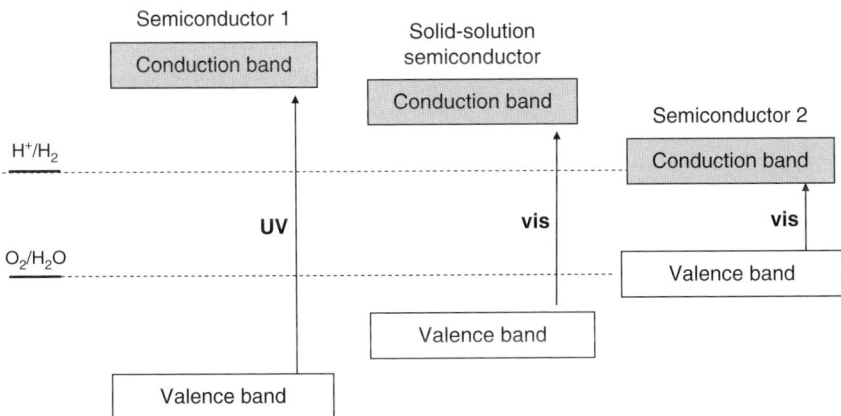

Figure 14 Band structure of photocatalysts made from the solid solution of wide and narrow band-gap photocatalysts (Kudo et al., 2004).

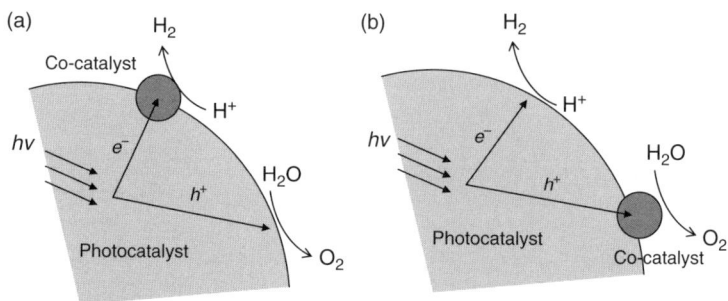

Figure 15 Schematic diagram of photocatalyst surface modification by the addition of cocatalyst to facilitate the hydrogen (a) or oxygen (b) evolution in water splitting.

of electrons and holes to surface water molecules, thereby reducing the activation energy for the reduction/oxidation of water (Maruthamuthu and Ashokkumar, 1989). The activity of the cocatalysts is found to be strongly dependent on the quantity of cocatalysts deposited on the photo-catalyst surface. When the amount exceeds a critical limit, the cocatalysts act as electron–hole recombination centers, reducing the efficiency of the host photocatalyst.

3.3. Photocatalyst development

Several types of materials, over 130 including oxides, nitrides, sulfides, and others have been reported to act as efficient photocatalysts for hydrogen evolution via water splitting.

Among these photocatalysts, the higher QYs are reported for Ba-doped $Sr_2Nb_2O_7$ (50% QY, pure water, UV light) (Kim et al., 2005) and $NiO/NaTaO_3$ (56% QY, pure water UV light) (Kato et al., 2003). Unfortunately, these exciting developments have only a limited value for practical hydrogen production because UV light accounts for only about 3–4% of solar radiation energy (Table 1). Therefore, regarding the use of solar energy, it is essential to develop photocatalysts that split water efficiently under visible light ($\lambda \sim 600$ nm, Section 1.4). The materials for visible-light-driven photocatalysts had been quite limited. However, many oxides, sulfides, oxynitrides, and oxysulfides have recently been found to be active for H_2 and O_2 evolution under visible-light irradiation (Table 2). So far, the maximum quantum efficiency over visible-light-driven photocatalysts achieves only a few percent at wavelengths as long as 500 nm (Cr/Rh—GaN/ZnO 2.5% QY, pure water, visible light) (Maeda et al., 2006a). This value is still far from the QY of 10% marked as the initial starting point for practical application (Turner et al., 2008). Hence, the development of new photocatalyst materials is still a major issue. As commented in the previous section, several approaches have been adopted in

Table 2 Overview of photocatalysts developed in last years for water splitting reaction under visible light

Photocatalysts (reference)	Band-gap energy (eV)	Cocatalyst	Sacrificial reagent	Activity μmol h^{-1} g^{-1} H$_2$	Activity μmol h^{-1} g^{-1} O$_2$
Titanium oxide and titanates					
TiO$_2$—Cr—Sb (Kato and Kudo, 2002)	2.2	—	AgNO$_3$ (0.05 M)		89
TiO$_2$—N (Kim et al., 2004)	2.73	Pt	CH$_3$OH (0.1 M) / AgNO$_3$ (0.05 M)	0	221
SrTiO$_3$—Cr—Ta (Ishii et al., 2004)	2.3	Pt	CH$_3$OH (0.1 M)	140	
SrTiO$_3$—Cr—Sb (Kato and Kudo, 2002)	2.4	Pt	CH$_3$OH (0.1 M)	156	
La$_2$Ti$_2$O$_7$—Cr (Hwang et al., 2005)	1.8–2.3	Pt	CH$_3$OH (0.3 M)	30	
La$_2$Ti$_2$O$_7$—Fe (Hwang et al., 2005)	1.8–2.3	Pt	CH$_3$OH (0.3 M)	20	
Sm$_2$Ti$_2$S$_2$O$_5$ (Ishikawa et al., 2002)		Pt / —	CH$_3$OH (0.3 M) / AgNO$_3$ (0.01 M)	40	16
Tantalates and niobiates					
TaON (Hitoki et al., 2002)	2.5	Pt / —	CH$_3$OH (0.06 M) / AgNO$_3$ (0.01 M)	50	3,300
CaTaO$_2$N (Yamashita et al., 2004)	2.4	Pt / —	CH$_3$OH (0.06 M) / AgNO$_3$ (0.01 M)	37	0
SrTaO$_2$N (Yamashita et al., 2004)	2.1	Pt / —	CH$_3$OH (0.06 M) / AgNO$_3$ (0.01 M)	50	0
BaTaO$_2$N (Yamashita et al., 2004)	1.9	Pt / —	CH$_3$OH (0.06 M) / AgNO$_3$ (0.01 M)	37	0
Sr$_2$Nb$_2$O$_{7-x}$N$_x$ (Ji et al., 2005)	2.1	Pt / —	CH$_3$OH (0.14 M) / AgNO$_3$ (0.01 M)	80	8

Table 2 (Continued)

Photocatalysts (reference)	Band-gap energy (eV)	Cocatalyst	Sacrificial reagent	Activity μmol h⁻¹ g⁻¹	
				H_2	O_2
Other transition metal oxides					
$BiVO_4$ (Kudo et al., 1998)	2.4	–	CH_3OH (0.1 M)	0	
			$AgNO_3$ (0.05 M)		421
Ag_3VO_4 (Konta et al., 2003)	2.0	–	CH_3OH (0.1 M)	0	
			$AgNO_3$ (0.05 M)		17
Metal oxynitrides					
$(Ga_{1-x}Zn_x)(N_{1-x}O_x)$ (Maeda et al., 2006)	2.4–2.8	Cr/Rh	–	930	466
$(Zn_{1+x}Ge)(N_2O_x)$ (Lee et al., 2007)	2.7	RuO_2	–	65	35
Metal sulfides					
CdS (Navarro et al., 2008)	2.4	–	S^{2-} (0.1 M)/SO_3^{2-} (0.04 M)	9.8	
CdS-CdO-ZnO (Navarro et al., 2008)	2.3	–	S^{2-} (0.1 M)/SO_3^{2-} (0.04 M)	11.6	
$Cd_{0.7}Zn_{0.3}S$ (del Valle et al., 2008)	2.68	–	S^{2-} (0.1 M)/SO_3^{2-} (0.04 M)	350	
CdS (Bühler et al., 1984)	2.4	Pt	S^{2-} (0.42 M)	34	
ZnS—Cu (Kudo and Sekizawa, 1999)	2.5	–	SO_3^{2-} (0.5 M)	450	
ZnS—Ni (Kudo and Sekizawa, 2000)	2.3	–	S^{2-} (0.005 M)/SO_3^{2-} (0.5 M)	280	

Photocatalyst					
ZnS—Pb (Tsuji and Kudo, 2003)	2.3	–	S^{2-}(0.005 M)/SO_3^{2-} (0.5 M)	40	
(AgIn)$_x$Zn$_{2(1-x)}$S$_2$ (Kudo et al., 2002)	2.4	Pt	S^{2-} (0.35 M)/SO_3^{2-} (0.25 M)	3,133	–
(CuIn)$_x$Zn$_{2(1-x)}$S$_2$ (Tsuji et al., 2005)	2.35	Pt	–	116	
(CuAg In)$_x$Zn$_{2(1-x)}$S$_2$ (Tsuji et al., 2005)	2.4	Ru	S^{2-}(0.35 M)/SO_3^{2-} (0.25 M)	4,000	
		Ru	S^{2-}(0.5 M)	7,666	
Na$_{14}$In$_7$Cu$_3$S$_{35}$ (Zheng et al., 2005)	2.0	–		18	

the search for photocatalysts for splitting water under visible-light irradiation: (i) find new materials, (ii) tune band-gap energy by modifying UV-active photocatalysts with cation/anion doping, and (iii) manufacture a multicomponent photocatalyst by forming solid solutions. The following sections will review the above approaches for the development of active visible-light photocatalysts for water splitting.

3.3.1. Titanium oxide and titanates

Titanium oxide (TiO_2) was the first material described as a photochemical water-splitting catalyst (Fujishima and Honda, 1972). However, TiO_2 oxide photocatalyst is unable to split water under visible light. This is mainly due to the large band gap of TiO_2 (3.1 eV), which allows solely for the use of a small fraction of the solar spectrum (UV fraction). Intensive studies have been carried out to improve the visible-light sensitivity of Ti oxide-based catalysts. One of the strategies for inducing visible-light response in TiO_2 was the chemical doping of TiO_2 with metal ions with partially filled d-orbitals (V, Cr, Fe, Co, Ni, etc.) (Hwang et al., 2005; Konta et al., 2004;). Although TiO_2 chemically doped with metal ions induces a visible-light response, no significant reactivity for water splitting under visible light was described for these doped TiO_2 photocatalysts. However, Kato and Kudo (2002) reported that TiO_2 codoped with a combination of Sb^{5+} and Cr^{3+} is active for O_2 evolution under visible-light irradiation ($\lambda > 440$ nm) from an aqueous solution using $AgNO_3$ as sacrificial reagent. In this system, doped Cr^{3+} ions form an electron donor level within the band gap of TiO_2 that enables this semiconductor to absorb visible light. Codoping with Sb^{5+} was necessary to maintain the charge balance, resulting in the suppression of the formation of Cr^{6+} ions and oxygen defects in the lattice.

The physical doping of transition metal ions into TiO_2 by the advanced ion-implantation technique has been shown to enable modified TiO_2 to work under visible light (Anpo, 2000; Anpo and Takeuchi, 2003). Thin TiO_2 films implanted with metal ions such as Cr^{3+} or V^{5+} are photoactive under visible light for H_2 evolution from an aqueous solution involving methanol as sacrificial reagent with a QY of 1.25 (Matsuoka et al., 2007). Although the ion-implantation method provides a way of modifying the optical properties of TiO_2, it was impractical for mass production due to the high cost of the ion-implantation apparatus used to develop these TiO_2-modified photocatalysts.

Another strategy followed to improve the visible-light response of TiO_2 involves the doping of anions such as N (Asahi et al., 2001), S (Umebayashi et al., 2002), or C (Khan et al., 2002) as substitutes for oxygen in the TiO_2 lattice. For these anion-doped TiO_2 photocatalysts, the mixing of the p states of doped anion (S, N or C) with the O 2p states shifts the VB edge upward and narrows the band-gap energy of TiO_2. Of these materials, only N-doped TiO_2 has been tested for photocatalytic water splitting (Kim et al., 2004).

Under visible light, this Pt-modified photocatalyst evolves O_2 from aqueous $AgNO_3$ as sacrificial electron acceptor and traces of H_2 from aqueous solutions of methanol as sacrificial electron donor.

When TiO_2 is fused with metal oxides (SrO, BaO, La_2O_3, Sm_2O_3), metal titanates with intermediate band gaps are obtained. Among these titanates, attention has been paid to $SrTiO_3$, $La_2Ti_2O_7$, and $Sm_2Ti_2O_7$. $SrTiO_3$ crystallizes in the perovskite structure and has a band gap of 3.2 eV, while La_2TiO_7 is a layered perovskite with a band gap of 3.8 eV. Neither of these titanates is active under visible light because of their wide band gap. The doping of foreign metal ions into $SrTiO_3$ and La_2TiO_7 is a conventional method for the development of visible-light-driven photocatalysts based on these titanates. A survey of dopants for $SrTiO_3$ revealed that the doping of Rh or the codoping of Cr^{3+}—Ta^{5+} or Cr^{3+}—Sb^{5+} were effective in making $SrTiO_3$ visible-light-responsive (Ishii et al., 2004; Kato and Kudo, 2002; Konta et al., 2004). These doped $SrTiO_3$ samples loaded with Pt cocatalysts have recorded activity for H_2 production from aqueous methanol solutions under visible-light irradiation. On the other hand, the doping of $La_2Ti_2O_7$ with Cr^{3+} or Fe^{3+} ions allows for visible-light absorption above 400 nm of the doped-$LaTi_2O_7$ samples by exciting the electrons in the Cr 3d or Fe 3d band to the CB (Hwang et al., 2004, 2005). However, these doped $La_2Ti_2O_7$ samples have no activity for pure water splitting under visible-light irradiation. The doped $La_2Ti_2O_7$ samples are active for H_2 evolution under visible light solely in the presence of methanol as sacrificial electron donor. Another strategy followed to improve the visible-light response of layered $Ln_2Ti_2O_7$ is related to the partial substitution of oxygen by sulfur anions in the $Ln_2Ti_2O_7$ lattice. For example, promising results have been reported for $Sm_2Ti_2S_2O_5$, which has been proven to be responsive to visible-light excitation at wavelengths up to 650 nm (Ishikawa et al., 2002). S 3p atomic orbitals constitute the upper part of the VB of $Sm_2Ti_2S_2O_5$ and make an essential contribution to lowering the band-gap energy (2.0 eV) from that of the corresponding $Sm_2Ti_2O_7$ (3.5 eV). Under visible-light irradiation, the $Sm_2Ti_2S_2O_5$ works as a stable photocatalyst for the reduction of H^+ to H_2 or oxidation of H_2O to O_2 in the presence of a sacrificial electron donor (Na_2S—Na_2SO_3 or methanol) or acceptor (Ag^+).

3.3.2. Tantalates and niobates

Oxides with structural regularities such as layered and tunneling structures are considered promising materials for efficient water photodecomposition. According to this fact, tantalate and niobate oxides with corner-sharing of MO_6 (M = Ta, Nb) octahedral structure have been considered as photocatalysts for water splitting. Tantalate and niobate oxides are highly active for pure water splitting even without cocatalysts but only under UV light because of their high-energy band gap (4.0–4.7 eV) (Kato and Kudo 1999, 2001; Kato et al., 2003; Kim et al., 2005; Takata et al., 1998). The high activity of these layered compounds is related to the easy migration and separation of the photogenerated electron–holes through the corner-shared framework of MO_6 units (Takata et al., 1998).

One way of increasing the visible response of photocatalysts based on tantalates and niobates is to form oxynitride compounds to reduce the band gap of the materials. O atoms in these compounds are partially substituted by N atoms in metal oxide causing the VB to shift to higher potential energy as a result of the hybridization of the O 2p orbitals with the N 2p orbitals. Following this strategy, $MTaO_2N$ (M = Ca, Sr, Ba), TaON, and $Sr_2Nb_2O_{7-x}N_x$ ($x = 1.5$–2.8) have been studied as photocatalysts for water splitting under visible-light irradiation (Hitoki et al., 2002; Ji et al., 2005; Yamashita et al., 2004). TaON and MTa_2N (M = Ca, Sr, Ba) have small band energies (TaON:2.5 eV, $MTaO_2N$: 2.5–2.0 eV) and are capable of absorbing visible light at 500–630 nm (Yamashita et al., 2004). TaON samples have photocatalytic activity under visible light for H_2 production from aqueous solutions using methanol as sacrificial reagent and also for O_2 production from aqueous solutions with $AgNO_3$ as electron acceptor. On the other hand, $MTaO_2N$ samples are capable of producing H_2 under visible light in the presence of methanol as sacrificial electron donor, whereas they did not have sufficient potential to oxidize water using $AgNO_3$ as electron acceptor.

Nitrogen substitution on $Sr_2Nb_2O_7$ oxides also enables the absorption of the oxynitride in the visible range as a result of the mixing of N 2p with O 2p states near the VB. $Sr_2Nb_2O_{7-x}N_x$ samples with different N substitution ($x = 1.5$–2.8) have photocatalytic activity under visible light for H_2 evolution from aqueous methanol solutions (Ji et al., 2005). The most active photocatalysts were those that achieve the higher nitrogen substitution while maintaining the original layered structure of $Sr_2Nb_2O_7$. An excess of nitrogen substitution facilitates the collapse of the layered structure of the parent oxide and turns into the less active unlayered $SrNbO_2N$.

3.3.3. Other transition metal oxides

$BiVO_4$ with scheelite structure and Ag_3VO_4 with perovskite structure were found to record photocatalytic activities for O_2 evolution from an aqueous silver nitrate solution under visible-light irradiation (Kudo et al., 1998, 1999; Konta et al., 2003). These oxides have steep absorption edges in the visible-light region. The steep edges indicate that the visible-light absorption of these oxides is due to a band–band transition. In contrast to other oxides, the VBs of $BiVO_4$ and Ag_3VO_4 consist of Bi and Ag orbitals mixed with O 2p states that result in an increase in VB potentials and a decrease in band-gap energy. This enables $BiVO_4$ and Ag_3VO_4 to absorb visible light. However, these catalysts had no potential for H_2 production because the CB potentials of the photocatalysts do not have sufficient overpotential for the reduction potential of water.

3.3.4. Metal (oxy) nitrides

The nitrides and oxynitrides of transition metal-cations with d^{10} electronic configuration (Ga^{3+}, Ge^{4+}) are a new type of photocatalysts that can split pure water under visible light without sacrificial reagents. In the development of oxynitride with d^{10} electronic configuration, the solid solution between GaN and ZnO (($Ga_{1-x}Zn_x$)($N_{1-x}O_x$)) was tested first (Maeda et al., Maeda et al., 2005a; Maeda et al., 2005b). GaN and ZnO have wurtzite structures with similar lattice parameters and can therefore form solid solutions. While GaN and ZnO have band-gap energies greater than 3 eV and therefore do not absorb visible light, the ($Ga_{1-x}Zn_x$)($N_{1-x}O_x$) solid solution has absorption edges in the visible region with band energies of 2.4–2.8 eV. Density functional calculations indicated that the visible-light response of the solid solution originates from the presence of Zn 3d and N 2p electrons in the upper VB that provide p–d repulsion for the VB, resulting in the narrowing of the band gap. The ($Ga_{1-x}Zn_x$)($N_{1-x}O_x$) solid solution has low photocatalytic activity even under UV irradiation. However, its activity under visible light increases remarkably with the modification of ($Ga_{1-x}Zn_x$)($N_{1-x}O_x$) solid solution by superficial deposition of cocatalyst nanoparticles. Different transition metals and oxides have been examined as cocatalysts to promote the activity of ($Ga_{1-x}Zn_x$)($N_{1-x}O_x$) solid solutions (Maeda et al., 2006a, b). Among the various cocatalysts examined, the largest improvement in activity was obtained when ($Ga_{1-x}Zn_x$)($N_{1-x}O_x$) was loaded with a mixed oxide of Rh and Cr. From this sample, it is observed that H_2 and O_2 evolve steadily and stoichiometrically under visible light from pure water without sacrificial reagents. The quantum efficiency of the Rh-Cr-loaded ($Ga_{1-x}Zn_x$)($N_{1-x}O_x$) photocatalyst for overall water splitting reaches ca. 2.5% at 420–440 nm (Maeda et al., 2005). These photocatalysts based on ($Ga_{1-x}Zn_x$)($N_{1-x}O_x$) solid solutions were the first particulate photocatalyst systems capable of performing overall water splitting by one-step photoexcitation under visible light.

The solid solution between ZnO and $ZnGeN_2$ (($Zn_{1+x}Ge$)(N_2O_x)) has also been found to be an active oxynitride photocatalyst for pure water splitting under visible light (Lee et al., 2007). The solid solutions ($Zn_{1+x}Ge$)(N_2O_x) present absorption in the visible region with a band-gap energy of ca. 2.7 eV, which is smaller than the band gaps of Ge_3N_4 (3.8 eV), $ZnGeN_2$ (3.3 eV), and ZnO (3.2 eV). The visible-light response of this material originates from the p–d repulsion between Zn 3d and N 2p and O 2p electrons in the upper part of the VBs that narrows the band gap. Neither $ZnGeN_2$ nor ZnO alone exhibits photocatalytic activity for overall water splitting under UV irradiation. However, the solid solution ($Zn_{1+x}Ge$)(N_2O_x) loaded with nanoparticulate RuO_2 cocatalyst becomes active under visible-light irradiation generating H_2 and O_2 stoichiometrically from pure water.

3.3.5. Metal sulfides

Sulfide photocatalysts, which have a narrow band gap and VBs at relatively negative potentials compared to oxides, are good candidates for visible-light-driven photocatalysts. Metal sulfide photocatalysts, however, are not stable in the water-oxidation reaction under visible light because the S^{2-} anions are more susceptible to oxidation than water, thereby causing the photodegradation of the photocatalyst (Ellis et al., 1977; Williams, 1960). For this reason, sulfide photocatalysts are not suitable for water splitting unless appropriate strategies are designed to minimize photodegradation. A common method for reducing the photocorrosion of the sulfides under irradiation is by means of the use of suitable sacrificial reagents. Photocorrosion may be effectively suppressed by using a Na_2S/Na_2SO_3 mixture as electron donor (Reber and Meier, 1986). Using this mixed solution, the photocatalytic reaction should proceed as follows, avoiding the degradation of the sulfide photocatalyst:

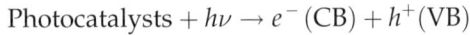

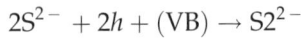

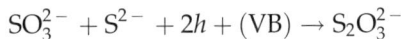

$$\text{Photocatalysts} + h\nu \rightarrow e^-(CB) + h^+(VB)$$

$$2H_2O + 2e^-(CB) \rightarrow H_2 + 2OH^-$$

$$SO_3^{2-} + H_2O + 2h^+(VB) \rightarrow SO_4^{2-} + 2H^{++}$$

$$2S^{2-} + 2h + (VB) \rightarrow S2^{2-}$$

$$S_2^{2-} + SO_3^{2-} \rightarrow S_2O_3^{2-} + S^{2-}$$

$$SO_3^{2-} + S^{2-} + 2h + (VB) \rightarrow S_2O_3^{2-}$$

Among the available sulfide semiconductors, nanosized CdS is an interesting photocatalyst material, since it has a narrow band gap (2.4 eV) and a suitable CB potential to effectively reduce the H^+ (Darwent and Porter, 1981; Matsumura et al., 1983; Reber and Meier, 1986). CdS loaded with Pt cocatalyst records a very high efficiency in light absorption and hydrogen production under visible light (QE = 25%, Bühler et al., 1984). However, the photocatalytic properties of CdS are limited as a consequence of its toxicity and photocorrosion under visible-light irradiation. In spite of the drawbacks associated with CdS, considerable efforts are still being made to improve its photocatalytic properties. The strategies reported in the technical literature to improve the activity of CdS include (i) changes in the structural characteristics of CdS and (ii) the combination of CdS with different elements or semiconductors to form composite photocatalysts with tuned band-gap size.

Considering the importance of the structural characteristics (crystalline phase, crystalline size, and geometrical surface area) in the control of band structure and in the concentration and mobility of photocatalyst charges, studies have been conducted on the influence of preparation methods on the photophysical properties of CdS (Arora et al., 1998; Jing and Guo, 2006). Improvement in CdS photoactivity is observed from preparation methods that lead to CdS phases with good crystallinity and few crystal defects.

Changes in the photoactivity of CdS can also be achieved by combining the CdS with other semiconductors with different energy levels: ZnO (Navarro et al., 2008; Spanhel et al., 1987) or TiO_2 (Fujii et al., 1998). In these composite systems, photogenerated electrons move from CdS to the attached semiconductors, while photogenerated holes remain in CdS. This charge-carrier separation stops charge recombination, thereby improving the photocatalytic activity of CdS. Improvements in CdS activity are reported for samples mixed with CdO and ZnO (Navarro et al., 2008). This improvement was linked to the better charge separation associated with the diffusion of photoelectrons generated in CdS toward surrounding CdO and ZnO. The incorporation of elements in the structure of CdS to make a solid solution is another strategy for improving the photocatalytic properties of CdS. ZnS is interesting as a semiconductor for combining with CdS. CdS and ZnS form a continuous series of solid solutions ($Cd_{1-x}Zn_xS$) where metal atoms are mutually substituted in the same crystal lattice (Fedorov et al., 1993; Nayeem et al., 2001). Valle et al. (2008) investigated the photophysical and photocatalytic properties of $Cd_{1-x}Zn_xS$ solid solutions with different Zn concentration ($0.2 < x < 0.35$). The solid solution between CdS and ZnS showed a blue shift of the absorption edge with the increase in Zn concentration. The photocatalytic activity of samples increases gradually when the Zn concentration increases from 0.2 to 0.3. The change in activity for H_2 production for these samples arises mainly from the modification of the energy level of the CB as the concentration of Zn increased in the solid solution photocatalyst.

ZnS is another sulfide semiconductor that has been investigated for photochemical water splitting. ZnS is unable to split water under visible light because of its large band gap (3.66 eV), which restricts light absorption to the UV region. Studies have been carried out to improve the visible-light sensitivity of ZnS-based photocatalysts. One of the strategies for inducing visible-light response in ZnS was the chemical doping of ZnS with metal ions: Cu^{2+} (Kudo and Sekizawa, 1999), Ni^{2+} (Kudo and Sekizawa, 2000), and Pb^{2+} (Tsuji and Kudo, 2003). The doped ZnS materials are capable of absorbing visible light as a result of the transitions from M^{n+} (M = Cu, Ni, Pb) levels to the CB of ZnS. These doped ZnS photocatalysts had high photocatalytic activity under visible light for H_2 production from aqueous solutions using SO_3^{2-}/S^{2-} as electron donor reagents.

A combination of ZnS with $AgInS_2$ and $CuInS_2$ to produce solid solutions $(CuAgIn)_xZn_{2(1-x)}S_2$ is another strategy followed to improve the optical absorption of ZnS in the visible range (Kudo et al., 2002; Tsuji et al., 2005a, b). The optical adsorption of these materials can be adjusted between 400 and 800 nm depending on solid solution composition. $(CuAgIn)_xZn_{2(1-x)}S_2$ solid solutions recorded high photocatalytic activities for H_2 evolution from aqueous solutions containing sacrificial reagents, SO_3^{2-} and S^{2-}, under visible-light irradiation. Loading solid solutions with cocatalysts improved photocatalytic activity. Pt loaded on $(AgIn)_{0.22}Zn_{1.56}S_2$ recorded the highest activity for H_2 evolution with an apparent QY of 20% at 420 nm.

Other ternary sulfides comprising In^{3+} and one type of transition metal cation (Cd^{2+}, Zn^{2+}, Mn^{2+}, Cu^+) are also investigated as photocatalysts for water splitting under visible light. However, the efficiency for water splitting under visible light achieved over these photocatalysts has so far been very low. For example, a quantum efficiency of only 3.7% at 420 nm was reported for the most active $Na_{14}In_{17}Cu_3S_{35}$ photocatalyst (Zheng et al., 2005).

4. CONCLUDING REMARKS AND FUTURE DIRECTIONS

1. Solar-hydrogen production via the water-splitting reaction on photocatalyst surfaces is one of the most promising technologies for the generation of energy in a clean and sustainable manner.
2. The success of this technology will be determined by the development of efficient photocatalysts that must satisfy very specific semiconducting and electrochemical properties.
3. Since the pioneering work by Fujishima and Honda, research has made significant progress in the development of efficient photocatalysts under visible light. From these studies, it is readily apparent that the energy conversion efficiency of water splitting is determined principally by the properties of the semiconductors used as photocatalysts.
4. In spite of these progresses, current results still record low efficiencies for visible-light-to-hydrogen conversion (2.5% QY) for practical purposes (10% QY).
5. To improve the efficiency of photocatalysts, developments in the future must be based on an understanding of the sophisticated factors that determine the photoactivity of the water-splitting reaction: (i) molecular reaction mechanisms involved in the oxidation and reduction of water on photocatalyst surfaces, (ii) structure and defect chemistry of photocatalyst surfaces, and (iii) charge transfer mechanisms between

semiconductor surfaces and cocatalysts. Nowadays, these factors have not been elucidated in sufficient detail and should be investigated as a way of refining the materials to maximize efficiency. On the other hand, the search for new photocatalytic materials with improved semiconducting and electrochemical properties is still likely to be the key to success. High throughput screening or combinational chemistry approaches, as well as more rational research based on fundamental calculations/predictions, would be useful in the search for new photocatalytic materials. Finally, the control of the synthesis of materials for customizing the crystallinity, electronic structure, and morphology of photocatalysts at nanometric scale also provide significant opportunities for improving water-splitting photocatalysts, as these properties have a major impact on photoactivity.

6. Considering the advances made in UV photocatalysts since the pioneering work of Fujishima and Honda in 1972 through to the present day, technically and economically viable visible-light photocatalysts for water splitting could become available in the near future.

ACKNOWLEDGMENTS

We are grateful to many of our colleagues for stimulating discussions and to our research sponsors CICyT and CAM (Spain) under grants ENE2007-07345-C03-01/ALT and S-0505/EN/0404, respectively. JAVM acknowledges financial support from the Autonomous Community of Madrid (Spain). MCAG acknowledges financial support from the Ministry of Science and Education (Spain) through the R&C Program.

REFERENCES

Abe, R., Sayama, K., and Sughihara, H. *J. Phys. Chem. B* **109**(33), 16052 (2005).
Anpo, M. *Pure Appl. Chem.* **72**, 1265 (2000).
Anpo, M., and Takeuchi, M. *J. Catal.* **216**, 505 (2003).
Archer, M.D., and Bolton, J.R. *J. Phys. Chem.* **94**, 8028 (1990).
Arora, M.K., Shinha, A.S.K., and Updhyay, S.N. *Ind. Eng. Chem. Res.* **37**(10), 3950 (1998).
Asahi, R., Morikawa, T., Ohwaki, T., Aoki, K., and Taga, Y. *Science* **293**, 269 (2001).
Ashokkumar, M. *Int. J. Hydrogen Energy* **23**(6), 427 (1998).
Balziani, V., Moggi, L., Manfrin, M.F., Bolleta, F., and Gleria, M. *Science* **189**, 852 (1975).
Bard, A.J. *J. Photochem.* **10**, 59 (1979).
Bard, A.J. *Science* **207**, 139 (1980).
Bird, R.E., Hulstrom, R.L., and Lewis, L. *J. Solar Cell* **15**, 365 (1985).
Blasse, G., Dirksen, J., and de Korte, P.H.M. *Mater. Res. Bull.* **16**, 991 (1981).
Bockris, J.O.M. *Int. J. Hydrogen Energy* **27**, 731 (2002).
Bolton, J.R., Strickler, S.J., and Conolly, J.S. *Nature* **316**, 495 (1985).
Bolton, J.R. *Solar Energy* **57**(1), 37 (1996).
Bühler, N., Meier, K., and Reber, J.F. *J. Phys. Chem.* **88**(15), 3261 (1984).
Choi, W.Y., Termin, A., and Hoffman, M.R. *J. Phys. Chem.* **84**, 13669 (1994).
Darwent, J.R., and Porter, G. *J. Chem. Soc. Chem. Commun.* **4**, 145 (1981).

142   R.M. Navarro et al.

Domen, K., and Yashima, M. *J. Phys. Chem. C* **111**, 1042 (2007).
Ellis, A.B., Kaiser, S.W., Bolts, J.M., and Wrighton, M.S. *J. Am. Chem. Soc.* **99**, 2839 (1977).
Fedorov, V.A., Ganshing, V.A., and Norkeshko, Y.U.N. *Mater Res. Bull.* **28**, 50 (1993).
Fonash, S.J. "Solar Cell Device Physics". Academic Press, San Diego (1981).
Fujihara, K., Ohno, T., and Matsumura, M. *J. Chem. Soc. Faraday Trans.* **94**, 3705 (1998).
Fujii, H., Ohtaki, M., Eguchi, K., and Arai, H. *J. Mol. Catal: A Chem.* **129**(1), 61 (1998).
Fujishima, A., and Honda, K. *Nature* **238**, 37 (1972).
Fujishima, A., Kohaakawa, K., and Hond, K. *J. Electrochem. Soc.* **122**, 1487 (1975).
Funk, J. *Int. J. Hydrogen Energy* **26**, 185 (2001).
Hitoki, G., Takata, T., Kondo, J.N., Hara, M., Kobayashi, H., and Domen, K. *Chem. Commun.* **16** 1698 (2002).
Hwang, D.W., Kim, H.G., Jang, J.S., Bae, S.W., Ji, S.M., and Lee, J.S. *Catal. Today* **93**, 845 (2004).
Hwang, D.W., Kim, H.G., Lee, J.S., Kim, J., Li, W., and Oh, S.H. *J. Phys. Chem. B* **109**, 2093 (2005).
Ikeda, S. Tanaka, A., Shinohara, K., Hara, M., Kondo, J.N., Maruya, K., and Domen, K. *Microp. Mater.* **9**, 253 (1997).
Ishii, T., Kato, H., and Kudo, A. *J. Photochem. Photobiol. A* **163**, 181 (2004).
Ishikawa, A., Takata, T., Kondo, J.N., Hara, M., Kobayashi, H., and Domen, K. *J. Am. Chem. Soc.* **124**, 13547 (2002).
Ji, S.M., Borse, P.H., Kim, H.G., Hwang, D.W., Jang, J.S., Bae, S.W., and Lee, J.S. *Phys. Chem. Chem. Phys.* **7**, 1315 (2005).
Jing, D.W., and Guo, L.J. *J. Phys. Chem. B*. **110**(23), 11139 (2006).
Kakuta, N., Park, K.H., Finlayson, M.F., Ueno, A., Bard, A.J., Campion, A., Fox, M.A., Webber, S.E., and White, J.M. *J. Phys. Chem.*, **89**, 732 (1985).
Kambe, S., Fujii, M., Kawai, T., and Kawai, S. *Chem. Phys. Lett.* **109**, 105 (1984).
Kato, H., Asakura, K., and Kudo, A. *J. Am. Chem. Soc.* **125**, 3082 (2003).
Kato, H., Hori, M., Konta, R., Shimodaira, Y., and Kudo, A. *Chem. Lett.* **33**, 1348 (2004).
Kato, H., Kobayashi, H., and Kudo, A. *J. Phys. Chem. B* **106**, 12441 (2002).
Kato, H. and Kudo, A. *Chem. Lett.*, **11**, 1207, (1999).
Kato, H., and Kudo, A. *J. Phys. Chem. B* **105**, 4285 (2001).
Kato, H., and Kudo, A. *J. Phys. Chem. B* **106**, 12441 (2002).
Khan, S.U.M., Al-Shahry, M., and Ingler, W.B. *Science* **297**, 2243 (2002).
Kim, J., Hwang, D.W., Kim, H.G., Lee, J.S., Li, W., and Oh, S.H. *Topics Catal.* **35**(3–4), 295 (2005).
Kim, H.G., Hwang, D.W., and Lee, J.S. *J. Am. Chem. Soc.* **126**(29), 8912 (2004).
Kogan, A., Splieger, E., and Wolfshtein, M. *Int. J. Hydrogen Energy* **25**, 739 (2000).
Kominami, H., Matsuura, T., Iwai, K., Ohtani, B., Nishimoto, S., and Kera, Y. *Chem. Lett.* 693 (1995).
Konta, R., Ishii, T., Kato, H., and Kudo, A. *J. Phys. Chem. B* **108**, 8992, (2004).
Konta, R. Kato, H., Kobayashi, H., and Kudo, A. *Phys. Chem. Chem. Phys.* **5**, 3061 (2003)
Kudo, A. *Catal. Survey Asia* **7**(1), 31 (2003)
Kudo, A., Kato, H., and Tsuji, I. *Chem. Lett.* **33**(12), 1534 (2004).
Kudo, A., Omori, K., ana Kato, H. *J. Am. Chem. Soc.* **121**, 11459 (1999).
Kudo, A., and Sekizawa, M. *Catal. Lett.* **58**, 241 (1999).
Kudo, A., and Sekizawa, M. *Chem. Commun.* **15**, 1371 (2000)
Kudo, A., Tsuji, I., and Kato, H. *Chem. Commum.* **17**, 1958 (2002).
Kudo, A., Ueda, K., Kato, and Mikami, I. *Catal. Lett.* **53**, 229 (1998).
Levene, J., and Ramsden, T. "*Summary of Electrolytic Hydrogen Production*". National Renewable Energy Laboratory, Golden, CO, MP-560-41099, (2007).
Lewis, N.S., Crabtree, G., Nozik, A.J., Wasielewski, M.R., and Alivisatos, A.P. "Basic Reseach Needs for Solar Energy Utilization". U.S. Department of Energy, Washington, DC, (2005).
Lewis, N.S., and Nocera, D.G. *Proc. Natl. Acad. Sci. USA* **103**(43), 15729 (2006).
Maeda, K., and Domen, K. *J. Phys. Chem. C* **111**, 7851 (2007).
Maeda, K., Teramura, K., Takata, K., Hara, M., Saito, N., Toda, K., Inoue, I., Kobayashi, H., and Domen, K. *J. Phys. Chem. B* **109**(43), 20504 (2005a).

Maeda, K., Takata, T., Hara, M., Saito, M., Inoue, Y., Kobayashi, H., and Domen, K. *J. Am. Chem. Soc.* **127**, 8286 (2005b).

Maeda, K., Teramura, K., Lu, D., Takata, T., Saito, N., Inue, Y., and Domen, K. *Nature* **440**(7082), 295 (2006a).

Maeda, K., Teramura, K., Saito, N., Inoue, Y., and Domen, K. *J. Catal.* **243**, 303 (2006b).

Maruthamuthu, P., and Ashokkumar, M. *Int. J. Hydrogen Energy* **13**, 677 (1988).

Maruthamuthu, P., and Ashokkumar, M. *Int. J. Hydrogen Energy* **14**, 275 (1989).

Matsumura, M., Saho, Y., and Tsubomura, H. *J. Phys. Chem.* **87**, 3807 (1983).

Matsuoka, M., Kitano, M., Takeuchi, M., Tsujimaru, K., Anpo, M., and Thomas, J.M. *Catal. Today* **122**, 51 (2007).

Mills, A., and Le Hunte, S. *J. Photochem. Photobiol. A Chem.* **108**, 1 (1997).

Miyake, J., Miyake, M., and Adsada, Y. *J. Biotechnol.* **70**, 89 (1999).

Morrison, S.R. "Electrochemistry at Semiconductor and Oxidized Metal Electrodes". Plenum Press, New York (1980).

Navarro, R.M., del Valle, F., and Fierro, J.L.G. *Int. J. Hydrogen Energy* **33**, 4265 (2008).

Nayeem, A., Yadaiah, K., Vajralingam, G., Mahesh, P., and Nagabhooshanam, M. *Int. J. Mod. Phys. B* **15**(7), 2387 (2001).

Nozik, A.J. *Appl. Phys. Lett.* **29**, 150 (1976).

Ogden, J.M. "Testimony to the Committee on Science" US House of Representatives; Washington, DC, (2003).

Reber, J.F., and Meier, K. *J. Phys. Chem.* **88**, 5903 (1984)

Reber, J.F., and Meier, K. *J. Phys. Chem.* **90**, 824 (1986).

Sato, S., and White, J.M. *Chem. Phys. Lett.* **72**, 83 (1980).

Sayama, K., Yoshida, R., Kusama, H., Okabe, K., Abe, Y., and Arakawa, H. *Chem. Phys. Lett.* **277**, 387 (1997).

Serpone, N., Borgarelo, E., and Grätzel, M. *Chem. Comm.* 342 (1984).

Serrano, M., and de Lasa, H. *Ind. Eng. Chem. Res.* **36**, 4705 (1997)

Service, R.F. *Science* **309**, 548 (2005).

Spanhel, L., Weller, H., and Henglein, A. *J. Am. Chem. Soc.* **109**, 6632 (1987).

Subramanian, V., Wolf, E., Kamat, P.V. J. of *Phys. Chem. B.* **105**(46), 11439 (2001).

Sung Lee, J. *Catal. Survey Asia* **9**(4), 217 (2005).

Takata, T., Tanaka, A., Hara, M., Kondo, J.N., and Domen, K. *Catal. Today* **44**, 17 (1998).

Tennakone, K., and Wickramanayake, S.J. *J. Chem. Soc. Faraday Trans.* **92**(1) 1475 (1986).

Tsuji, I., Kato, H., Kobayashi, H., and Kudo, A. *J. Am. Chem. Soc.* **126**, 13406 (2004).

Tsuji, I., Kato, H., Kobayashi, H., and Kudo, A. *J. Phys. Chem. B.* **109**, 7329 (2005a).

Tsuji, I., Kato, H., and Kudo, A. *Angew. Chem. Int. Ed.* **44**, 3565 (2005b).

Tsuji, I., and Kudo, A. *J. Photochem. Photobiol. A* **156**(1–3), 249 (2003).

Turner, J., Sverdrup, G., Mann, M.K., Maness, P-C., Kroposki, B., Ghirardi, M., Ewans, R.J., Blake, D. *Int. J. of Energy Res.* **32**(5), 379 (2008).

Umebayashi, T., Yamaki, T., Itoh, H. and Asai, K. *Appl. Phys. Lett.* **81**, 454 (2002).

Valle, F., Navarro, R.M., Ishikawa, A., Domen, K., and Fierro, J.L.G. "International Symposium on Catalysis for Clean Energy and Sustainable Chemistry". Madrid, Spain (2008).

Williams, R., *J. Chem. Phys.* **32**, 1505 (1960).

Xu, Y., and Schoonen, M.A.A., *Am. Mineral.* **85**, 543 (2000).

Yamashita, H., Harada, M., Misaka, J., Takeuchi, M., Ikeue, K., and Anpo, M. *J. Photochem. Photobiol. A* **148**, 257 (2002).

Yamashita, D., Takata, T., Hara, M., Kondo, J.N., and Domen, K. *Solid State Ionics* **172**, 591 (2004)

Yoshimura, J., Ebina, Y., Kondo, J., Domen, K., and Tanaka, A. *J. Phys. Chem.* **97**, 1970 (1993)

Yoshimura, J., Kudo, A., Tanaka, A., Domen, K., Maruya, K., and Onishi, T. *Chem. Phys. Lett.* **147**, 401 (1988).

Zheng, N., Bu, X., Vu, H., and Feng, P. *Angew. Chem. Int. Ed.* **44**, 5299 (2005).

CHAPTER 5

Photocatalytic Reactor Configurations for Water Purification: Experimentation and Modeling

Ajay K. Ray

Contents   1. Introduction   145
2. Macrokinetic Studies   148
     2.1   A novel kinetic reactor   152
     2.2   Distinguishing different regimes for slurry systems   154
3. Major Challenges in the Design and Development of Large-Scale Photocatalytic Reactors for Water Purification   159
     3.1   Multiple tube reactor   164
     3.2   Tube light reactor (TLR)   168
     3.3   Taylor vortex reactor   169
     3.4   Experimental details   172
4. Conclusions   181
Acknowledgments   183
References   183

1. INTRODUCTION

Wastewater treatment by low-energy UV-irradiated titanium dioxide, generally known as heterogeneous photocatalysis, has great potential as an alternative method for water purification and has become a subject of increasing interest over the last 15 years. The potential of this new technique has attracted numerous researchers to work in this area. The process couples low-energy ultraviolet light with semiconductors acting as photocatalysts. In this process, electron–hole pairs that are generated by

Department of Chemical and Biochemical Engineering, University of Western Ontario, ON N6A 5B9, Canada.
E-mail address: aray@eng.uwo.ca

Advances in Chemical Engineering, Volume 36
ISSN 0065-2377, DOI: 10.1016/S0065-2377(09)00405-0

the band-gap excitation carry out in situ degradation of toxic pollutants. The holes act as powerful oxidant to oxidize toxic organic compounds while electrons can reduce toxic metal ions to metals, which can subsequently be recovered by solvent extraction (Chen and Ray, 2001; Wang et al., 2004a, b). The first clear recognition and implementation of photocatalysis as a method of water purification was conducted by Bard (1980) and Pruden and Ollis (1983) in the photomineralization of halogenated hydrocarbon contaminants in water, including trichloroethylene, dichloromethane, chloroform, and carbon tetrachloride, sensitized by TiO_2. Since then, numerous studies have shown that a great variety of dissolved organic compounds could be oxidized to CO_2 by heterogeneous photocatalysis technique. Many reviews (Chen et al., 2000b; Fox, 1993; Hoffmann et al., 1995) and books (de Lasa et al., 2005) have also been devoted to this area of research. In the review article by Blake (1997), more than 1,200 references on the subject have been reported and an exhaustive list of chemicals that can be treated by heterogeneous photocatalysis process has also been provided.

The reason for this huge interest in photocatalysis is the significant number of advantages this process has over the traditional methods and other advanced oxidation processes of water treatment, particularly, (a) complete mineralization of many organic pollutants to environmentally benign effluents such as carbon dioxide, water, and mineral acid; (b) there is no need for use of expensive and dangerous oxidizing chemicals (such as O_3 or H_2O_2) since dissolved oxygen (or air) is sufficient; (c) TiO_2 as catalyst is active, inexpensive, nonhazardous, stable, and reusable; (d) the light required to activate the catalyst is low-energy UV-A ($\lambda < 380$ nm) of intensity around $1-5\,W\,m^{-2}$ for photoexcitation, and it is also possible to use solar light as an alternative; and (e) possibility of simultaneous oxidation of toxic organics as well as reduction of toxic metals ions.

In spite of the potential of this promising technology, development of a practical water treatment system has not yet been successfully achieved. In the last decade, a large number of publications have appeared based on laboratory-scale studies with generally positive results for very diverse categories of toxic compounds in water. However, technical development to pilot-scale level has not been successfully achieved although there are numerous patents approved worldwide.

An important impediment in the development of efficient photocatalytic reactors is the establishment of effective reactor designs for large-scale use as demanded by industrial and commercial applications. In order to achieve a successful commercial implementation, several reactor design parameters must be optimized, such as the photoreactor geometry, the type of photocatalyst, and the utilization of radiated energy. In this type of reactors, besides conventional reactor complications such as reactant–catalyst contacting, flow patterns, mixing, mass transfer, reaction kinetics, catalyst installation, and temperature control, an additional engineering factor related to illumination of catalyst becomes relevant. A fundamental issue regarding the successful implementation of photocatalytic reactors in contrast to traditional chemical reactors is the transmission of irradiation in a highly scattering and absorbing medium composed of water and fine TiO_2 particles. It is also necessary to achieve efficient exposure of the catalyst to light irradiation. Without photons of appropriate energy content, the catalyst shows no activity. The problem of photon energy absorption has to be considered regardless of reaction kinetics mechanisms. The illumination factor is of utmost importance since the amount of catalyst that can be activated determines the water treatment capacity of the reactor. The high degree of interaction between the transport processes, reaction kinetics, and light absorption leads to a strong coupling of physicochemical phenomena and no doubt, it is the major obstacle in the development of a photocatalytic reactor.

The assessment of irradiation and its distribution inside photocatalytic reactors is essential for the extrapolation of bench-scale results to large-scale operations and the comparison of the efficiencies of different installations. The successful scaling-up of photocatalytic reactors entails increasing the number of photons absorbed per unit time and per unit volume of reactor as well as efficient utilization of the electron holes created during the photo-catalytic transformations. While some of the physicochemical principles of photocatalysis are relatively well understood, reactor design and reactor engineering of photocatalytic units still require consideration. Cassano et al. (1995) emphasized the importance of the selection of radiation sources and the design of reactor geometry with respect to the irradiation source including mirrors or reflectors.

The central problem in a photocatalytic reactor is focused on a uniform distribution of light to a large surface area of catalyst. For particular photo-reactor geometry, scale-up in the axial and/or radial directions is con-strained by the phenomenon of opacity, light scattering, depth of radiation penetration, and local volumetric light absorption. The arrangement of light source–reactor system influences the reactor design in such a strong way that independent consideration is not possible. Moreover, the need for at least one of the reactor walls to transmit the chosen radiation imposes the utilization of transparent materials, such as glass for the reactor construc-tion, and thus imposes size limitations, sealing problems, and breakage risks.

The overall reaction rate of photocatalytic processes is usually slow compared to conventional chemical reaction rates due to low concentration level of the pollutants, and therefore, there is a need to provide large amounts of active catalyst inside the reactor. Although the effective surface area of the porous anatase catalyst coating is high, there can only be a thin coating (about 1 µm thick) applied to a surface. Thus, the amount of active catalyst in the reactor is limited, and even if individual degradation processes can be made relatively efficient, the overall conversion efficiency will still be low. This problem severely restricts the processing capacity of the reactor, and the necessary time required to achieve high conversions are measured in hours, if not days. These aspects are discussed in detail in the section "Major challenges in the design and development of large-scale photocatalytic reactors."

The mechanism of photocatalysis reaction has been discussed in many books (de Lasa et al., 2005; Serpone and Pelizetti, 1989) and articles (Chen et al., 2000; Dutta et al., 2005). The individual steps that drive the above reaction are still not well understood due to the complex processes involved. The semiconductor photocatalytic oxidation of organic species dissolved in water by gaseous oxygen poses a number of questions with respect to the three-phase system and its interaction with light. It can, therefore, be referred to as a *four-phase* system, the fourth phase being the *UV light-electronic* phase. These questions present a great challenge, and the first step of the investigation must be a systematic study and understanding of the macrokinetic factors affecting the photocatalytic degradation rate.

2. MACROKINETIC STUDIES

TiO$_2$ catalyst has been used in two forms: freely suspended particles in aqueous solution and immobilized particles onto rigid inert surfaces. In the former case, a high ratio of illuminated catalyst surface area to the effective reactor volume can be achieved for a small, well-designed photocatalytic reactor (Chen and Ray, 1998), and almost no mass transfer limitation exists since the diffusional distance is very small, resulting from the use of ultrafine (<30 nm) catalyst particles. In large-scale applications, however, the catalyst particles must be filtered prior to the discharge of the treated water, even though TiO$_2$ is harmless to the environment. Hence, a liquid–solid separator must follow the slurry reactor (SR) (e.g., Photo-Cat$^®$ reactor of Purifics Environmental Canada Technologies). The installation and operation of such a separator will raise the cost of the overall process, as the separation of the ultrafine catalyst particles is a slow and expensive process. Besides, the penetration depth of the UV light is limited due to the strong absorption by TiO$_2$ and dissolved organic species, particularly for dyes. All these disadvantages render the scale-up of a SR very difficult (Mukherjee and Ray, 1999).

Many macrokinetic studies have been done to date to understand the effects of various parameters in the degradation of compounds through heterogeneous photocatalysis. The past works have analyzed various macrokinetic factors, namely (a) catalyst loading, (b) initial concentration of solute, (c) light intensity, (d) circulation rate, (e) pH of solution, (f) partial pressure of oxygen, and (g) temperature, which affect the degradation rate (Table 1). References used in Table 1 are typical references, and it should be

Table 1 Important macro kinetic factors affecting photocatalytic degradation rate as reported by few investigators in open literature

Investigators	Chen and Ray, 1998	Rideh et al., 1997	Inel and Okte, 1996	Matthews, 1990
Pollutant	4-Nitro phenol	2-Chloro phenol	Malonic acid	Phenol, 4-CP, benzoic acid
Experimental setup	Swirl-flow reactor with suspended catalyst and recirculation	Cylindrical annular Pyrex reactor with external recirculation	Annulus reactor with a pump to recirculate the suspension	Spiral reactor, with a pump to recirculate the suspension
Macro kinetic factors	Results			
1 Initial concentration	Initial observed rate is pseudofirst-order with respect to initial concentration	Rate decreases with increase of initial concentration	Rate increases and reaches a plateau at high concentration, Langmuir-type dependence	Rate is zero-order at high concentration and first-order at low concentration
2 Catalyst loading	Rate increases and reaches a plateau at a catalyst loading of $2\,g\,L^{-1}$	Rate increases and reaches a plateau at a catalyst loading of $0.2\,g\,L^{-1}$	Rate increases and reaches a plateau at a catalyst loading of $1\,g\,L^{-1}$,	At low solute concentration lowest loading gives greater rate at high solute concentration, the more the better

Table 1 (*Continued*)

	Macro kinetic factors	Results		
3	Light intensity	Rate $\propto I^\beta$ where β is between 0.5 and 1.0	Rate is proportional to light intensity	Rate $\propto I^{0.5}$	Not studied
4	Circulation rate	Not studied	Not studied	Rate increases with circulation rate and reaches a plateau	No significant effect of circulation rate and assumed no mass transfer limitation
5	pH	At both high and low pH values, degradation rate is quite slow, the best pH is near point of zero charge, that is, between 5.6 and 6.4	Low rate in acidic region, constant rate in neutral region and increasing rate in basic region	Rate increase with pH and reaches a plateau at pH = 9. Surface charges of TiO_2 is strongly influenced by pH	Not studied
6	Partial pressure of oxygen, P_{O_2}	Rate increases with P_{O_2} and reaches a plateau at about 0.4 atm	Rate increases nonlinearly with P_{O_2}	Not studied	Not studied
7	Temperature	Not significant	Not significant	Rate increases with T	Not studied

noted that many other authors have reported similar results but are not included here for brevity. The following general conclusions can be drawn from the past macrokinetic studies:

1. The reaction rate depends on the initial concentration of the solute. Although at high concentration, the observed rate is independent of solute concentration (zeroth-order dependence), at low concentration it shows pseudofirst-order dependence while at some intermediate concentration it follows Langmuir-type dependence (Ray and Beenackers, 1997).
2. The reaction rate increases with catalyst loading and reaches a maximum at some optimum catalyst loading and then decreases. The researchers attribute the decrease of rate at high catalyst loading to the obstruction of light transmission and defined it as the "shielding effect" (Chen et al., 2000).
3. The reaction rate is proportional to the light intensity, but at higher intensities the reaction rate becomes proportional to the square root of light intensity. It is explained that as the light intensity increase, so does the recombination rate of hole and electron (Ray and Beenackers, 1997).
4. The effect of circulation rate on the degradation rate has also been studied. Typically, no influence of circulation rate on the degradation rate has been found in slurry systems (Chen and Ray, 1998), although increases in the degradation rate due to circulation rate has been reported under some specific conditions (Inel and Okte, 1996; Serrano and de Lasa, 1997, 1999).
5. In most cases, it is found that the pH has direct impact on the degradation rate. It has been observed that the pH affects the surface properties of TiO_2 catalyst, and hence the degradation rate gets affected (Mehrotra et al., 2005).
6. The photocatalytic conversion rate increases with the increase of the oxygen partial pressure. The typical explanation is that oxygen acts as an electron scavenger, thus reducing the rate of electron–hole recombination (Chen and Ray, 1999).
7. The effect of temperature is not so significant within the studied interval, probably due to the counter effect of a decrease in the adsorption constant of solute on the catalyst with increasing thermal levels. It is also to be noted that reactions carried out in heterogeneous photocatalysis display low activation energies. Moreover, most of the photocatalytic experiments are conducted at isothermal conditions (usually at room temperature), and therefore, temperature changes do not play a major role (Chen and Ray, 1999).

One of the most important issues of macrokinetic studies is distinguishing the "kinetic" region in which the observed rate is governed by kinetic dependences. Obviously, intrinsic kinetic data are necessary for catalyst characterization and reactor design. Traditionally, in the kinetic studies of heterogeneous catalysis, to distinguish the "kinetic" regime from the "transport-limited" regime, data related to the substance transformation over catalyst must be presented using the dependence mol per unit mass of catalyst per unit time instead of mol per unit volume of solution per unit time. This kind of data representation with dimension of rate of reaction based on per unit mass of catalyst can be found in the literature of multiphase slurry reactions with suspended catalyst. However, in the literature devoted to the photocatalytic degradation of organic solutes such representation of data is not a common feature. Typically, kinetic data are reported using the dimension mol per unit volume of solution per unit time and the "kinetic" region, rigorously to say, cannot be identified. In the next section, a novel kinetic reactor design is described that can be used for identifying different kinetic as well as transport-limited regimes for both slurry and immobilized systems. This reactor also allows determining pure kinetic rate constants and its dependence as a function of various parameters such as light intensity, amount of catalysis, pH, and oxygen concentration.

2.1. A novel kinetic reactor

Figure 1 shows a semibatch swirl-flow monolith-type reactor that has been designed to study purely the kinetics of photocatalytic reactions for any model components (Ray and Beenackers, 1997). Monoliths are unique

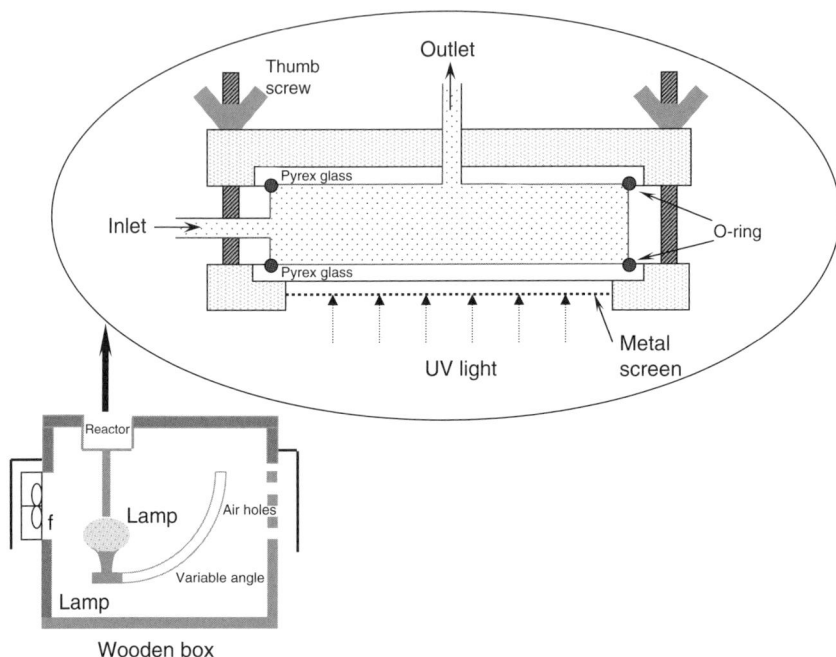

Figure 1   Schematic diagram of the novel kinetic reactor (Mehrotra et al., 2003).

catalyst supports that provide a high surface to volume ratio and allow high flow rates with low pressure drop. Catalyst has been immobilized for continuous use and to eliminate the need of separation for post-process treatment. Experiments were performed to determine the various rate constants, in particularly, the effect of light intensity and catalyst layer thickness on reaction rate constant, effect on overall reaction rate when light has to travel through the scattering and absorbing heterogeneous medium to activate the catalyst, and the effect of angle of incidence of light falling on catalyst particles. As it is apparent in most fixed-catalyst systems, mass transfer limitations do exist. However, this unique reactor can be used to determine the mass transfer effect and the correct kinetic rate constants and this to extract pure kinetic data. The reactor consists of two circular glass plates and is placed between soft padding housed within stainless steel and aluminium casing separated by 3–10 mm (Figure 1). The catalyst is (note: to use the "present" tense) consistently better deposited either on the top of the bottom plate or on the bottom of the top plate. The polluted water is introduced tangentially between the two glass plates through an inlet tube and exited from the center of the top plate through an outlet tube. The tangential introduction of liquid created a swirl-flow and thereby helped in obtaining better mixing and residence time distribution inside the reactor.

The lamp (Philips HPR 125 W high pressure mercury vapor) is placed underneath the bottom glass plate on a holder that can be moved to create different angle of incidence of light on the catalyst plate. The lamp and reactor are placed inside a wooden box painted black so that no stray light can enter into the reactor. The lamp is constantly cooled by a fan to keep the temperature down and protect the lamp from overheating. The lamp has a spectral energy distribution with a sharp peak at $\lambda = 365.5$ nm of 2.1 W and thereby the incident light intensity of our interest is 213 W m^{-2}. Provision is made for placement of several metal screens of different mesh size between the lamp and bottom glass plate to obtain variation in light intensity.

The flow pattern inside the reactor can be calculated by solving momentum equations using commercial software package, FLUENT. Figure 2 reports the particle trajectory for both one inlet and two inlets in opposite directions. The figure clearly shows that tangential introduction of fluid indeed creates swirl-flow inside the reactor. It reveals the presence of a dead zone in upper right corner and bypassing of fluid when one inlet stream is used. The location of dead zone can vary depending on the inlet velocity. In order to eliminate the dead zone and bypassing of fluid, two inlets placed in opposite directions were incorporated and the Figure 2b clearly shows improved flow patterns inside the reactor thereby resulting in uniform residence time distribution of the fluid.

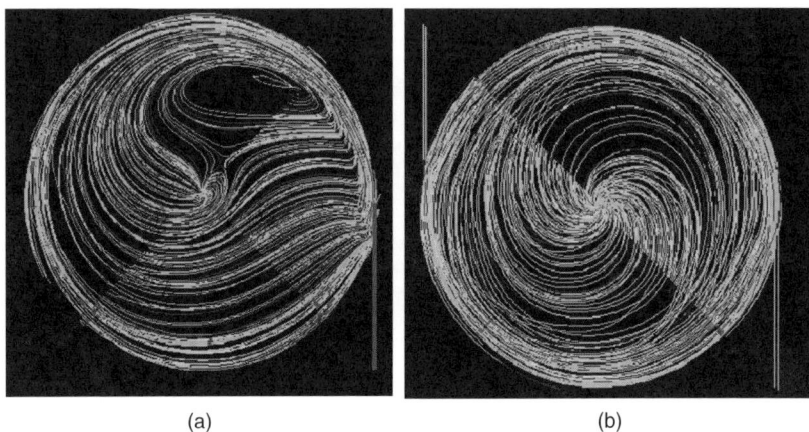

(a)   (b)

Figure 2 Flow pattern inside the reactor. (a) one inlet, (b) two inlets. Figure clearly shows use of two inlets can eliminate presence of dead volume and bypassing of fluids that is present in reactor with one inlet (Chen et al., 2001).

2.2. Distinguishing different regimes for slurry systems

Figure 3 represents experimental data of the photocatalytic degradation rate of benzoic acid for different catalyst loading while Figure 4 represents rate versus circulation rate. The experimental studies have been done with catalyst loading in the range of 0.01–$2.0\,g\,L^{-1}$ and at circulation rate between 50 and $1,050\,mL\,min^{-1}$. The figure reveals that the degradation rate is almost constant (independent of the catalyst loading) for catalyst loading between 0.01 and $0.05\,g\,L^{-1}$. The rate then drops gradually as the catalyst loading is increased from 0.05 to $2.0\,g\,L^{-1}$. The rate is not affected by the circulation rate at the same catalyst loading, but the magnitude of rates with respect to circulation rates is different at different catalyst loading. Clearly, Figure 3 depicts the "kinetic" regime for catalyst loading between 0.01 and $0.05\,g\,L^{-1}$. In this range, the overall rate is solely controlled by kinetics as when we double the amount of catalyst (in g-cat), the conversion (in $mmol\,min^{-1}$) also doubles, thereby the resultant rate (in $mmol\,g^{-1}$-cat min^{-1}) remains constant. When catalyst loading is increased above $0.05\,g\,L^{-1}$, the rate decreases gradually. Hence, in this region, the overall rate is not controlled entirely by kinetics and must be influenced by the transport of either pollutant or light to the catalyst surface.

In this transport-limited region, the conversion increases at a slower rate than the increase of amount of catalyst, and thereby, the overall rate decreases instead of remaining constant as in the "kinetic" regime. In other words, in this region conversion is influenced by different factors: (a) the additional catalyst surface area provided does not come in full contact with the pollutant due to external mass transfer resistance, and/or

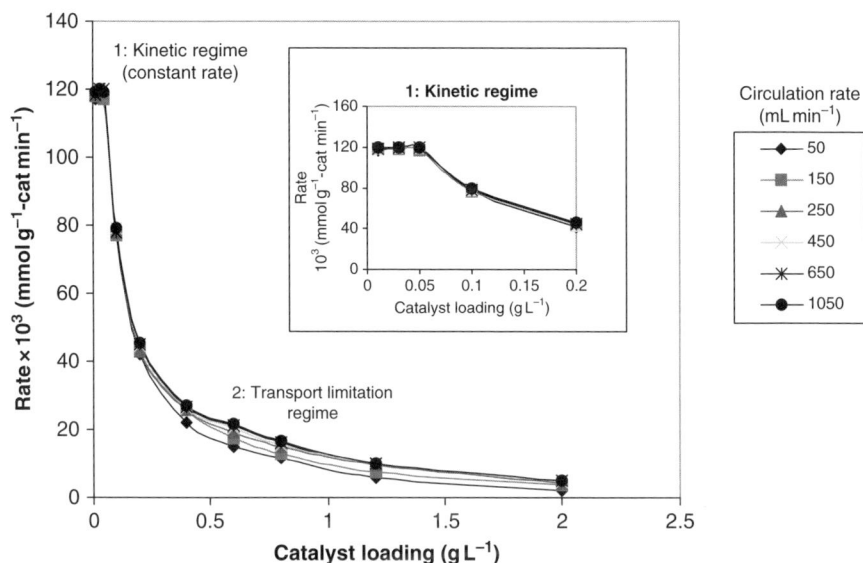

Figure 3 Degradation rate (mmol g^{-1}-cat min^{-1}) of benzoic acid for different catalyst loading showing the kinetic regime. Experimental conditions: $C_{SO} = 0.2$ mM, pH $= 3.7 \sim 3.9$, $T = 303$ K, $I = 9.90$ mW cm^{-2}, and O_2 saturated (Mehrotra et al., 2005).

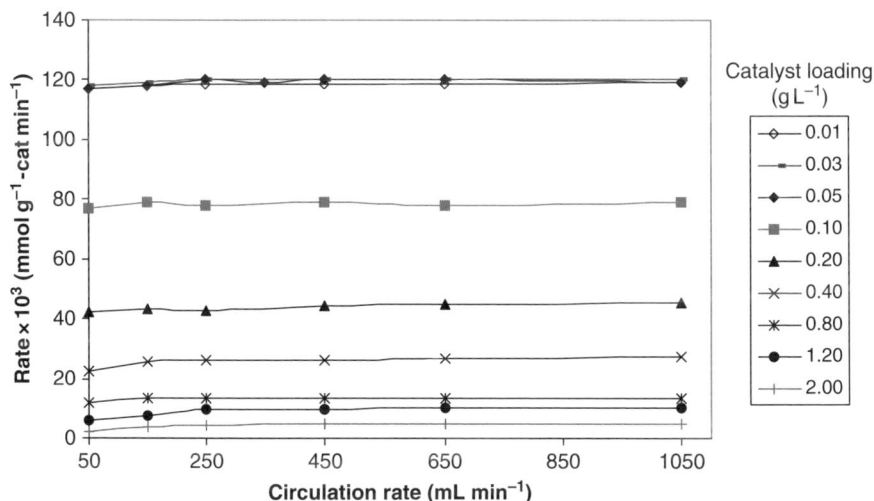

Figure 4 Degradation rate (mmol g^{-1}-cat min^{-1}) of benzoic acid against circulation rates at different catalyst loading. Experimental conditions: $C_{SO} = 0.2$ mM, pH $= 3.7 \sim 3.9$, $T = 303$ K, $I = 9.90$ mW cm^{-2}, and O_2 saturated (Mehrotra et al., 2005).

(b) the pollutant molecules cannot reach some of the available catalyst surface area due to agglomeration of the catalyst particles (internal mass transfer resistance), and/or (c) the irradiation provided cannot reach some catalyst surface area due to absorption and scattering (shielding effect) of light, and/or (d) the light cannot penetrate the agglomerates and activate the inner surfaces, and/or (e) there is a combination of all of the above. Thus, Figure 3 reveals two regimes: (1) "kinetic" regime at the low catalyst range ($<0.05\,g\,L^{-1}$) and (2) "transport (mass or light)" limitation regime at the high catalyst loading ($0.05\,g\,L^{-1}$–$2.0\,g\,L^{-1}$). In the transport limitation regime, the problem lies in understanding of the factor responsible for the above described transport limitations.

Figure 4 reveals that the overall reaction rate remains constant with the increase of circulation rate at any particular catalyst loading. Such results corresponding to the experimental data obtained previously (Mehrotra et al., 2005) mean that there is no external mass transfer limitation under the experimental conditions. The specific features related to our experiments are small volume of the reactor (63.5 mL), low residence time (11 s), internal stirring, and small catalyst loading (0.0125–$0.05\,g\,L^{-1}$). The difference is evident as our results clearly distinguish data related to the "kinetic" regime. In order to analyze the cause of such discrepancy in more detail, a special experiment was performed (Mehrotra et al., 2005). A beaker containing 70 mL of the slurry mixture of 0.2 mM benzoic acid and desired amount of TiO_2 catalyst was taken. The mixture was stirred in the dark for half an hour to allow for adsorption (dark reaction) of benzoic acid on the catalyst surface to reach equilibrium. The beaker was then placed over a glass plate with UV lamp underneath. The reaction mixture was bubbled with oxygen and stirred very slowly. The resultant initial rate at different catalyst loading is shown in Figure 5.

Comparing typical rate dependence on the catalyst loading from the previous studies with results reported in Figures 3 and 5, it can be concluded that at very slow internal stirring, high residence time and absence of external circulation (conditions in Figure 5) diminishes the region of rate constancy, and under certain conditions the "kinetic regime" could not be observed. Hence, proper mixing, small catalyst loading, small volume of the reactor, and low residence time, which are characteristics of our experiments shown in Figure 3, are the necessary conditions to obtain data under a well-distinguished "kinetic" regime.

To analyze the transport (mass or light) limitation regime in more detail, one can represent in Figure 6 the same experimental results of apparent rate reported in Figures 3–4 using another dimension $mmol\,L^{-1}$-solution min^{-1}, which is most commonly used in open literature by researchers working in photocatalysis. Figure 6 shows three different domains: (a) the increase of rate at the low catalyst loading, (b) the constancy of rate at the intermediate catalyst loading ("optimal catalyst loading" domain), and (c) the decrease of rate at the high catalyst loading.

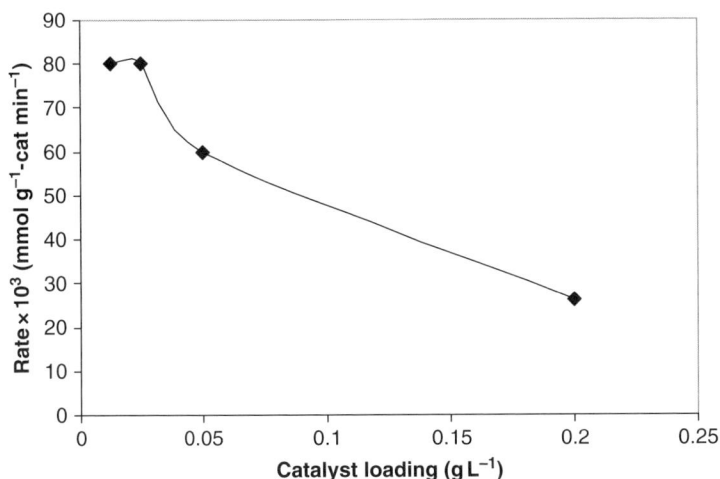

Figure 5 Degradation rate (mmol g^{-1}-cat min^{-1}) of benzoic acid for different catalyst loading without external circulation. Experimental conditions: $C_{SO} = 0.2$ mM, pH $= 3.7 \sim 3.9$, $T = 303$ K, $I = 8.0$ mW cm^{-2}, and O_2 saturated (Mehrotra et al., 2005).

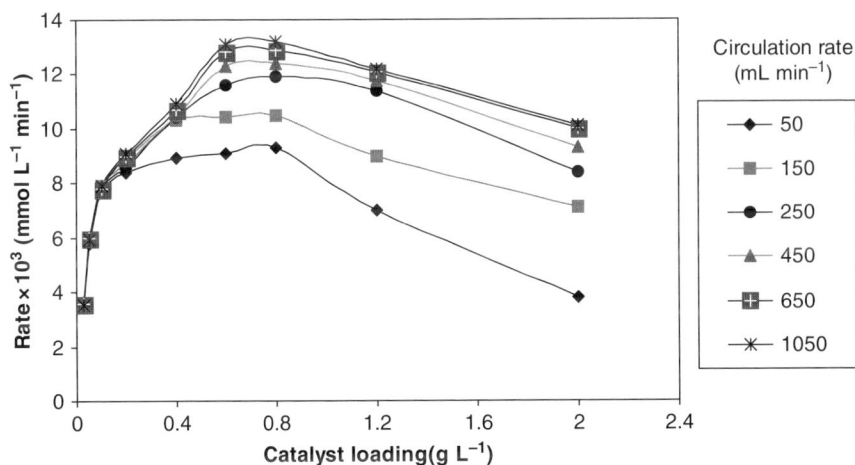

Figure 6 Degradation rate (mmol L^{-1}-solution min^{-1}) of benzoic acid against catalyst loading at different circulation rates. Experimental conditions: $C_{SO} = 0.2$ mM, pH $= 3.7 \sim 3.9$, $T = 303$ K, $I = 9.90$ mW cm^{-2}, and O_2 saturated (Mehrotra et al., 2005).

This plateau in the optimal loading (working regime) domain becomes smaller and even insignificant with the increase of the circulation rate. We have already reported in Figure 4 that external mass transfer is negligible as rate (in mmol g^{-1}-cat min^{-1}) was not affected by the circulation rate at any given catalyst loading. When same experimental data are plotted in

Figure 6, it shows that as the circulation rate is increased, the plateau of constant rate (in $mmol\,L^{-1}$-solution min^{-1}) not only diminishes but also shifts toward higher catalyst loading. Obviously, the circulation rate changes the transport rate, but it is not a case of the external transport limitation. It is very likely that in this region (catalyst loading between 0.4–$0.85\,g\,L^{-1}$), the circulation rate influences agglomeration of catalyst particles. Internal mass transfer resistance caused by the agglomerate primarily controls the degradation rate. At higher circulation rates, the average size of agglomerated catalyst particles decreases, and therefore, the observed degradation rate increases. Another important aspect to note is the difference between pure "kinetic" regime values and values corresponding to the "optimal catalyst-loading" regime. There is a significant difference between these rates, and such difference has not been presented in the previous studies. The ratio of the rate at "optimal catalyst loading" to the rate in the "kinetic" regime can be estimated as about 0.166 irrespective of circulation rate using the dimension of rate as $mmol\,g^{-1}$-cat min^{-1} (Figure 4) for "optimal catalyst loading" of $0.8\,g\,L^{-1}$ (Figure 6).

The three factors that can be considered to be responsible for the transport limitation of overall rate can be distinguished as (a) increase of catalyst loading per unit solution volume causes an increase of apparent rate, while combination of (b) mutual influence of particle (agglomeration of particles) hinders the organic substance to reach the catalyst surface, and (c) the ability of photons to penetrate agglomerates and activate the inner surfaces causes decrease of apparent rate. The first and the last two factors compensate each other in the second region (domain), resulting in a region of "optimal catalyst loading." When catalyst loading is increased further, the third factor dominates, thus reducing the apparent rate, which Figure 6 clearly demonstrates.

A sound explanation for the optimal catalyst-loading regime is the appearance of particle agglomerates under some sufficient catalyst loading. Generally, the limitation domain can be interpreted as the internal diffusion region in which the agglomerate plays the role of the porous particle. An estimation of such agglomerate size can be easily done using the known theoretical relationship for the Thiele modulus, according to which the effectiveness factor η is given by the following equation:

$$\eta = \frac{3}{R}\sqrt{\frac{D_e}{k\rho_p S_a}} \tag{1}$$

where k is the kinetic constant, D_e is the diffusion coefficient, and R is the radius of the agglomerate. The value of kinetic rate constant can be obtained using the rate value in the "kinetic" regime $\approx 4 \times 10^{-3}$ $mmol\,L^{-1}$-solution min^{-1} (Figure 6) with initial concentration, $C_{SO} = 0.2$ mmol L-solution, and specific surface area of the reactor as $100\,m^2\,m^{-3}$. Using the above values, the degradation rate constant k for benzoic acid is equal to $3.33 \times 10^{-6}\,\mathrm{m\,s^{-1}}$. In the "optimal catalyst-loading" domain, $\eta = 0.166$, $D_e = 1 \times 10^{-10}\,\mathrm{m^2\,s^{-1}}$ (Chen et al., 2000), $\rho_p = 2.4\,\mathrm{g\,cm^{-3}}$, and $S_a = 50\,\mathrm{m^2\,g^{-1}}$, therefore, the characteristic size of the agglomerate, R, is given by

$$R = \frac{3}{\eta}\sqrt{\frac{D_e}{k\rho_p S_a}} = 9\,\mu\mathrm{m} \qquad (2)$$

The agglomerate size is in close agreement with data obtained independently and reported elsewhere (Hoffmann et al., 1995) compared to average primary particle size of Degussa P25 as 30 nm.

When the catalyst loading is increased even further (greater than $1.2\,\mathrm{g\,L^{-1}}$), the overall rate is limited primarily by limitation of transport of light to the catalyst surface, which is described in literature as shielding effect. In this region, addition of more catalyst is like adding inert materials and therefore, meaningless. The most important point to note is that constancy of rate while expressing rate data as mol/mass of catalyst/time reveals the "kinetic" region whereas constancy of rate data when rate is expressed as mol/liter of solution/time reveals the region of "optimal catalyst loading." Determination of all the three regions is important, and it is possible to accomplish by representing same set of experimental data in two different rate dimensions. Intrinsic kinetics can be obtained from the "kinetic" region, which is necessary for catalyst characterization and reactor design. "Optimal catalyst loading" region is essential from a practical application point of view while the shielding region states that there is no point of adding any more semiconductor particles, which acts as an inert instead of acting as a photocatalyst.

3. MAJOR CHALLENGES IN THE DESIGN AND DEVELOPMENT OF LARGE-SCALE PHOTOCATALYTIC REACTORS FOR WATER PURIFICATION

Photocatalytic reactions are promoted by solid photocatalyst particles that usually constitute the discrete phase distributed within a continuous fluid phase in the reactor. Therefore, at least two phases, that is, liquid and solid, are present in the reactor. The solid phase could be dispersed (SPD) or stationary (SPS) within the reactor. SPD photoreactors may be operated with the catalyst particles and the fluid phase(s) agitated by mechanical or other means. Depending on the means of agitation, the photoreactor resembles that of slurry or fluidized bed reactors. In numerous investigations, an aqueous suspension of the catalyst particles in immersion or annular-type photoreactors has been used. However, the use of suspensions requires the separation and recycling of the ultrafine (submicron size) catalyst from the treated liquid and can be an inconvenient, time-consuming expensive process. In addition, the depth of penetration of UV light is limited because of strong absorption by both catalyst particles and dissolved organic species. Above problems could be avoided in SPS photoreactors in which photocatalyst particles are immobilized onto a fixed transparent surface, such as the reactor wall or a fiber mesh, or supported on particles, such as glass or ceramic beads, that are held in fixed positions in the photoreactor. However, immobilization of catalyst on a support generates a unique problem. The reaction occurs at the liquid–solid interface and usually only a part of the catalyst is in contact with the reactant. Hence, the overall rate may be limited to mass transport of the pollutant to the catalyst surface. In addition, the rate of reaction is usually slow because of the low concentration level of the pollutant, and therefore, there is a need for a reactor whose design provides a high ratio of illuminated immobilized catalyst to illuminated surface and provides the possibility of total reactor illumination.

The volume of photocatalytic reactor can be expressed as

$$V_R = \frac{Q\,C_{in}\,X}{\kappa\,\mathfrak{R}} \tag{3}$$

where Q is the volumetric flow rate ($m^3\,s^{-1}$), C_{in} is the inlet pollutant concentration ($mol\,m^{-3}$), X is the fractional conversion desired, κ is illuminated catalyst surface area in contact with reaction liquid inside the reactor volume ($m^2\,m^{-3}$), and $\mathfrak{R}$ is the average mass destruction rate ($mol\,m^{-2}\,s^{-1}$). Hence, smallest reactor volume will result when κ and $\mathfrak{R}$ are as large as possible for specified values of Q, C_{in}, and X. $\mathfrak{R}$ is a *reaction-specific* parameter as it expresses the performance of catalyst for the breakdown of a specific model component, while κ is a *reactor-specific* parameter representing the amount of catalyst inside a reactor that is sufficiently illuminated so that it is active and is in contact with the reaction liquid. An increase in $\mathfrak{R}$ can be accomplished by modifying the physical nature of the catalyst in terms of its structure and morphology or by the addition of additional oxidizing agents (thereby increasing active hydroxyl radical concentrations) or band-engineering of the catalyst in which photocatalyst is chemically modified to reduce band gap and increasing catalyst activity (Zhou et al., 2006a, b, 2007a, b). Improving the breakdown rates would lead to the need of a less demanding amount of catalyst to be illuminated, and therefore, a smaller reactor volume. The parameter κ, namely illuminated specific surface area, helps to compare design efficiency of different photocatalytic reactors as it defines the efficacy to install as much active catalyst per unit volume of reaction liquid in the reactor.

One major barrier to the development of a photocatalytic reactor is that the reaction rate is usually slow compared to conventional chemical reaction rates, due to low concentration levels of the pollutants. Other crucial hurdle is the need to provide large amounts of active catalyst inside the reactor. Even though the effective surface area of the porous catalyst coating may be high, there can only be a thin coating (about 1 μm thick) applied to a surface. Thus, even if individual degradation processes can be made relatively efficient, the overall conversion efficiency will still be low. This problem severely restricts the processing capacity of the reactor, and the time required to achieve high conversions are measured in hours, if not days.

A number of photocatalytic reactors have been patented in recent years but none has so far been developed to pilot-scale level. Based on the manner in which catalyst is used, and the arrangement of the light source and reactor vessel, all photocatalytic reactor configurations fall under four categories. They are *slurry type* in which catalyst particles are in suspension form (Chen and Ray, 1999), *immersion* type with lamp(s) immersed within the reactor (Ray and Beenackers, 1996), *external* type with lamps outside the reactor (Assink et al., 1993), and *distributive* type with the light distributed from the source to the reactor by optical means such as reflectors and light conductors (Ray, 1999) or optical fibers (Peill and Hoffman, 1995). Majority of reactors patented are variation of the SR, and classical annular reactor (CAR) of immersion or external type in which catalyst is immobilized on reactor wall (Sato, 1992; Taoda, 1993), on pipes internally (Matthews, 1990a), on ceramic membranes (Anderson et al., 1991), on glass wool matrix between plates (Cooper, 1989), on semipermeable membranes embedded (Miano and Borgarello, 1991; Oonada, 1994) in water permeable capsules (Hosokawa and Yukimitsu, 1998), on a mesh of fiberglass (Henderson and Robertson, 1989), on beads (Heller and Brock, 1993), on fused silica glass fibers (Hofstadler et al., 1994), on porous filter pipes (Haneda, 1992), on glass fiber cloth (Masuda et al., 1994), and so on. The reactors are either helical (Ritchie, 1991), spiral (Matthews, 1988), shallow cross flow basins (Cooper and Ratcliff, 1991), or optical fiber (Wake and Matsunaga, 1994). However, all these reactor designs are limited to small scales by the low values of illuminated catalyst density. The only way to apply these systems for large-scale applications is by using large numbers of multiple units.

The parameter κ (Ray and Beenackers, 1998a), namely illuminated specific surface area, helps to compare design efficiency of different photocatalytic reactors as it defines the efficacy to install as much active catalyst per unit volume of reaction liquid in the reactor. The parameter κ represents the amount of catalyst inside a reactor that is sufficiently illuminated so that it is active and is in contact with the reaction liquid. Table 2 lists κ values for the four different classes of photocatalytic reactors. In a SR, small catalyst particles could provide large surface area for reaction but essentially most of the catalyst surface area will be inactive, particularly for large reactor dimensions as the catalyst particles will not receive enough light from the external light source. This happens since the organics and the liquid medium itself are absorbing light. This is especially true for large reactor

Table 2 Comparison of κ (m^2 m^{-3}) for different reactors

Photocatalytic reactor	κ (m^2/m^3)	Parameters	κ (m^{-1})	Remarks
Slurry reactor [19]	$\left[\dfrac{6C_c}{\rho_c}\right]\dfrac{1}{d_p}$	$d_p = 0.3\,\mu m$ $C_c = 0.5\,kg\,m^{-3}$	2,631[a]	Scale-up not possible
External type – annular reactor [21]	$\dfrac{4d_0}{d_0^2 - d_i^2}$	$d_0 = 0.2\,m$ $d_i = 0.1\,m$	27	Scale-up not possible
Immersion type – with classical lamps [16]	$\left[\dfrac{4\varepsilon}{1-\varepsilon}\right]\dfrac{1}{d_0}$	$d_0 = 0.09\,m$ $\varepsilon = 0.75$	133	Scale-up possible but large V_R
Immersion type – with novel lamps [18]	$\left[\dfrac{4\varepsilon}{1-\varepsilon}\right]\dfrac{1}{d_0}$	$d_0 = 0.0045\,m$ $\varepsilon = 0.75$	2,667	Scale-up possible with small V_R
Distributive type – with hollow tubes [22]	$\left[\dfrac{4\varepsilon}{1-\varepsilon}\right]\dfrac{1}{d_0}$	$d_0 = 0.006$ $\varepsilon = 0.75$	2,000	Scale-up possible with small V_R

[a]The value will be much lower than 2,631 m^{-1} as all the suspended catalyst particles will not be effectively illuminated. Catalyst concentration, $C_c = 0.5\,kg$ m^{-3}, is normally used. $\rho_c = 3800\,kg\,m^{-3}$.

dimensions, resulting in low efficiencies and the drawback of impossibility for scale-up to commercial use applications. In addition, use of suspension requires separation and recycling of the ultrafine catalyst from the treated effluent by filtration, centrifugation or coagulation, and flocculation. These add various levels of complexity to an overall treatment process and clearly decrease the economical viability of the SRs. An external-type reactor will always be limited by low values of κ. An immersion-type reactor could be scaled-up to any dimension but when classical lamps of diameter between 0.07 and 0.1 m are used, the κ value is very low even if it is assumed that the lamps occupy 75% of the reactor volume.

Many other new *innovative* type of reactor designs exist in literature, addressing specific problems and applications or others being designed especially for treating specific types of pollutants. These are in the form of treatment agents containing novalak resins, photoelectrochemistry or elec-trophotographic methods, use of magnetic or sound waves, photocatalysis and acoustics, catalyst particles coated with polymers as well as ion exchange process (Mukherjee and Ray, 1999). Others include the use of optical fibers as the source of illumination catalyst contained in removable filter units and water-permeable capsules containing the catalyst. Although, innovative in approach, the main problem associated with all these reactor configurations is again the issue of scale-up for commercial use purposes.

Of particular interests are several Photo-CREC-Water photocatalytic reac-tors developed by Professor de Lasa's research group. Their devices involve unique designs with respect to both suspended and supported catalyst. The Photo-CREC Water I reactor is an annular reactor with a lamp placed at the reactor centerline. The arrangement allows high catalyst loading of the glass mesh, good catalyst irradiation, and uniform contact of the catalyst with the circulating water. The Photo-CREC-Water I reactor was originally proposed by de Lasa and Valladares (1997). Subsequently, modifications were introduced to the original design by Serrano and de Lasa (1997). The Photo-CREC-Water II (Salaices et al., 2001) is an annular vessel with a lamp placed in the center of the reactor. In the upper section, there is a slurry distribution system ensuring intense mixing of the slurry suspension at the reactor entry. The unit is equipped with quartz windows and accessory collimator tubes. This configuration allows the measurement of photon absorption and the quantification of back and forward reflection, and it is of particular value to establish energy and quantum efficiencies in photo-catalytic reactor units. Their Photo-CREC-Water III is an annular vessel with external illumination designed to simulate a solar-irradiated reactor. The reactor is irradiated externally by a set of eight UV lamps. They measured the radiation field distribution using a fiber optic sensor placed inside a concentric inner tube with the sensor device connected to a spectrophotor-adiometer. The external illumination permitted the simulation of solar irradiation, whereas the combination of internal and external illumination allowed the smoothing of the radiation field radial distribution and the increase of irradiation efficiency.

The scale-up has been severely limited by the fact that reactor configurations have not been able to address the three most important parameters, namely, light distribution inside the reactor through the absorbing and scattering liquid to the catalyst, providing high surface areas for catalyst coating per unit volume of reactor, and mixing inside the reactor (Ray and Beenackers, 1996). The new reactor design concepts must provide a high ratio of activated immobilized catalyst to illuminated surface and must have a high density of active catalyst in contact with the liquid to be treated inside the reactor. Few new design of photocatalytic reactors, addressing the above issues have been developed by our research group and are described in the next few sections.

3.1. Multiple tube reactor

In order to overcome some of these deficiencies inherent in conventional photocatalytic reactor designs, a distributive-type photocatalytic reactor design in which catalyst is fixed to a structure in the form of glass slabs (plates), rods, or tubes inside the reactor has the greatest potential for scale-up. This will allow for high values of κ and will eliminate light passage through the reaction liquid. This is advantageous because when light approaches the catalyst through the bulk liquid phase, some radiation is lost due to absorption in the liquid. In particular, this effect is more pronounced for highly colored dye pollutants as they are strong UV absorber and will therefore significantly screen the TiO_2 from receiving UV light.

Our research group considered scale-up configurations that contain both high surface areas to volume and efficient light distribution to the catalyst phase. Two ways above deficiencies inherent in conventional photocatalytic reactor designs can be overcome. First, by using a distributive type of photocatalytic reactor design in which catalyst is fixed to a structure in the form of hollow glass tubes, and second, by using an immersion-type reactor with very narrow diameter tube lamps. The design based on hollow glass tubes allows for a much higher illuminated surface area per unit reactor volume, while the other design provides not only much higher values for active catalyst surface area but also the catalyst can be activated uniformly at its highest possible level. Furthermore, the designs of the reactors are flexible enough to be scaled-up for commercial scale applications.

The limitation to the size of the reactor with light conductors is the UV transparency of the material and the light distribution to the catalyst particle. The critical and probably the most intricate factor is the distribution of the available light in the conductors to the catalyst particles and to ensure that each particle receives at least the minimum amount of light necessary for activation. The reactor configuration conceptually applicable for photo-catalysis, satisfying most of the above-mentioned requirements, is a rectan-gular vessel in which light conductors as glass slabs (or rods) coated on its outside surface with catalysts are embedded vertically. The lamps together with reflectors are placed on two sides of the reactor while liquid enters and exits from the other two sides. Light rays entering the conductor through one end are repeatedly reflected internally down the length and at each reflection come in contact with the catalyst present around the outer surface of the conductors. Thus, conducting materials might be considered as a means of light carrier to the catalyst. Since the ratio of the surface area on which catalyst is present to the light entering area could be as high as 500, evidently a very large catalyst area can be illuminated. Moreover, with a large number of such light conducting material packed inside the reactor, the configuration provides a high total light transfer area and allows for a higher illuminated catalyst area per unit reactor volume. Densely packing the reactor with light conducting object not only increases surface to volume ratio but also reduces the effective mass transfer diffusion length for the pollutant to catalyst surface.

The vital issue in the distributive type of reactor concept is how to introduce light from the external source efficiently into the light conductors, and likewise, how to get it out again at the proper location and in the apropos amount. The predominant obstacle we came across in the use of glass slabs (or rods) as the light-conducting object is the occurrence of total internal reflection. It transpires when light travels from denser to rarer medium and is determined by the critical angle given by

$$\theta_c = \sin^{-1}\left[\frac{n_2}{n_1}\right] \tag{4}$$

where n_1 and n_2 are the refractive indices of the denser and rarer medium, respectively. In the case of light travelling from air, to glass to air (or water), the angle θ will always exceed the critical angle, θ_c, for the interface between glass and air (or water) irrespective of angle of incidence, α (0–90°). In other words, all the light rays that are entering through the top surface will experience the phenomena of total internal reflection and will come out axially rather than emerging from the lateral surface. However, refractive index of TiO_2 (between 2.4 and 2.8) is higher than that of glass (about 1.5) in the wavelength range of 200–400 nm, and it is likely that total internal reflection would not take place when the glass surface is coated with titania. Nevertheless, if coating consists of small spheres of catalyst particles dispersed along the surface, the actual glass–titania interface will be small, as most of the glass surface will still be in contact with water. Therefore, it is best, if possible, to avoid the occurrence of total internal reflection entirely.

One way of avoiding total internal reflection is by surface roughening. Moreover, surface roughening assists in achieving better catalyst adhesion to the substrate. Both are indeed found out to be the case experimentally by us. In fact, when lateral surface was roughened by sand blasting, most of the light emerged within few centimeters and hardly any light remained thereafter in the axial direction. This is not only because roughening desists total internal reflection phenomena but also UV transparency of most light conducting material is very poor. Although the use of quartz as light conductors will naturally help to overcome light transmission problem, it will certainly make the overall reactor set-up more expensive.

The total internal reflection problem can also be effectively avoided when the surface's light that has to pass through is parallel instead of perpendicular. One such configuration is a hollow glass tube coated on its surface with semiconductor catalysts. The hollow tube might be considered as a pore carrying light to the catalyst. In this novel configuration, light rays entering through one end of the hollow tube are repeatedly internally reflected down the length of the tube and at each reflection come in contact with the annular catalyst coating present around the outer surface of the tube. Although total internal reflection could be avoided completely in this configuration, the angle of incidence of light will be a critical factor. When light falls on the glass surface, a part of it is reflected and the rest is transmitted. The ratio between the reflection and transmission of light is a strong function of angle of incidence. When the light beam is nearly parallel with the surface (α close to $0°$), most of the light is reflected and exits axially rather than laterally while for light rays with α close to $90°$ most of the light will emerge laterally within few centimeters and barely any light will remain thereafter as reflection is only 4% for a glass–air interface. Hence, it is important that in the design of reactor based on hollow tubes, light must be guided into the conductors at a very precise angle through a combination of optical lenses and reflectors.

Computer simulation of the reactor was performed using a commercial CFD package FLUENT/UNS to determine effects of flow rates, diffusion coefficients, reaction rate constants and inlet species concentrations and intertube spacing. Application of CFD to reactors entails a number of tasks: (a) formulating the relevant transport equations, (b) establishing the necessary constitutive and closure equations, (c) formulating appropriate boundary conditions, (d) selecting the most suitable numerical techniques to solve the equations, (e) choosing (or developing) a suitable computer code to implement the numerical techniques, and (f) developing effective flow simulation strategies. In order to get accurate results, the 3D model of the reactor was divided into control volumes ranging from 120,000 to 600,000 cells distributed unevenly in the computational domain. Computation of each problem is very time consuming, as it requires at least 2,000 iterations to be performed to obtain accurate results. In the wake of this, a state-of-the-art computational technique, viz., distributed computing, was used to expedite the calculation and overcoming the memory bottleneck present in single workstations (Periyathamby and Ray, 1999). Initially, a reasonably converged solution was obtained for the velocity and pressure equations. The reaction part of the solver was then activated, and using the almost converged solution as the initial condition, a complete solution of the reacting flow model was obtained. The converged velocity profiles at various regions of the reactor were observed to determine the degree of mixing. Fluid mixing is one of the important criteria in efficient photocatalytic reactor design as the transport of reactants from the bulk of the liquid to the catalyst surface determines the extent of fluid-catalyst contacting and overall conversion.

In Figure 7, streamlines are used to show the mixing pattern in the reactor. The streamline pattern shows that maximum mixing takes place at the bottom of the reactor where fluid enters the reactor tangentially with a swirling motion. There is also a good degree of mixing just before the fluid exits at the top of the reactor. However, the flow is uniform without much mixing in the midregion between the inlets and the outlets.

Figure 7   Velocity vectors showing the degree of mixing inside the reactor.

The advantage of using computer simulations is that the length of the reactor required for complete degradation of a particular pollutant can be determined easily compared to time-consuming expensive experimental studies. In addition, the flexibility of determining the effects of various process parameters in computer simulations will be useful, particularly flow rates, as the degradation rate strongly depends on the residence time of the fluid inside the reactor. A typical case of degradation of a pollutant, monochlorobenzene, C_6H_5Cl, is shown in Figure 8.

3.2. Tube light reactor (TLR)

During the development of this new concept of photocatalytic reactor based on multiple hollow tubes [multiple tube reactor (MTR)], we developed a unique new lamp design. These are extremely narrow diameter fluorescent tube lamps of low wattage emitting lights in the wavelength of our interest ($\lambda < 365\,nm$). These new lamps address many of the solutions to the

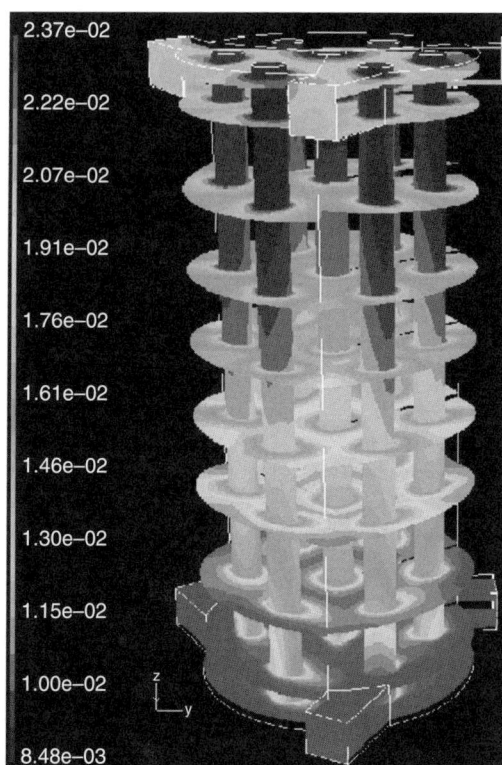

Figure 8 Photocatalytic degradation of monochlobenzene in MTR. The flow direction from bottom to top (Periyathamby and Ray, 1999).

problems that have restricted the development of technical scale photocatalytic reactors for water purification. These lamps are available in various shapes and lengths and can be placed inside a reactor to form a variety of different configurations. Development of a reactor using these new lamps will provide all the advantages of the MTRs, plus the additional advantage that catalyst could be activated at its highest level. In the reactor, catalyst was deposited on the outer surface of the low wattage lamps using a dip-coating apparatus. Thus, the main problem encountered in the development of a reactor based on multiple hollow tubes (MTR) was avoided. In the MTR concept, it is impossible to obtain a uniform light distribution along the length of the tubes and therefore, it will severely restrict the maximum length of tubes that can be used inside a reactor and thereby the overall performance of the reactor. The new lamps eliminated this drawback in the development of the MTR reactor concept, as the new design is capable of uniform light distribution over long tube lengths. Of course, this was possible with classical lamps too. However, the new lamps allow for a 50–100 times larger surface area for catalyst per unit reactor volume compared to a classical reactor design.

3.3. Taylor vortex reactor

In the design of fixed bed photocatalytic reactors, one must address two issues, namely, uniform distribution of light and mass transfer of pollutants to the catalytic surface. Earlier, our research group did experimental studies on reactors containing catalyst-coated tube bundles (Ray, 1999), catalyst-coated extremely narrow diameter immersion-type lamps (Ray and Bee-nackers, 1998), and catalyst-coated rotating tube bundles (Ray, 1998). The experimental as well as simulation results revealed that photocatalytic reaction is primarily diffusion (mass transfer) controlled when catalyst is fixed (Periyathamby and Ray, 1999). The photocatalytic reaction takes place at the fluid-catalyst interface, and in most cases, the overall rate of reaction is limited to the transfer of pollutants to the catalyst surface. In our earlier studies, we have enhanced mass transfer by increasing mixing through turbulence and/or use of baffles. This new reactor design uses flow instability to increase reaction yield throughout the reactor volume. We considered unsteady Taylor–Couette flow in between two coaxial cylinders where inner cylinder coated with TiO_2 catalyst is rotated at different speed to achieve the desired instability.

Taylor (1923) first observed the instability of fluid when the inner cylinder exceeds a critical speed between two coaxial cylinders in his established work. The laminar flow confined within the annulus region between two coaxial cylinders with the inner one differentially rotating with respect to the outer suffers centrifugal instability depending on the geometry and rotation rates. Taylor (1923) showed that an inviscid rotating flow to be unstable, if the energy of rotation associated with fluid particle decreases radially outward. Under such unstable configuration, one notices the appearance of circumferential toroidal vortices in between the two cylinders and is known as Taylor–Couette vortices. These vortices evolve due to the adverse gradient of angular momentum that creates potential unstable arrangement of flow. Such an unstable condition arises naturally if the outer cylinder is held stationary while the inner cylinder is rotated at a sufficiently high rotation rate. In such an unstable condition, the annulus is filled with pair of counter rotating vortices. Subsequently, significant research works have been published on the hydrodynamics, transport properties, and applications of Taylor–Couette flow. Researchers analyzed the Taylor–Couette vortex flow experimentally, mathematically, and by numerical simulation and reported different flow patterns at different speed, shape and size of vortices, role of Taylor vortex on radial mixing, effect of axial flow and axial dispersion in Taylor vortex flow, contribution of Taylor vortices in heat and mass transfer, and so on.

Computer simulation was performed to observe the flow pattern within the annular space for the same reactor configuration used in our experimental study. In computing, the flow of the three-dimensional Navier–Stokes equation was solved in primitive formulation by using the commercial computational fluid dynamics (CFD) software Fluent®. *Time accurate* solution was obtained by solving the governing equations in which no simplification was made regarding symmetry and reflection of the solution. Fluent preprocessor GAMBIT® was used to create geometry and generate grid. The computation of Navier–Stokes equation for Taylor–Couette geometry is expensive. The problem was solved in a mainframe supercomputer (SGI origin 2000). The details of the computational details of simulation runs were reported elsewhere (Sengupta et al., 2001). Figure 9 shows computer-simulated results of the vortex formation and flow patterns in the annular gap between the two coaxial cylinders where the inner cylinder rotates above the critical speed. The boundary layer oscillates periodically between almost zero thickness to a maxima in between the counter-rotating vortex pair, where the two shear layers approaching each other spews out a jet of fluid toward the outer wall. The fluid particles in their motion around the toroidal vortices come periodically in contact with the inner surface where the catalyst is immobilized, and in the presence of light the catalyst is activated and as a result the redox reaction takes place. The residence time in the illuminated region is thus a function of the angular velocity of the recirculating vortex as well as the size of the vortex. The latter, once again depends on the gap size and the number of vortices formed in a given length of the reactor. The critical speed of the inner cylinder at the onset of the Taylor instability depends upon the aspect ratio (ratio of length and the annular gap) and the kinematic viscosity of fluid. If the inner cylinder speed is increased further, the flow system

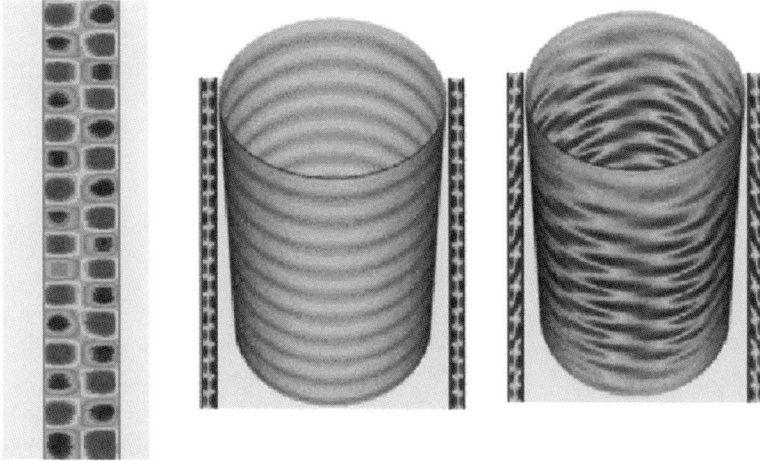

Figure 9 Flow configuration of Taylor vortices in the annular gap and contours of axial velocity in the annular space from simulation for vertical and cylindrical sections of Taylor-Couette ($Re = 177$) and wavy vortex flow ($Re = 505$) (Dutta and Ray, 2004).

exhibits a sequence of time-dependent stable vortex flow regimes as well as complicated patterns in the wide transition region between the laminar Couette flow and the turbulent vortex flow (TVF). Kataoka (1986) observed different flow patterns at different rotation speed and reported flow regions as laminar Taylor vortex flow (LTVF), singly periodic wavy vortex flow (SPWVF), doubly periodic wavy flow (DPWF), weakly turbulent wavy vortex flow (WTWVF), and TVF. The size and number of the vortices formed depended on the aspect ratio of the cylinder. However, the situation can be different for very short or infinitely long cylinder. Moreover, the effective wavelength depends on the initial and boundary conditions. This has been attributed to the existence of many stable solutions of the governing Navier–Stokes equation for flows far from equilibrium. Coles (1965) investigated systematically the wavy vortex flow and found that there are several distinct stable flow states depending upon the path through which final (steady state) Reynolds number is reached. This suggests nonuniqueness of the flow pattern or the possibility of existence of multiple solutions. Rayleigh (1920) was first to deduce the criteria of centrifugal instability where he stated that an inviscid rotating flow is unstable if the energy of the rotating fluid particles decreases radially outward. An unstable condition arises when the outer cylinder is held at stationary and the inner cylinder is rotated, the fluid near the inner cylinder experiences a centrifugal force while the fluid in the outer cylinder experiences the presence of stationary wall. If the rotation is increased further, the centrifugal force overcomes the stabilizing viscous flow and these two opposing forces create instability by creating series of pair of counter rotating vortices where the diameter of an individual vortex is approximately equal to the annular gap. In other words, stability is ensured if

$$\frac{d}{dr}[rv_\theta]^2 > 0 \tag{5}$$

According to Taylor (1923), the flow instability is observed when the Taylor number exceeds a critical value where Taylor number is defined by geometrical parameters and the speed of rotation:

$$Ta = Re^2\left[\frac{d}{r_i}\right] \tag{6}$$

where $Re = r_i \omega d/\nu$.

Sczechowski et al. (1995) was the first to study experimentally Taylor–Couette flow instability in photocatalytic reactor to enhance the photoefficiency. They observed that when catalyst particles were used as suspension, the useful reaction took place only periodically. Taylor–Couette flow allows catalyst particles to get into the continuous periodic illumination since only part of the reactor is illuminated due to optical dense fluid. Vortices created in Taylor–Couette flow move the catalyst particles into and out of the illumination area and thus allow pollutants to come into periodic contact with light and darkness. The residence time of the particles in the illuminated area is thus a function of the angular velocity of the recirculating vortices as well as the size of the vortex. They observed 30% higher photoefficiency at 300 rpm by using an unusually high catalyst loading of $10\,g\,L^{-1}$. However, slurry system raises the question of separating the submicron-size catalyst particles after degradation of pollutants. In addition, photoefficiency can only be increased at higher rotation speed and at unusual higher catalyst loading, which might not be feasible in practical application. Based on the above considerations, our research group has designed a new Taylor Vortex photocatalytic Reactor (TVR) where the outer surface of the inner cylinder is coated with catalyst and the fluorescent lamp is placed inside of the inner cylinder. The immobilization of catalyst eliminates the need for separation of submicron-size particles after treatment.

3.4. Experimental details

3.4.1. The reactors
Figure 10 shows schematic drawing of the novel bench-scale reactor system based on hollow tubes. The reactor (MTR) consists of a cylindrical vessel of diameter 0.056 m within which 54 hollow quartz glass tubes of diameter 0.006 m coated on its peripheral surface with catalyst were placed. The tubes were held securely within the reactor by two teflon end plates on which

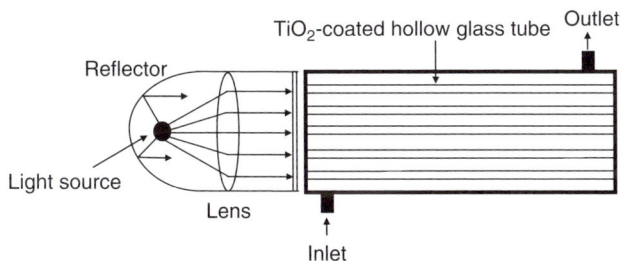

Figure 10 Schematic diagram of multiple tube reactor (MTR) (Ray, 1999).

54 holes were drilled. The reactor resembles that of a shell- and tube-heat exchanger with reaction liquid flowing through the shell-side over the outside surfaces of the coated tubes while light travels through the inside of hollow tubes. The tubes were arranged in triangular pitch of 0.007 m, thereby achieving a very high surface area per unit volume. The feed was introduced through four equally distributed ports at one end of the vessel, thereby minimizing formation of any dead zones. Similarly, the exit flow from the reactor was collected through four ports distributed at the other end of the reactor. One end of each tube was closed to prevent any reaction liquid entering the inside of the tubes. The closed ends were also coated with aluminum for better utilization of axially exiting light. The glass material used was quartz instead of Pyrex. Although quartz is more expensive than Pyrex glass when glass slabs or rods are considered, the price difference is not so appreciable when glass test tubes are considered. The quartz was used in the setup particularly for two reasons: (a) transmission of light in Pyrex is very poor compared to quartz and therefore, the length of hollow glass tubes that can be used in the reactor will be restricted and (b) when using large number of 5–6-mm test tubes, use of quartz tubes will increase the strength of the reactor and it will be much easier to handle bundle of long but narrow diameter hollow test tubes without worrying about breakage of the tubes. Of course, if one uses Pyrex tubes, reactor will be cheaper but then the length of the reactor has to be reduced. The light source (Philips GBF 6436, 12 V, 40 W) used in MTR was a low-voltage halogen lamp optically positioned in a lightweight highly glossy anodized aluminium reflector spanning 0.056 m for a clearly defined beam spread. In addition, a condenser lens of focal length 0.04 m was placed between the lamp and the reactor to obtain light beam at a half intensity beam angle between 2 and 4°.

The TLR (Figure 11) consists of a stainless steel flat top plate (0.132 × 0.016 m) with 21 holes onto which another plate (0.248 × 0.132 m) was welded. Twenty-one U-shaped lamps were placed around the latter plate and its end extended through the holes for electrical connections.

Figure 11 Schematic diagram of tube light reactor (TLR) (Ray and Beenackers, 1998).

Electrical wires were connected to the novel lamps through copper holders that are screwed around the lamp end. This part acts as a clamp for the lamps. The assembly was put in a rectangular stainless steel reactor vessel. Feed is introduced at the top of the vessel and is equally distributed over the width of the reactor through five inlet ports thereby minimizing formation of any dead zones. Similarly, the flow exits the reactor through five ports at the other end. The effective illuminated surface areas of the catalyst and the volume of the reactor are $0.15\,m^2$ and $5.36 \times 10^{-4}\,m^3$, respectively. The parameter κ, defined as total illuminated catalyst surface area that is in contact with reaction liquid per unit volume of liquid treated in the reactor volume, is equal to $618\,m^2\,m^{-3}$. The novel lamps (Philips NDF-U2 49-6W) used in TLR were specially developed by Philips Lighting for our experiments. The U-shaped lamps are $0.498\,m$ long and have a diameter of $0.0045\,m$ only. These operate at $1020\,V$, produce $6\,W$ of which 15% is in the UV-A region. The light intensity ($\lambda < 380\,nm$) on catalyst particles is $127.8\,W\,m^{-2}$.

In Figure 12, a schematic view of the Taylor Vortex Reactor (TVR) used is shown, which consists of two coaxial cylinders (with $r_i/r_o = 0.796$ and aspect ratio, $L/d = 38.24$) made of Perspex in which the inner one rotates while the outer one remains stationary. Perspex is used as it makes it easy for handling, and moreover, it can cut off light in UV-B and UV-C ranges and thereby eliminates direct photolysis of organic compounds. The catalyst was coated on the outer surface of the inner cylinder, which can be rotated at variable speed achieved through gear coupling, a stepper motor together with a frequency generator. A UV lamp (Philips, TLK 40W/10R) was mounted inside of the inner cylinder. The lamp has a spectral distribution

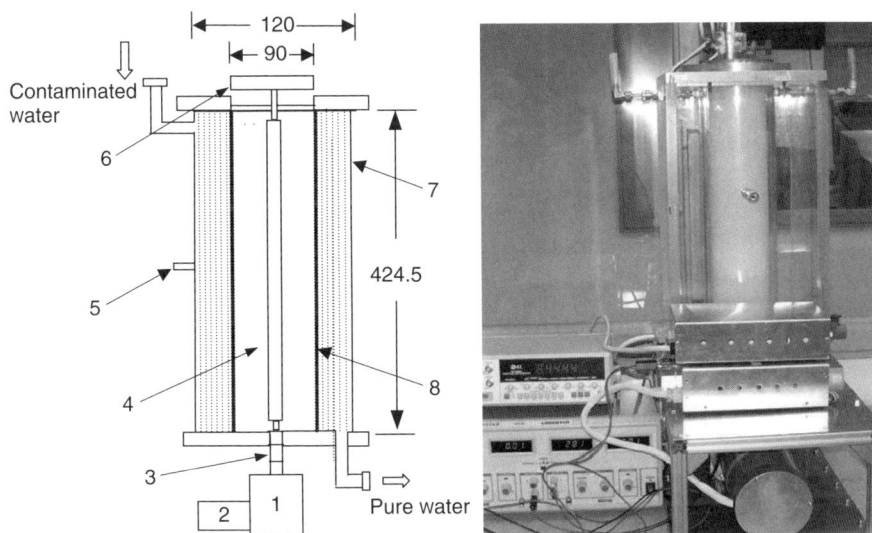

Figure 12 Experimental set-up of Taylor vortex photocatalytic reactor (Dutta and Ray, 2004): (1) motor, (2) speed controller, (3) gear coupling, (4) UV lamp, (5) sample collection point, (6) lamp holder, (7) outer cylinder, (8) catalyst-coated inner cylinder.

energy with a sharp (primary) peak at $\lambda = 365$ nm with an incident light intensity 13 W m^{-2}. The volume of the annular region (in which the reaction liquid is present) is equal to 1.45 L and the reactor was operated in batch mode. A sampling port was made near the middle portion of the reactor through which samples were collected by a syringe. A UV radiometer (Cole-Parmer Instrument Series 9811) was used to measure the intensity of the light around 365 nm wavelength. The rpm of inner cylinder was determined with a tachometer.

3.4.2. Catalyst
Degussa P25 grade TiO_2 was used as catalyst for all the experiments. The crystalline product is nonporous primarily in the anatase form (70:30 anatase to rutile) and is used without further treatment. It has a BET surface area of $(5.5 \pm 1.5) \times 10^4$ m^2 kg^{-1} and crystallite sizes of 30 nm in 0.1–0.3 μm diameter aggregates.

3.4.3. Catalyst immobilization
For better catalyst fixation and its durability, the glass surface of the tubes and the lamps on which titania was deposited were roughened by sand blasting. This makes the catalyst surface uneven but increases the strength and amount of catalyst per unit area that could be deposited. It is known that adherence of TiO_2 on quartz is poor than on Pyrex. However, when the

glass surface was roughened, the adherence improved appreciably. The glass surface was carefully degreased, cleaned with 5% HNO_3 solution overnight, washed with water, and then dried. A 5% aqueous suspension of the catalyst was prepared with water out of Millipore Milli-Q water purification system. The suspension was mixed in an ultrasonic cleaner (Branson 2200) bath to obtain a milky suspension that remained stable for weeks. The lamp's surfaces were coated with catalyst by dip-coating apparatus designed for coating catalyst. This is completely automated equipment capable of immobilizing catalyst onto a variety of different shaped and sized substrates to any desired thickness by successive dipping of the objects into a suspension at controlled speed that can be varied between 0.4 and $4.0 \times 10^{-4}\,\mathrm{m\,s^{-1}}$. Four 250-W infrared lamps were attached to a clamp that can be moved both vertically and horizontally for instant drying of the coating. After coating, it was dried and then fired at a temperature of 300°C. Catalyst coating thus obtained on a roughened glass surface was very stable. Hardly any catalyst washes away under running water. In addition, roughening of glass surface helped in achieving better light distribution due to reduction in total internal reflection of light.

3.4.4. Model component and analysis
Laboratory grade Orange II (Acid orange 7, MW 350.3, dye content 85%) was used as model component. This is an excellent model component for the characterization of a photocatalytic reactor as the dye is reactive only in presence of both TiO_2 and UV light and biologically not degradable. Changes in Orange II concentration were measured using Shimadzu UV-1601PC UV-Visible-Spectrophotometer at wavelength of 485 nm. A shimadzu 5000A total organic carbon (TOC) analyzer with an ASI-5000 autosampler was used to analyze the TOC for the model compounds.

3.4.5. Experimental setup
A gear pump (Verder model 2036) was used for pumping the reactant between the reactor and the reservoir via a flow-through cuvet placed inside a universal photometer (Vitatron 6000) for continuous on-line measurement of the model component (Figure 13). Two three-way glass valves were used between the water and specially designed reactant reservoir for initial zero setting of the analytical instrument before the start of an experiment, introduction of the reactant into the system, elimination of bubbles formed during experiment, and final flushing of the entire system.

3.4.6. Experimental procedure
At the start of every experiment, the reactor was rinsed with Milli-Q water before zero-setting the analytical instrument. The reactor was then filled with the dye solution and it was ensured that no air bubbles remained in the system. The change in the dye concentration was continuously recorded.

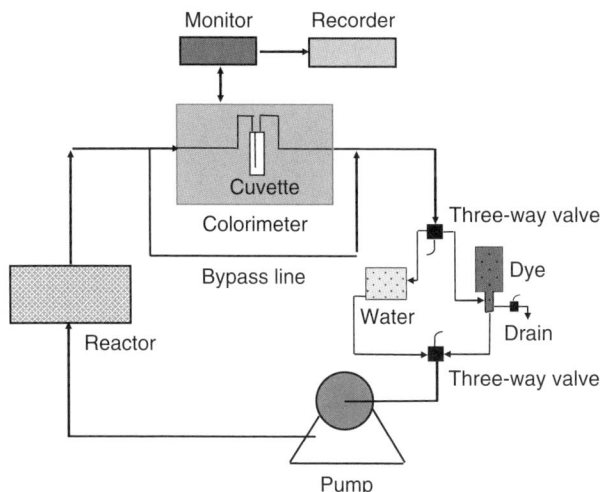

Figure 13   Experimental setup. Different reactors described in this chapter were used in the "Reactor" block.

New silicon connecting tubes and fresh catalysts were found to adsorb the dye for about an hour, but no noticeable adsorption by the entire system was observed afterward. Light was turned on only when the colorimeter reading was stabilized. For the analysis of Taylor vortex reactor, various visualization fluids (AQ-RF, AQ-1000, ST-1000, AQ-Red Dye) from Kalliroscope Corporation were used to observe the flow pattern in the annular gap for Taylor vortex reactor. AQ-RF rheoscopic fluid was used as received while AQ-1000 rheoscopic concentrate, ST-1000 bacterial stabilizer, and AQ-Red dye were mixed in required proportion to get proper visualization at different speeds. Degussa P25 grade TiO_2 was used as photocatalyst as received. The outer surface of the inner cylinder was coated with TiO_2 using a fully automated dip-coating apparatus described elsewhere (Ray and Beenackers, 1998).

Varied catalyst film thickness can be obtained by controlling the number of times coated and the speed of coating. For the Taylor vortex reactor, the experiments were carried out in two phases. In the first phase, only Taylor-Couette flow development was studied at different speeds of rotation (Reynolds number) of the inner cylinder using different flow visualization fluid. Time needed for vortex formation and its movement toward center from both ends, vortex pattern, number of vortices within the reactor length, changes of flow pattern upon increasing the speed of rotation, and the effect of mode of starting (slow or sudden increase) of rotation of the inner cylinder were recorded. In the second phase, photocatalytic degradation experiments were carried out for the three model compounds.

3.4.7. Results and discussion

The reaction rate was found out to be a function of flow rate indicating that reaction is mass transfer controlled. Experimental results reveal that 90% of the pollutant was degraded in about 100 min for MTR (Ray, 1999) and in about 30 min for TLR (Ray and Beenackers, 1998). This was achieved even though the reactor was far from optimum with respect to mass transfer, flow distribution, and efficiency of packing of the tubes in the reactor. In fact, performance of the reactor can be instantly improved by decreasing the length of the hollow tubes as it is likely that catalyst is almost inactive near the end of the 0.5 m long tube away from the light source. It is apparent that MTR design idea creates great opportunities for building much more efficient photocatalytic reactor. The main problem in the development of MTR design concept is that it is impossible to obtain uniform light distribution along the length of the tubes, and thereby, restricting the maximum length of tubes that can be used. One way of avoiding this is to place one extremely narrow diameter novel lamps inside each of the hollow tubes. In this way, all the advantages of the MTR can be retained while eliminating the basic drawback of uniform light distribution dilemma. Moreover, this will also eliminate the main problem experienced in the TLR with the prolonged use of the novel lamps in contact with reaction liquid.

Time-dependent LTVF was observed when the rotation of the inner cylinder exceeds 2.2 rpm ($Re > 111$). The vortex starts at the bottom of the cylinder at the critical Reynolds number and moves toward center of the cylinder with time. The flow was monitored using the Kalliroscope fluid (AQ 1000) and in our system, it took about 12 min to reach steady state as shown in Figure 14.

At steady state, about 21 counter rotating vortex pairs were observed and the dimension of the each counter rotating pair was almost twice the gap of the annular space. It was also observed that

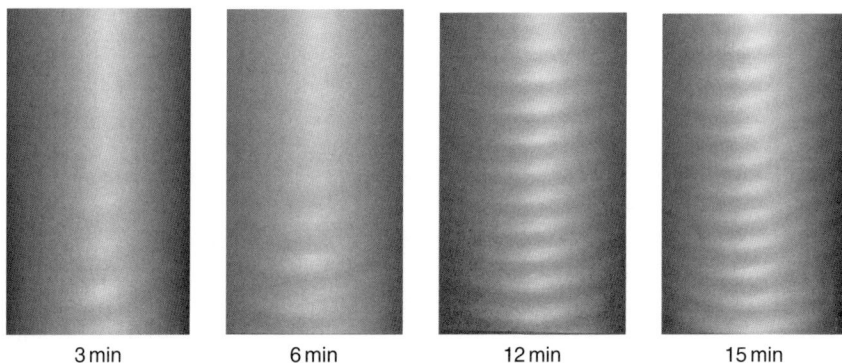

| 3 min | 6 min | 12 min | 15 min |

Figure 14 Progress of time-dependent Taylor vortex flow around critical Reynolds number, $Re_c = 111$. (Dutta and Ray, 2004).

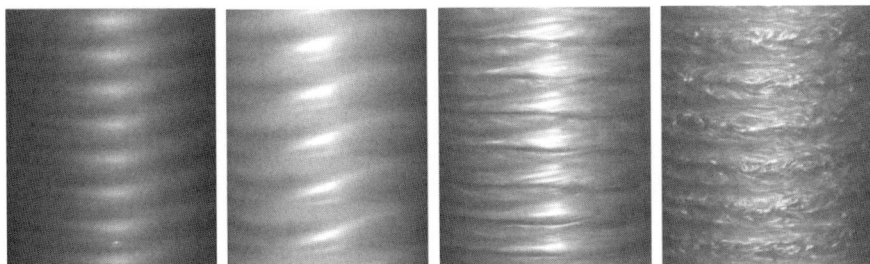

Figure 15 Photograph of flow pattern at different Reynolds number. (a) Taylor vortex flow ($Re = 177$), (b) wavy vortex flow ($Re = 505$), (c) weakly turbulent vortex flow ($Re = 3027$), and (d) turbulent vortex flow ($Re = 8072$) (Dutta and Ray, 2004).

the outflow boundary is easily distinguishable while the inflow boundary is not so easily distinguishable with bare eye at this critical Reynolds number. When the Reynolds number was increased beyond a secondary critical value of about 228 ($Re/Re_c = 2.05$), Taylor vortex flow takes the shape of azimuthally traveling waves. This wavy vortex flow exists until the Reynolds number reaches about 1,770 ($Re/Re_c \approx 16$) at which the flow transforms into turbulent flow. In the turbulent region, we observed at first weakly TVF, which changes to fully TVF at even higher Reynolds number. Figure 15 shows photographs of the flow patterns at different Reynolds number observed using the flow visualization fluid.

In Table 3, reactor specifications and experimental conditions used and efficiency obtained for the different reactors are compared. A more practical "engineering" definition for efficiency is used instead of more scientific quantum efficiency. The efficiency of each of the reactors, expressed in terms of 50% pollutant converted per unit time per unit reactor volume per unit electrical power consumed, is compared for the same model component (Orange II dye) and same initial concentration ($C_0 = 0.024$ mol m^{-3}). The SR consists of 20 tubes each of volume 7×10^{-5} m^3 containing 3×10^{-5} m^3 of liquid (with TiO$_2$ concentration of 0.5 kg m^{-3}), placed on a holder that rotates around a magnetic stirrer and is surrounded by 24 Philips TLK 40 W/10 R lamps. The CAR was of 0.099 m outside diameter and 0.065 m inside diameter and 0.77 m long, surrounded externally with 10 Philips TLK 40 W/10 R lamps. When the efficiencies of these test reactors are compared (Table 3) with the experimental results of CAR, an increase of about 229% for SR, 1157% for MTR, 1668% for TLR, and 872% for TVR. This increase in efficiency was despite the fact that the design of these test reactors was far from optimum. In our laboratory, we have used Perspex as material (to avoid breakage), which reduces light intensity significantly. Moreover, the lamp used in this study had UV-A intensity of only 13 W m^{-2},

Table 3 Comparison of reactor specifications, experimental conditions, and reactor performance efficiency for photocatalytic decomposition of Orange II dye

Photocatalytic reactor	CAR[c]	SR[c]	MTR[d]	TLR[c]	TVR[e]
Volume of reactor (m^3)	3.5×10^{-3}	3.4×10^{-3}	1.2×10^{-3}	5.4×10^{-4}	3.7×10^{-3}
Catalyst surface area (m^2)	0.18	3.7	0.51	0.15	0.12
Parameter κ (m^{-1})	69	6139^b	1087	618	102
Flow rate ($m^3 \, s^{-1}$)	8.4×10^{-5}	Batch operation	3.0×10^{-5}	1.7×10^{-5}	Batch operation
Volume of liquid treated (m^3)	2.6×10^{-3}	6.0×10^{-4}	4.7×10^{-4}	2.4×10^{-3}	1.1×10^{-3}
Electrical energy input (W)	400	960	40	126	40
Efficiency[a] ($s^{-1} \, m^{-3} \, W^{-1}$)	9.43×10^{-4}	3.1×10^{-3}	1.18×10^{-2}	1.67×10^{-2}	9.17×10^{-3}
% increase in efficiency	0	229	1157	1668	872

[a]Efficiency is defined as 50% pollutant converted per unit time per unit reactor volume per unit electrical energy used.
[b]The value will be lower than 6,139 m^{-1} as all suspended catalyst particles will not be effectively illuminated and the assumption of average particle diameter of 0.3 μm may be too low.
[c]Ray and Beenackers, 1998.
[d]Ray, 1999.
[e]Dutta and Ray 2004.

and the overall rate could be increased significantly if a higher wattage lamp is used. It is apparent that TVR design idea creates great opportunities for building much more efficient photocatalytic reactor.

Both the MTR and the TLR design concept creates great opportunities for building much more efficient photocatalytic reactors for water purification as the reactors will be more economical. From Tables 2 and 3, it can be seen that higher values for the parameter κ can be achieved for both MTR and TLR than other reactor configurations. It is expected that the performance of TLR will surpass that of MTR because of superior catalyst activation, but the overall reactor efficiency may be much lower due to the application of a large excess of light energy than is required for catalyst activation. It is apparent that MTR design idea creates great opportunities for building much more efficient photocatalytic reactor for water purification, as the reactor most likely will be economical. We believe that MTR reactor will be cost effective compared to other photocatalytic reactors since it consists of inexpensive hollow glass tubes, cheap catalyst, and requires low-wattage lamps. It needs a reflector, which usually comes with the lamp and, of course, a lens to direct the light entry at proper angle. Moreover, the hollow test tubes could easily be replaced. It is well known that water purification by photocatalysis will not be cheaper than, for example, by biotreatment. However, if one is interested in purifying water-containing toxic chemicals, the best method may be to break open the benzene ring first by photocatalysis to eliminate the toxic chemicals and then send the water for biotreatment. It would then not be necessary to completely mineralize the pollutants present in water by a photocatalytic reactor. A combination of the two methods could be best suited for water purification and may be more economical. A problem for TLR is still the burning stability and lifetime of the lamps, particularly when the lamps are used immersed in water containing toxic chemicals. The main obstacle to the development of MTR design concept is that it is impossible to obtain uniform light distribution along the length of the tubes and thereby severely restricting the maximum length of tubes that can be used. One way of avoiding both the problems is to place one extremely narrow diameter novel tubelight lamp inside each of the hollow tubes. In this way, all the advantages of the MTR concept can be utilized while eliminating the basic drawback of uniform light distribution dilemma. Moreover, this will also eliminate the main problem experienced in the TLR with the prolonged use of novel lamps immersed in polluted water.

4. CONCLUSIONS

Heterogeneous photocatalysis on semiconductor particles has been shown to be an effective means of removing toxic organic pollutants as well as toxic metal ions from water. In the first part of this chapter, systematic kinetic studies of parameter influence (catalyst loading, circulation rate, substance initial concentration, light intensity) on the reaction rate of photocatalytic degradation of benzoic acid in TiO_2 slurry systems has been presented. Operating with the data "reaction rate versus catalyst loading," it was observed that when rate data is plotted per unit mass of catalyst, one could identify "kinetic" region from the constancy of rate data. In contrast, when the same experimental data was plotted with rate data expressed as per unit volume of reaction liquid, one could identify the region of "optimal catalyst loading" from the constancy of the rate data where rate is limited primarily by internal mass transfer due to agglomeration of catalyst particles. "Kinetic limitation" domain occurs at low catalyst loading and intrinsic (true) kinetic dependences, particularly "reaction rate – light intensity" can only be obtained from the data in the "kinetic" region. "Optimal catalyst loading" domain is a distinctive part of the "transport" limitation domain (which occurs at intermediate values of catalyst loading and is a general feature of the multiphase catalytic process) in which reaction rate is likely governed by transport processes within the catalyst particle agglomerate (estimated size of the agglomerate was found to be 9 μm), that is, internal mass transport limitation. It was observed that in the "optimal catalyst loading" region, there is no external mass transport limitation, but circulation rate influences agglomeration of catalyst particles as the rate plateau shifts toward higher value of catalyst loading when circulation rate is increased. Significant difference in the rate values between the "optimal catalyst loading" regime and the "kinetic regime" was observed. Determination of this region is very important as one can readily obtain from the "kinetic" regime the intrinsic reaction rate constant, which is required for catalyst characterization, elucidation of mechanism of the photocatalytic reaction and in the reactor design, while determination of "optimal catalyst loading" is important for practical applications. It is, however, difficult to distinctively distinguish "internal mass transfer" limitation regime and "shielding effect" domain as they are most likely overlapping. The analysis presented in the first part for slurry system can be used for any heterogeneous catalysis systems.

Several challenges related to photocatalysis and reactor designs are described in the later part of the chapter. The central problems in the development of a photocatalytic reactor, namely, light distribution inside the reactor, providing high surface areas for catalyst per unit reactor volume, and mixing inside the reactors are discussed. Several new reactor concepts, namely, a distributive-type fixed-bed reactor system that employs hollow glass tubes as a means of light delivery to the catalyst particles, an immersion-type reactor where new extremely narrow diameter artificial fluorescent lamps coated with catalyst, and a reactor based on utilization of flow instability to increase mixing, are discussed. All reactors result in a 100- to 150-fold increase in surface area per unit volume of reaction liquid inside the reactor relative to a CAR design and a 10- to 20-fold increases relative to an immersion-type reactor using classical lamps. The design of all the reactors increases the surface-to-volume ratio while eliminating the prospect of light loss by absorption and scattering in the reaction medium. Simulation as well as experiments studies performed to understand the degradation of an Orange II dye showed promising results. All reactor configurations are flexible enough to be scaled-up for commercial applications.

ACKNOWLEDGMENTS

The author acknowledges research contributions made by research collaborators Professor Virender K. Sharma (Florida Institute of Technology), Professor Gregory Yablonsky (Washington University at St. Louis), Professor Tapan K. Sengupta (Indian Institute of Technology, Kanpur), Professor George Zhao (National University of Singapore), and graduate students (Dingwang Chen, Jinkai Zhou, Yuxin Zhang, Vidyaniwas Khetawat, Preety Mukherjee, Uvaraj Periyathamby, Mohammad Kabir, Shuhua Zhou, Kanheya Mehrotra, Xiaoling Wang, Paritam K Dutta, Yong Yong Eng, Atreyee Bhattacharya, Fengmei Li, Muthusamy Sivakumar, Debjani Mukherjee, Mahmoud Housyn, and Pankaj Chowdhury) without whom the research presented here would not be possible.

REFERENCES

Anderson, M.A., Tunesi, S., and Xu, Q. US 5035784 A 910730 (1991).
Assink, J.W., Koster, T.P.M., and Slaager, J.M. "Fotokatalytische oxydatie voor afvalwater behandeling". Internal report reference no. 93–137, TNO – Milieu en Energie, The Netherlands (1993).
Bard, A.J. *Science* **207**, 139 (1980).
Blake, D.M. "Bibliography of work on Photocatalytic Removal of Hazardous Compounds from Water and Air". NREL/TP-430-22197, National Renewable Energy Laboratory, Golden (1997).
Cassano, A., Martin, C, Brandi, R., and Alfano, O. *Ind. Eng. Chem. Res.* **34**, 2155 (1995).
Chen, D.W., Li, F., and Ray, A.K. *AIChE J.* **46**, 1034 (2000a).
Chen, D.W., Li, F., and Ray, A.K. *Catal. Today* **66**, 475 (2001).
Chen, D.W., and Ray, A.K. *Appl. Catal. B* **23**, 143 (1999).
Chen, D.W., and Ray, A.K. *Chem. Eng. Sci.* **56**, 1561 (2001).
Chen, D.W., and Ray, A.K. *Water Res.* **32**, 3233 (1998).
Chen, D.W., Sivakumar, M., and Ray, A.K. *Dev. Chem. Eng. Miner. Proces.* **8**, 505 (2000b).
Coles, D. *J. Fluid Mech.* **21**, 385 (1965).
Cooper, G.A.US patent 4888101 (1989).
Cooper, G.A., and Ratcliff, M.A.WO 9108813 A1 910627 (1991).
de Lasa, H., Serrano, B., and Salaices, M. "Photocatalytic Reaction Engineering". Springer, USA (2005).
de Lasa, H.I., and Valladares, J. US Patent No. 5, 683, 589 (1997).
Dutta, P.K., and Ray, A.K. *Chem. Eng. Sci.* **59**, 5249 (2004).
Dutta, P.K., Sharma, V., and Ray, A.K. *Environ. Sci. Technol.* **39**, 1827 (2005).
Fox, M.A., and Dulay, M.T. *Chem. Rev.* **93**, 341 (1993).

Haneda, K.Patent JP 04061933 A2 920227 (1992).

Heller, A., and Brock, J.R.Patent WO 9317971 A1 930916 (1993).

Henderson, R.B., and Robertson, M.K.Patent EP 3063301 A1 890308 (1989).

Hoffmann, M.R., Martin, S.T., Choi, W., and Bahnemann, D.W. *Chem. Rev.* **95**, 69 (1995).

Hofstadler, K., Bauer, R., Novallc, S., and Helsler, G. *Environ. Sci. Technol.* **28**, 670 (1994).

Hosokawa, M., and Yukimitsu, K.Patent JP 63042793 A2 880223 (1988).

Inel, Y., and Okte, A.N. *J. Photoch. Photobio. A.* **96**, 175 (1996).

Kataoka, K. Taylor vortices and instabilities in circular Couette flows. *in* N.P. Cheremisinoff (Ed.), "Encyclopedia of Fluid Mechanics", vol. 1. Gulf Publishing, Houston (1986), p. 236.

Masuda, R., Kawashima, K., Takahashi, W., Murabayashi, M., and Ito, K.Patent JP 06320010 A2 941122 (1994).

Matthews, R.W. Patent AU 600289 B2 900809 (1990a).

Matthews, R.W. Patent WO 8806730 A1 880907 (1998).

Matthews, R.W. *Water Res.* **24**, 663 (1990b).

Mehrotra, K., Yablonsky, G.S., and Ray, A.K. *Chemosphere* **60**, 1427 (2005).

Mehrotra, K., Yablonsky, G.S., and Ray, A.K. *Ind. Eng. Chem. Res.* **42**, 2273 (2003).

Miano, F., and Borgarello, E.Patent EP 417847 A1 910320 (1991).

Mukherjee, P.S., and Ray, A.K. *Chem. Eng. Technol.* **22**, 253 (1999).

Oonada, J.Patent JP 06071256 A2 940315 (1994).

Peill, N.J., and Hoffmann, M.R. *Environ. Sci. Technol.* **29**, 2974 (1995).

Periyathamby, U., and Ray, A.K., *Chem. Eng. Technol.* **22**, 881 (1999).

Pruden, A.L., and Ollis, D.F. *Environ. Sci. Technol.* **17**, 628 (1983).

Ray, A.K. *Catal. Today* **44**, 357 (1998).

Ray, A.K. *Des. Chem. Eng. Sci.* **54**, 3133 (1999).

Ray, A.K., and Beenackers, A.A.C.M. *AIChE J.* **43**, 2571 (1997).

Ray, A.K., and Beenackers, A.A.C.M. *AIChE J.* **44**, 477 (1998a).

Ray, A.K., and Beenackers, A.A.C.M. *Catal. Today* **40**, 73 (1998b).

Ray, A.K., and Beenackers, A.A.C.M. European patent, EP 96200942.9-2104 (1996).

Rayleigh, L. "Scientific Papers", vol. 6. Cambridge, England (1920), p. 447.

Rideh, L., Wehrer, A., Ronze, D., and Zoulalian, A. *Ind. Eng. Res.* **36**, 4712 (1997).

Ritchie, D.G. Patent US 5069885 A 911203 (1991).

Salaices, M., Serrano, B., and de Lasa, H. *Ind. Eng. Chem. Res.* 40, 5455 (2001).

Sato, K.Patent JP 04114791 A2 920415 (1992).

Sczechowski, J.G., Koval, C.A., and Noble, R.D. *Chem. Eng. Sci.* **50**, 3163 (1995).

Sengupta, T.K., Kabir, M.F., and Ray, A.K. *Ind. Eng. Chem. Res.* **40**, 5268 (2001).

Serpone, N., and Pelizetti, E. (Ed.), "Photocatalysis: Fundamentals and Application". Wiley, New York (1989).

Serrano, B., and de Lasa, H.I. *Chem. Eng. Sci.* **15**, 3063 (1999).

Serrano, B., and de Lasa, H.I. *Ind. Eng. Chem. Res.* **36**, 4705 (1997).

Taoda, H. "Water Treatment". Patent JP 05076877 A2 930330 (1993).

Taylor, G.I. *Trans. Roy. Soc. Lond. A* **223**, 289 (1923).

Wake, H., and Matsunaga, T.Patent JP 06134476 A2 940517 (1994).

Wang, X., Pehkonnen, S.O., andRay, A.K. *Electrochem. Acta* **49**, 1435 (2004a).

Wang, X., Pehkonnen, S.O., and Ray, A.K. *Ind. Eng. Chem. Res.* **43**, 1665 (2004b).

Zhou, J., Takeuchi, M., Ray, A.K., Anpo, M., and Zhao, X.S. *J. Colloid Interface Sci.* 311, 497 (2007a).

Zhou, J., Takeuchi, M., Zhao, X.S., Ray, A.K., and Apno, M. *Catal. Lett.* **106**, 67 (2006a).

Zhou, J., Zhang, Y., Zhao, X.S., and Ray, A.K. *Ind. Eng. Chem. Res.* **45**, 3503 (2006b).

Zhou, J., Zou, Z., Zhao, X.S., and Ray, A.K. *Ind. Eng. Chem. Res.* **46**, 745 (2007b).

CHAPTER 6

Development and Modeling of Solar Photocatalytic Reactors

Camilo A. Arancibia-Bulnes[*]**, Antonio E. Jiménez,** and **Claudio A. Estrada**

Contents

1.	Introduction	186
2.	Solar Photocatalytic Reactors	187
	2.1 Solar reactor types	187
	2.2 CPC photoreactors	192
	2.3 The dependence of reaction rates on radiation intensity	196
	2.4 Solar photoreactors comparisons	199
	2.5 Concentrating CPC reactors	202
3.	Radiation Transfer in Photocatalytic Reactors	206
	3.1 Optical properties of the medium	206
	3.2 Modeling dye degradation	208
	3.3 The radiation transfer equation	210
	3.4 Methods of solution of the RTE	211
4.	The P1 Approximation	213
	4.1 General solution of the P1 approximation for solar tubular reactors	215
	4.2 Solution for a parabolic trough reactor	217
	4.3 Solution for an annular lamp reactor	218
	4.4 Applicability of the P1 approximation	221
5.	Conclusions and Perspectives	222
	Acknowledgments	223
	List of Symbols	223
	Abbreviations	225
	References	225

Centro de Investigación en Energía, Universidad Nacional Autónoma de México, Privada Xochicalco s/n, Col. Centro, A. P. 34, Temixco, 62580 Morelos, México

[*] Corresponding author.
E-mail address: caab@cie.unam.mx

Advances in Chemical Engineering, Volume 36
ISSN 0065-2377, DOI: 10.1016/S0065-2377(09)00406-2

1. INTRODUCTION

Photocatalysis is an advanced oxidation process with a high potential for the treatment of water contaminated with persistent organic pollutants. In this process ultraviolet radiation is used for the excitation of a semiconductor catalyst (usually TiO_2) in contact with the polluted water. When photons are absorbed by the catalyst electrons are promoted to the conduction band, leaving holes in the valence band. These charges can diffuse to the surface of the material where they initiate oxidation/reduction reactions with molecules adsorbed in the semiconductor surface. Due to the high cost and environmental impact of the utilization of artificial UV light in photocatalytic processes, there has been a great deal of interest in the application of solar energy. Other advanced oxidation processes able to use solar energy, particularly photo-Fenton, are also very interesting but will not be considered in this review due to the space limitations.

In solar photocatalytic detoxification of water, as in any other solar energy application, the adequate design and sizing of systems is of the greatest importance. As the solar resource is relatively diluted, the need to invest in large collector fields is a major factor affecting the final cost per unit product. This is particularly true for photocatalysis, where only a fraction of the solar radiation spectrum can be utilized. Therefore, the design of solar photoreactors should be aimed at making the best use of the available UV solar radiation.

Solar photoreactors are often classified into nonconcentrating and concentrating, depending on whether they operate with solar radiation as it arrives to earth's surface or use a solar concentrator to augment the radiative flux impinging on the reactor. From the basic mechanisms of photocatalysis it is well known that the kinetics of reactions behaves linearly with the radiation intensity when this intensity is low, and as the square root when this intensity is high. As a consequence of this, nonconcentrating reactors make a more efficient use of solar radiation, although concentrating reactors are faster. Because of this, recent research has focused mostly on the former and on reactors with low concentration factors, particularly those based on compound parabolic collectors (CPC). These types of reactors have also been used for the disinfection of water.

Another area which has gained interest in recent years is the utilization of supported catalyst in solar photoreactors, as opposed to suspended catalyst. Different geometries have been proposed for supporting the catalyst, which are strongly linked to photoreactor geometry. Results have been in some cases very good, despite of the fact that traditionally suspended catalyst has been considered more effective.

The modeling of radiation absorption within solar photocatalytic reactors has been addressed by some researchers, but is one of the areas where less work has been carried out. Both heuristic and first principles methods, based on the radiative transfer equation (RTE), have been used. This is linked also with the utilization of kinetic models based on the radiation absorbed by the catalyst, as opposed to models based only on radiation arriving to the solar collector.

In some cases, evaluating a global reaction rate based on the total radiation absorbed by the reactor is not equivalent to evaluate rates based on the radiation absorbed locally, especially when mass transfer limitations occur. Thus, appropriate analysis of radiative transfer in reactors can be a valuable tool in making design decisions.

2. SOLAR PHOTOCATALYTIC REACTORS

In solar photocatalytic reactors the source of the photons necessary to carry out the excitation of the catalyst is near-UV solar radiation. This implies that two optical problems have to be dealt with in the design of these reactors: (1) to collect solar radiation, and (2) to distribute this radiation in the best possible manner inside the reaction volume. These tasks should be done with the highest possible efficiency, as the amount of solar energy available in the wavelength range utilizable by TiO_2 is small. The work in solar photocatalytic reactors has been reviewed in several works (Alfano et al., 2000; Bahnemann, 2004; Blanco-Galvez et al., 2007; Goslich et al., 1997; Goswami, 1995, 1997; Goswami et al., 2000; Malato et al., 2007; Malato Rodríguez et al., 2004), so our aim here is only to discuss the main ideas and to review some recent work.

2.1. Solar reactor types

Different types of solar photocatalytic reactors have been developed over time. Two main distinctions have been made: (i) concentrating versus non-concentrating reactors and (ii) slurry versus fixed catalyst reactors. These two classifications are not mutually exclusive, and other categories may also be established, for instance, a very important is the one of tubular solar photoreactors.

Let us start by pointing out some characteristics of solar radiation which affect the design of photoreactors, as one may try to make an optimal use of the available radiation. First of all, the usable solar radiation is restricted by two factors: on the one hand, the fact that solar radiation at the earth's surface starts at wavelengths larger than 300 nm. On the other hand, that the most common semiconductor used in photocatalysis is anatase TiO_2. Because of the magnitude of its energy band gap, small particles of this material absorb radiation very poorly for wavelengths above 370 nm. Many attempts to develop catalysts with a more extended spectral response have been done, but they have failed to date to provide a low-cost alternative

(Blanco-Galvez et al., 2007). Thus, the usable wavelengths barely coincide with the solar UV range (from 300 to 400 nm). This means that the radiation available to carry out photocatalytic reactions is around 5% of the global solar radiation; that is, at most 50 W m^{-2} (Hulstrom et al., 1985).

A second characteristic of UV solar radiation is that, even for very clear atmospheres, it is composed in similar amounts of both beam and diffuse radiation (Hulstrom et al., 1985). The first is defined as the radiation arriving directly from the sun, while the second is the solar radiation that has been scattered by gases and aerosols after entering the earth's atmosphere. This second type of radiation reaches the ground in a more or less diffuse manner; that is, with similar intensity from all directions in the sky. In this respect, the situation encountered in solar photocatalytic reactors is quite different from the one encountered in solar thermal collectors. The latter are able to use the whole solar spectrum, and in that case diffuse radiation accounts for a much smaller fraction of the global irradiance.

The solar concentration ratio of a solar collector is defined as the ratio of the aperture area of the reflector to the area of the receiver (Rabl, 1985). This quantity gives an approximation to the number of times the radiative flux density (W m^{-2}) is increased in the surface of the receiver, as compared to the incoming solar radiation. The concentration ratio is commonly expressed as a number of "suns"; for instance, a collector that increases the radiative flux density five times is said to have a five suns concentration ratio. Thus, the formula for the concentration ratio is

$$C_R = \frac{A_a}{A_r} \tag{1}$$

The first type of photocatalytic reactors employed were parabolic trough concentrators (PTC) with one- or two-axis tracking systems to follow the apparent movement of the sun in the sky. These photocatalytic reactors were just standard medium concentration ($C_R \approx 15$ suns) thermal collectors, which were readily adapted as photocatalytic reactors by replacing the absorber tube by a Pyrex glass tube (Alpert et al., 1991; Minero et al., 1993) (Figure 1). However, it was eventually found that PTC reactors had several disadvantages which made them energetically less efficient and more expensive than nonconcentrating reactors (Malato et al., 1997):

1. Concentrating collectors are only able to use a limited fraction of the diffuse solar radiation. This utilizable fraction can be estimated as the inverse of the concentration ratio of the collector ($1/C_R$) (Rabl, 1985). Because of this, parabolic trough photocatalytic reactors, which have a concentration ratio around 15 or higher, miss practically all diffuse radiation. This amounts to losing around half of the available UV solar irradiance.

Figure 1   Parabolic trough photocatalytic reactor at Plataforma Solar de Almería (reprinted from Malato et al., 2007 with permission from Elsevier).

2. Reaction rates in photocatalysis have been observed to be proportional to the square root of radiation intensity, at medium and high concentration levels (see Section 2.3). Thus, nonconcentrating reactors make a more efficient use of the collected photons. In the end, this diminishes the total collector area required as compared to concentrating reactors.
3. Medium and high concentrating reactors require tracking systems that are costly, require maintenance, and consume energy.
4. Concentration may lead to overheating of the water and evaporation, especially when dealing with nontransparent pollutants as dyes.

Nevertheless, PTC reactors do have some attractive features. First of all, they allow the degradation of pollutants at faster rates (although with larger collector areas), simply by delivering more photons per unit volume in the reaction space. Also, tracking helps to make a more efficient use of the beam UV radiation component along the entire day by keeping the collector aperture normal to the sun rays.

In parallel with the experiences with PTC reactors, the development of several types of nonconcentrating reactors was pursued by different research groups. Among nonconcentrating photoreactors we can mention, falling film reactors (Böckelmann et al., 1995; Gernjak et al., 2004) (Figure 2), the shallow pond reactor (Bedford et al., 1994), the flat tubular photoreactors (Goswami, 1997), flat plastic reactors (van Well et al., 1997), multistep

Figure 2   Thin film fixed bed reactor (reprinted from Bahnemann, 2004, with permission from Elsevier).

cascade falling-film reactors (Guillard et al., 2003; Pichat et al., 2004), and reactors based in CPC (Blanco et al., 1999) (Figure 3). On the other hand, some researchers have even used systems with higher concentration ratio than PTC (Oyama et al., 2004).

In particular, nonconcentrating CPC reactors are currently regarded as one of the best options for large-scale applications of solar photocatalysis, because they combine some advantages of both PTC and nonconcentrating reactors (Malato Rodriguez et al., 2004) as follows:

1. They are able to collect both beam and diffuse solar radiation.
2. They have low fabrication costs as compared to PTC, mainly because they do not require sun tracking systems.
3. They use a closed tubular reaction space as receiver, like PTC, and this helps avoiding the evaporation of volatile compounds.
4. They withstand more pressure than other reactors due to the tubular geometry of the flow space. This allows working in the turbulent regime.
5. They avoid water overheating.

It is important to point out that most of these advantages are also shared by other nonconcentrating tubular reactors. However, the optical efficiency of CPC is near to the theoretical maximum (Rabl, 1985), as will be discussed

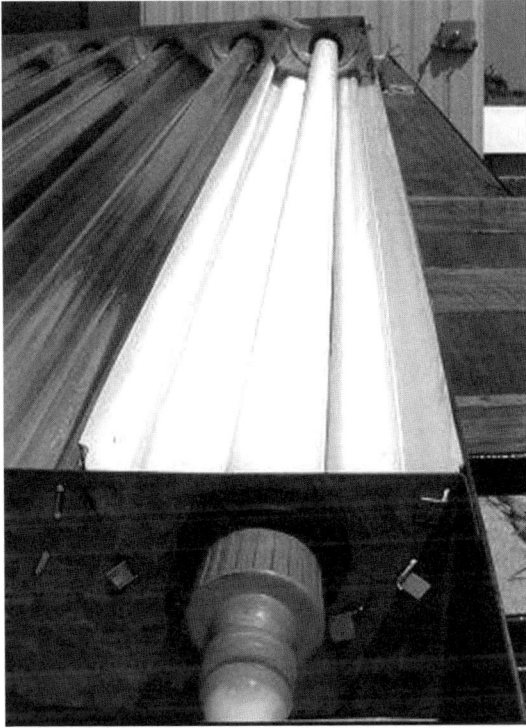

Figure 3   CPC photoreactor (developed by AoSol, Portugal).

in the next section. This is not so for other nonconcentrating tubular collectors (TC), which miss some amounts of diffuse and beam radiation depending on the incidence angle.

As stated previously, another distinction usually made is between slurry and supported catalyst reactors. In slurry photocatalytic reactors the catalyst is present in the form of small particles suspended in the water being treated. These reactors generally tend to be more efficient than supported catalyst reactors, because the semiconductor particles provide a larger contact surface area per unit mass. In fact, the state of the photocatalyst is important both to increase contaminant adsorption and to improve the distribution of absorbed radiation. In a slurry unit the photocatalyst has a better contact with the dissolved molecules and is allowed to absorb radiation in a more homogeneous manner over the reaction volume. Using suspended catalyst has been the usual practice in PTC, CPC, and other types of tubular reactors. The drawback of this reactor design is the requirement for separation and recovery of the very small particles at the end of the water treatment process. This may eventually complicate and slow down the water throughput.

Given this situation, the use of supported photocatalyst has attracted significant research efforts. Several ways have been proposed to support the semiconductor, such as coating of extended surfaces (Diaz et al., 2007; Gryglik et al., 2004) or glass beads (Gelover et al., 2004). In the first case, photocatalytic reactors are designed with shallow depths to ensure proper illumination of the supported semiconductor. For instance, this is the case of the thin film fixed bed reactor (Bockelmann et al., 1995) or multistep cascade reactor (Guillard et al., 2003).

Supporting the catalyst on beads gives more flexibility to reactor geometry than doing it on extended surfaces. It has been observed in some cases that reactors with TiO_2-coated glass beads can reach efficiencies similar to slurry photoreactors. For instance, Gelover et al. (2004) deposited anatase thin films over small cylindrical pieces of glass, using a sol–gel technique. They carried out the degradation of 4-chlorophenol and carbaryl in a PTC in which the tubular reaction space was filled with these glass pieces. The results obtained with this configuration were compared to results obtained in the same reactor when suspended catalyst was used. Similar degradation rates were observed in both cases.

Besides the applications for detoxification, in recent years the catalyst support geometry and its optimization have also been investigated for application in water disinfection (see, for instance, McLoughlin et al., 2004a; Meichtry et al., 2007; Navntoft et al., 2007).

2.2. CPC photoreactors

As discussed in the previous section, one type of solar collector geometry that has been found to be very advantageous for solar photocatalysis is based on CPC (Blanco et. al., 1999), which are a type of nonimaging optical devices.

A common classification for solar collectors is between imaging and nonimaging collectors. The first are based on optical designs originally intended for imaging applications; that is, they are able to produce an image of the radiation source (the sun) in their focal region. For instance, this is the case of PTC, which are able to produce a two-dimensional image of the sun over a flat receiver.

The requirement of producing an image of the source imposes serious limitations in the capacity of optical systems for collecting light. Nonimaging solar collectors are freed from these constraints in order to reach better energetic performance. In this kind of optical systems the goal is reaching the maximum possible capture of light over a given collection angle, which is called acceptance half angle θ_{max}. Thus, light rays arriving to the system's aperture within the acceptance angle are ensured to reach the receiver, regardless of the fact that they may not produce an image of the source (Figure 4).

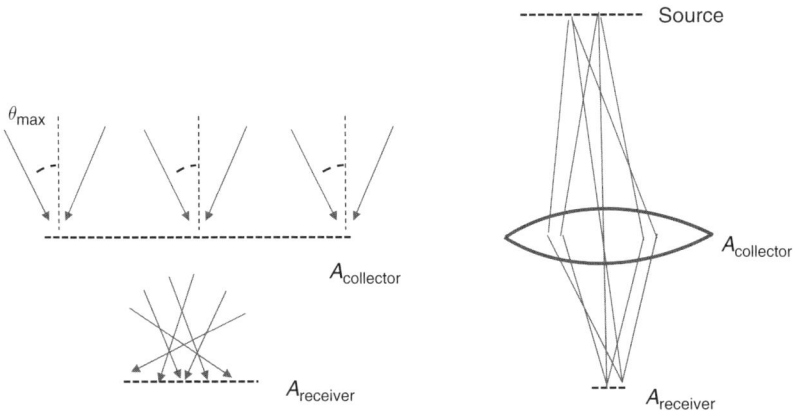

Figure 4   Nonimaging (left) versus imaging (right) optical systems. In the latter all light rays coming from a point in the source must be imaged to a point in the receiver.

Due to the relaxation of the requirement to form images, some nonimaging collectors, called ideal collectors, are able to reach the maximum possible collection of light for a given concentration ratio. In particular, the maximum concentration ratio of line focusing solar collectors is given by (Rabl, 1976, 1985; Winston et al, 2005)

$$C_{R,max} \leq \frac{1}{\sin \theta_{max}} \tag{2}$$

The interpretation of Equation (2) is that if a solar collector is to be able to capture all of the radiation incident within the acceptance cone defined by θ_{max} (Figure 4), at most it can have a concentration ratio given by C_{max}. This rule comes from the fundamental principles of thermodynamics (Rabl, 1985) and is obeyed by all optical systems.

The other way of looking at Equation (2) is that, for a given concentration ratio, an ideal collector is able to collect radiation over the widest possible angle. CPC may be designed in principle to have different concentration ratios, depending on their acceptance angle. In particular, nonconcentrating CPC ($C_R \approx 1$) used in photocatalysis are able to collect radiation from the whole hemisphere ($\theta_{max} = \pi/2$) (Blanco et al., 1999), being this their main advantage over other nonconcentrating tubular reactors.

Originally, the term CPC was coined to refer to a concentrator consisting of two parabolic segments rotated relative to each other. This kind of geometry was designed specifically to concentrate radiation in a planar absorber. However, the designation CPC has evolved to include other kinds of nonimaging solar concentrators (Winston et al., 2005). For instance, in the case of a tubular receiver the ideal reflector is usually called a CPC,

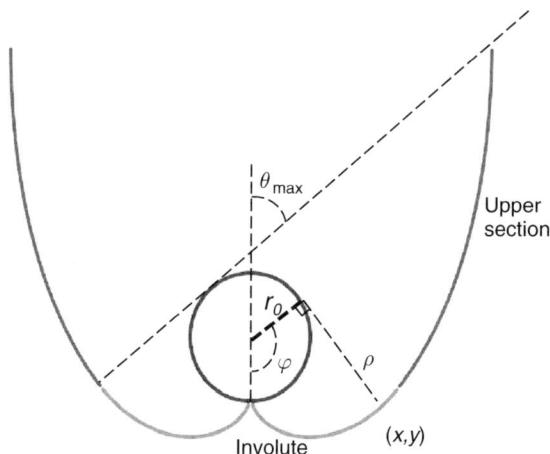

Figure 5   Geometry of a CPC for a tubular receiver.

but is formed by two involute sections instead of two parabolas. For concentrations higher than unity these involute sections are continued by segments that are also not parabolic (Figure 5).

The following are the equations of a CPC for a tubular receiver (Rabl, 1985): the coordinates of the points of the concentrator are given in a parametric form by

$$x = r_0 \sin \varphi - \rho \cos \varphi \tag{3}$$

$$y = -r_0 \cos \varphi - \rho \sin \varphi \tag{4}$$

with φ a polar angle around the tubular receiver of radius r_0, as presented in Figure 5. For the involute (lower) section of the concentrator this angle is bound as $0 \le \varphi \le \pi/2 + \theta_{max}$ and the distance ρ is given by

$$\rho = r_0 \varphi \tag{5}$$

For the upper section of the concentrator φ is restricted to the range $\pi/2 + \theta_{max} \le \varphi \le 3\pi/2 - \theta_{max}$, and

$$\rho = r_0 \frac{\pi/2 + \theta_{max} + \varphi - \cos(\varphi - \theta_{max})}{1 + \sin(\varphi - \theta_{max})} \tag{6}$$

In particular, in nonconcentrating CPC only the involute part of the reflector is used with φ up to π (see, for instance, Figure 3).

As it can be observed in Figure 5, the upper part of a CPC consists of a large reflector area that is nearly vertical, and which adds very little to the

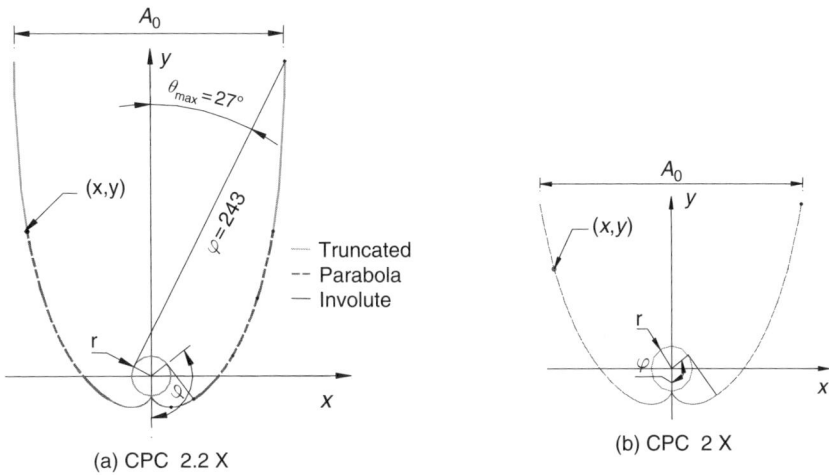

Figure 6   Truncation of a 2.2 suns CPC (a), to 2 suns (b) (Jiménez and Salgado, 2008).

aperture area A_a. Therefore, this part of the concentrator can be heavily truncated without reducing significantly the concentration ratio (Carvalho et al., 1985). For instance, Figure 6 illustrates the truncation of a CPC with 2.2 suns concentration ratio. With this operation, around 50% of the reflecting material is saved, but the collector loses only about 10% concentration ratio.

It is important to point out that Equations (3)–(6) are only valid for CPC with tubular receivers. Nevertheless, the methods of nonimaging optics (Winston et al., 2005) allow the design of CPC reflectors for receivers of varied geometry. For instance, Chaves and Collares Pereira (2007) proposed a design of a CPC with a vertical stripe of supported catalyst as receiver (fin). This fin is contained in the water phase (Figure 7) inside a glass tube. This illustrates how, as new geometries for catalyst support are studied, consideration should be paid to the optimal design of the required reflectors.

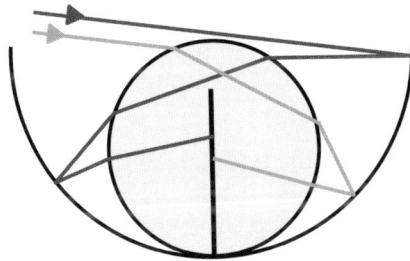

Figure 7   CPC reflectors for a tubular reactor with catalyst supported in a vertical stripe. Adapted from Chaves and Collares Pereira (2007), with permission from The American society of Mechanical Engineers.

2.3. The dependence of reaction rates on radiation intensity

Radiation absorption is the initiating step in photocatalytic processes. When a photon of energy greater than the band gap of the semiconductor catalyst arrives to its surface, it is absorbed and promotes an electron from the valence to the conduction band, creating an electron–hole pair. Therefore, the quantity of interest is the number of photons absorbed by the catalyst, characterized in a slurry photoreactor by the so-called local volumetric rate of photon absorption (LVRPA). This quantity is the number of photon mol (Einstein) absorbed in the suspension per unit volume and time. Taking into account that the energy of a photon of wavelength λ is given by $E_\lambda = hc/\lambda$, where h is Planck's constant and c is the speed of light, the LVRPA is obtained in terms of the power absorbed by the catalyst at each wavelength

$$e_L(\mathbf{r}) = \frac{1}{hcN_A} \int Q^m_{abs-cat,\lambda}(\mathbf{r})\,\lambda\,d\lambda \tag{7}$$

Here e_L is the LVRPA, N_A is Avogadro's number, and $Q^m_{abs-cat,\lambda}$ is the volumetric power absorbed by the catalyst (see Section 3.2) at a given point $\mathbf{r}$ in the reactor volume and per unit wavelength. This integral is carried out over the wavelength range defined by the intersection of the spectrum of the radiation source, with the absorption region of the catalyst. In the case of solar radiation and TiO$_2$ catalyst this region is usually considered to be from 0.3 to 0.4 µm.

The relationship between the local reaction rate and the LVRPA can be obtained from models of the reaction mechanism. By considering the steps proposed by Turchi and Ollis (1990), and taking into account the simplifications proposed by Bandala et al. (2004), it is obtained

$$\frac{dC_p(\mathbf{r})}{dt} = \frac{-k_\alpha k_F\,C_p(\mathbf{r})\,(-1+\sqrt{1+\gamma^2 e_L})}{k_\alpha C_p + k_{O,1}[R_1] + k_{O,2}[R_2] + \cdots + k_{O,n}[R_n]} \tag{8}$$

where

$$\gamma = \frac{4k_e k_R}{k_F k_E} \tag{9}$$

The local concentration of the pollutant is denoted by $C_p(\mathbf{r})$. This concentration depends on the position inside the reactor (denoted by vector $\mathbf{r}$), unless perfect mixing is attained. The quantities $[R_1]$, $[R_2]$, $[R_n]$ are the concentrations of possible intermediate compounds. k_α, k_F, $k_{O,1}$, $k_{O,2}$, $k_{O,n}$, k_e, k_R, k_F, and k_E are kinetic constants for the different steps involved in the photocatalytic process (Alfano et al., 1997; Bandala et al., 2004; Turchi and Ollis, 1990).

The above equation gives the local reaction rate as a product of two functions, one dependent on the concentration of the pollutant and the intermediates only, and a second one dependent only on the LVRPA. Note that all the concentrations are location dependent as well as e_{L}. To obtain the evolution of the average concentration of the pollutant it is necessary to average Equation (8) over the suspension volume V_{T}

$$\frac{dC_{\mathrm{p,av}}}{dt} = \frac{1}{V_{\mathrm{T}}} \int_{V_{\mathrm{T}}} F_1(C_{\mathrm{p}}, [R_i]) \, F_2(e_{\mathrm{L}}) \, dV \tag{10}$$

where, from Equation (8), the functions F_1 and F_2 are defined as

$$F_1(C_{\mathrm{p}}, [R_i]) = \frac{-k_\alpha k_{\mathrm{F}} \, C_{\mathrm{p}}(\mathbf{r})}{k_\alpha C_{\mathrm{p}} + k_{\mathrm{O},1}[R_1] + k_{\mathrm{O},2}[R_2] + \cdots + k_{\mathrm{O},n}[R_n]} \tag{11}$$

$$F_2(e_{\mathrm{L}}) = -1 + \sqrt{1 + \gamma^2 \, e_{\mathrm{L}}} \tag{12}$$

Note that while it is often possible to improve mixing in order to have pollutant concentrations that are not site dependent, it is generally not possible to have a homogeneous value for the LVRPA. Therefore the above equation can be expressed as

$$\frac{dC_{\mathrm{p,av}}}{dt} = F_1(C_{\mathrm{p,av}}, [R_i]) \frac{1}{V_{\mathrm{T}}} \int_{V_{\mathrm{R}}} F_2(e_{\mathrm{L}}) \, dV \tag{13}$$

In this equation the integration should be restricted to the irradiated section (the photoreactor) of volume V_{R}. In the dark parts of the system the second function is zero. However, it must be pointed out that when important mass transfer limitations occur Equation (13) is no longer valid. For instance, Ballari et al. (2008) found that this may be the situation under conditions of high catalyst loading and high irradiation rates.

The determination of the distribution of the LVRPA requires the use of some type of radiative transfer model. In the case of transparent pollutants, it can be considered that e_{L} depends on TiO$_2$ concentration (C_{catal}) only, and not on the concentration of the pollutant, since it is the former component which absorbs and scatters radiation. This allows uncoupling the radiation problem from the degradation kinetics when Equation (13) is solved; that is, one can first evaluate e_{L} and then, independently of the value of the pollutant concentration, integrate $F_2(e_{\mathrm{L}})$ over the reactor volume. Once this quantity has been calculated, its numerical value is taken as a constant in Equation (13), which can now be solved to obtain the evolution of $C_{\mathrm{p,av}}$.

For Equation (12) there are two limit cases at high and low solar radiation concentrations, respectively

$$F_2(e_L) \cong \gamma\sqrt{e_L} \quad \text{if} \quad \gamma^2 e_L \gg 1 \tag{14}$$

$$F_2(e_L) \cong \left(\frac{\gamma^2}{2}\right) e_L \quad \text{if} \quad \gamma^2 e_L \ll 1 \tag{15}$$

These equations mean that a linear dependence of the reaction rate with light intensity (LVRPA) is observed when intensities are small, while square root dependence is observed when intensity is high. This latter dependence occurs for high intensities because the recombination of electron–hole pairs starts to limit the efficient use of the available photons (Alfano et al., 1997). The intensity at which the crossover between these two types of behavior occurs depends on the value of the lumped kinetic parameter γ, which in turn depends on the specific reaction under consideration.

The above theoretical expressions coincide with the experimental observations. Bahnemann et al. (1991) carried out the photocatalytic degradation of chloroform, varying radiation intensities by five orders of magnitude. They found a nonlinear correlation between the degradation rate and radiation intensity. Similarly, studies of phenol degradation carried out by Okamoto et al., (1985) showed a square root relationship between degradation rate and intensity. Nevertheless, at small intensities a linear relationship is observed. For instance, Román Rodríguez (2001) studied the solar photocatalytic degradation of carbaryl with concentration ratios up to five suns and observed this linear relationship.

For nonabsorbing substrates, a factor can be defined that takes into account the effect of all the optical processes in the reaction rate. This factor has been called reaction rate optical factor (RROF) (Arancibia-Bulnes and Cuevas, 2004)

$$R_{\text{opt}} = \int_{V_R} \left(-1 + \sqrt{1 + \gamma^2 e_L}\right) dV \tag{16}$$

For small values of γ, when the kinetics is linear with respect to the LVRPA, this integral increases asymptotically with catalyst concentration (Arancibia-Bulnes and Cuevas, 2004). However, when square root dependence dominates, this factor reaches a maximum and then starts to decrease with increasing catalyst concentration. This means that too much catalyst has a negative optical effect at high radiation intensity leading to smaller degradation rates. This behavior is illustrated in Figure 8.

These conclusions, however, have to be considered in the proper context, given that in many cases the optimal catalyst concentration is determined

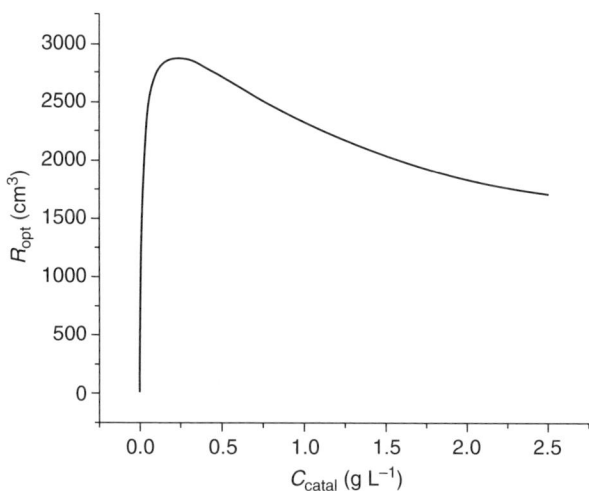

Figure 8   Reaction rate optical factor as a function of catalyst concentration for a parabolic trough solar photocatalytic reactor. Adapted from Arancibia-Bulnes and Cuevas 2004, with permission from Elsevier.

not by the optimal value of the RROF but by the requirement of having enough sites available for pollutant adsorption. Therefore, concentrations are often much higher than optimal from the optical point of view.

2.4. Solar photoreactors comparisons

Comparison of the performance between different photocatalytic reactors is not an easy task due to the diversity of experimental conditions and variables considered in each case. Work has been carried out by evaluating the reactor performance for the degradation of different model pollutants (Bahnemann et al., 1991; Bockelmann et al., 1995; Curcó et al., 1996a; Giménez et al., 1999; Goslich et al., 1997; Malato et al., 1997). The emphasis has been, however, on ranking concentrating versus nonconcentrating solar reactors. For instance, Bockelmann et al. (1995) carried out evaluation study of the degradation of dichloroacetic acid, as well as samples of wastewaters, with a parabolic trough (concentrating) and a thin film fixed bed reactor (1 sun). They observed that both concentrating and nonconcentrating collectors had inherent advantages and disadvantages, and it was not possible to decide unequivocally which system performed better.

More recently, reactors based on nonimaging collectors, like the CPC, have attracted interest (Blanco et al., 1999). These reactors share some of the advantages of both parabolic troughs and nonconcentrating reactors

(Malato et al., 1997) as mentioned previously. These features were confirmed in several studies using CPC and other nonconcentrating reactors as well as parabolic troughs (Curcó et al., 1996b; Giménez et al., 1999; Malato et al., 1997).

When comparing different solar collectors it is important to take into account the different quantities of radiation collected in each case. For a given receiver, the radiative power collected increases with aperture size (i.e., with concentration). However, this is not a linear effect and, for instance, a CPC with a two suns concentration ratio provides twice the aperture area of a nonconcentrating CPC. Nevertheless the power received by the tubular absorber is not doubled, because the former CPC misses around half of the diffuse radiation. Other important consideration is the fact that concentrating reactors are faster simply because they collect more radiation, which is associated with larger collector area.

For all the above reasons, when comparing different collectors it is necessary to consider the pollutant degradation against collected energy and not against irradiation time (Curcó et al., 1996b). One possibility for this assessment is to use the accumulated available energy, defined as the time integral of the UV power impinging on the collector's aperture area, per unit volume of the reactor

$$E_{\text{available}} = \int G_{\text{UV,global}}(t)\frac{A_{\text{a}}}{V_{\text{T}}}\,\mathrm{d}t \qquad (17)$$

where $E_{\text{available}}$ is the accumulated available energy (W m^{-3}), A_{a} is the collector's aperture area, V_{T} treated volume, and $G_{\text{UV,global}}(t)$ is the global UV irradiance (W m^{-2}) at time t. This latter quantity is the radiative flux impinging on the collector surface due to both beam and diffuse solar rays, as measured for instance with a global UV radiometer. When working with experimental data the integral of Equation (17) has to be approximated as a sum over discrete radiation measurements taken on a prescribed sampling interval.

It is important to note that the accumulated available energy does not take into account the actual energy collected by each reactor. In fact, the actual collected energy is a function of concentration ratio, as concentrating collectors are able to capture only a $1/C_{\text{R}}$ fraction of the diffuse radiation. In order to reflect this, another quantity can be defined called the accumulated collected energy

$$E_{\text{collect}} = \int G_{\text{UV,collect}}(t)\frac{A_{\text{a}}}{V_{\text{T}}}\,\mathrm{d}t \qquad (18)$$

where

$$G_{\text{UV,collect}} = (G_{\text{UV,global}} - G_{\text{UV,diffuse}}) + \frac{1}{C_{\text{R}}}G_{\text{UV,diffuse}} \qquad (19)$$

Here $G_{UV,diffuse}(t)$ is the diffuse UV irradiance ($W m^{-2}$) at time t. The first term of the right-hand side of Equation (19) accounts for the contribution of beam solar rays (hence, global minus diffuse irradiance), and the second one for the diffuse solar rays captured by the system.

Bandala et al. (2004) and Bandala and Estrada (2007) carried out comparisons, for the degradation of oxalic acid and carbaryl, between tubular solar photocatalytic reactors with different mirror types: V-trough collector (VTC), CPC, PTC, and directly illuminated tubes (TC) (see Figure 9). All of the collectors tested were coupled to the same azimuthal solar tracking system to ensure the same irradiation conditions over them. This comparison gave results relatively similar for both CPC and VTC, with important gains observed over the PTC and the TC (Figure 10). While the PTC was able

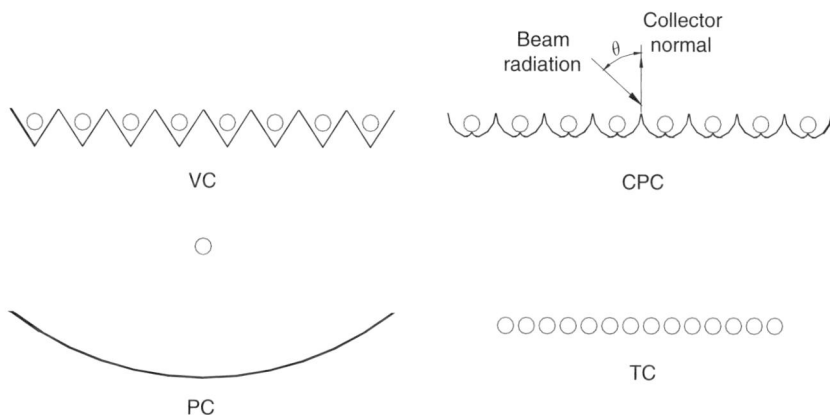

Figure 9   Different photoreactor geometries compared by Bandala et al. 2004, reprinted with permission from Elsevier.

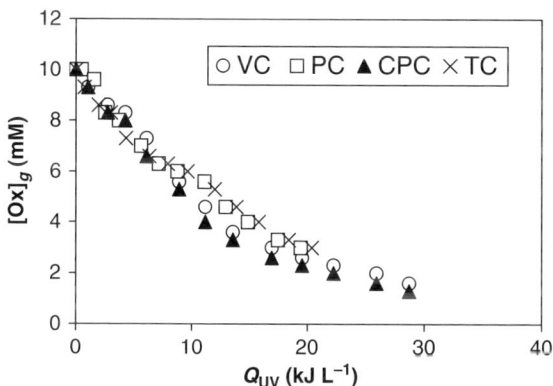

Figure 10   Photocatalytic degradation of oxalic acid in four different solar reactors as a function of accumulated energy (reprinted from Bandala et al., 2004, with permission from Elsevier).

to use less of the available solar UV radiation (only the beam part), it had a high optical efficiency. The same happened for TC, where the absence of a reflector reduces reflection losses. Nevertheless, it appears that the ability of the VTC and the CPC to distribute the radiation more homogeneously in the reactor walls is an important consideration and produces better results in the pollutant photodegradation process, with the CPC being ranked in the first place.

McLoughlin et al. (2004b) have also made a similar comparison between a VTC, a PTC, and a CPC for disinfection of water heavily contaminated with *Escherichia coli*, both by photocatalysis and by UV irradiation without catalyst. In this case the collectors did not track the sun but were inclined at local latitude with reactor tubes running east-west. It was also found that the CPC had the best performance, followed by the PTC and the VTC, which showed comparable results. It is necessary to point out that the PTC studied was of the nonconcentrating type with a very different configuration with respect to the one used by Bandala et al. (2004).

2.5. Concentrating CPC reactors

Photocatalysis is a rather inefficient process at moderate and high solar concentrations. This is due to the square root dependence of the photoconversion rate with radiation intensity (see Section 2.3). However, at low concentrations up to five suns, no such dependence has been observed (Román Rodríguez, 2001), because saturation conditions of the electron–hole generation in the catalyst are not reached. Actually, information about photocatalysis with low concentration solar reactors is scarce. This type of reactors has not been considered in the past probably because they do not make use of diffuse radiation. Jiménez and Salgado (2009) have studied the use of concentrating CPC photocatalytic reactors, with different concentration ratios between 1 and 2 suns. Four different CPC reactors with tubular receivers of the same size and having 1, 1.5, 1.75, and 2 suns concentrations were compared. Degradation of carbaryl, a pesticide widely used in Latin America, was studied. The catalyst was deposited by the sol–gel technique in long glass tubes that were fixed inside the reaction space (Figure 11).

The deposit was carried out in Duran glass tubes of 70 cm length and 0.6 cm outer diameter. The sol–gel technique (Brinker and Scherer, 1990) allows immobilizing the TiO_2 catalyst in the form of thin films over metallic and ceramic substrates, in particular glass. This deposition technique has the flexibility to permit modification of properties like particle size, surface area, crystalline structure, surface texture, photosensitivity, and chemical reactivity by controlling the process parameters, like the used reactives, solvent, drying temperature, and doping with different metal traces (see Gelover et al., 2004, and references therein). In the case considered, the films

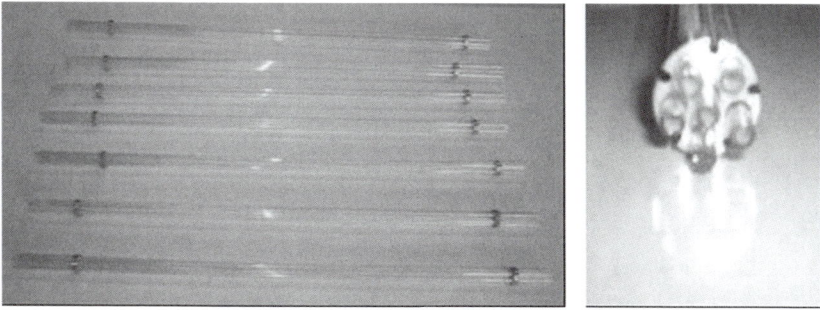

Figure 11   Glass tubes coated with sol-gel TiO$_2$ as support for catalyst to be used inside CPC reactors.

were 80 nm thick and had anatase crystalline structure (Gelover et al., 2004; Jiménez González and Gelover Santiago, 2007), with 15 nm average particle size.

Figure 12 presents the cross section of the four constructed CPCs. The collectors were mounted in parallel on a single structure to facilitate the comparison. This structure was oriented facing south with inclination equal to the local latitude (19°) and tubes oriented east-west (Figure 13). Each reactor processed 3 L of water in the recirculation batch mode.

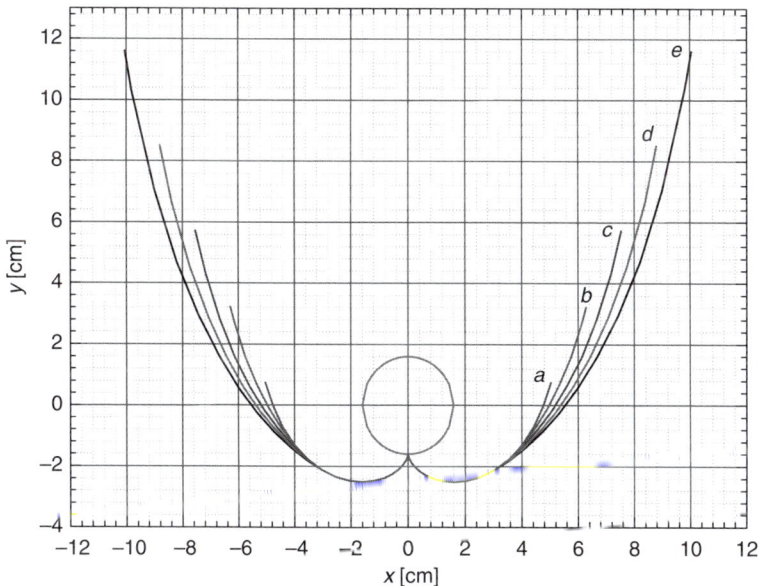

Figure 12   Cross section of the four CPC with 1, 1.5, 1.75, and 2 suns concentration.

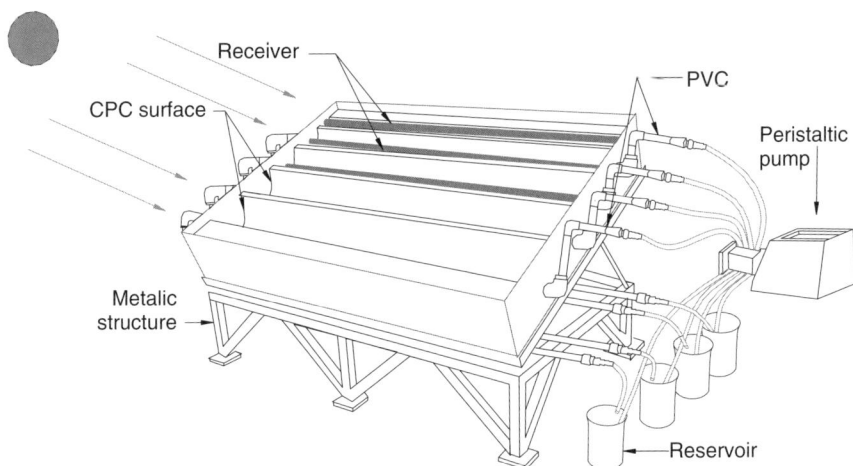

Figure 13 Multiple CPC test bed and experimental scheme. Each reactor tube contains five small glass tubes inside, supporting the TiO_2 catalyst.

As it has been discussed previously, for the 2 suns CPC the collected UV radiation is about 75% of the available solar radiation, while a nonconcentrating CPC captures in principle all the available radiation. However, as the aperture area of the 2 suns CPC is twice the aperture of the one sun system, the net result is that it collects 1.5 more energy. In the same way, the 1.5 and 1.75 suns reactors are able to capture 1.4 and 1.25 times more energy than the 1 sun system, respectively. The effect of these differences in the energy accepted by each system can be clearly seen in the degradation of carbaryl (Figure 14), where degradation proceeds faster as the concentration ratio increases.

When results are analyzed as a function of accumulated collected radiation (Equation (19)), as shown in Figure 15, it is found that the four collectors described utilize the captured energy with similar efficiency. This seems reasonable, considering that the four collectors have very similar designs, with concentration ratios being the only difference. It is also apparent that none of the collectors operates under a regime where catalyst electron–hole generation is saturated. Nevertheless, it seems as if the 2 suns collector performs slightly better than the others, at least at the beginning of the reaction period.

The four reactors were also compared on the basis of cost. Taking into account the cost of the aluminum reflectors it is found that the increase in the total cost of the 2 suns against the 1 sun collector is only of 15%. This is more than compensated by the fact that the same quantity of supported catalyst is used in both systems, so the overall cost per unit collector area is reduced.

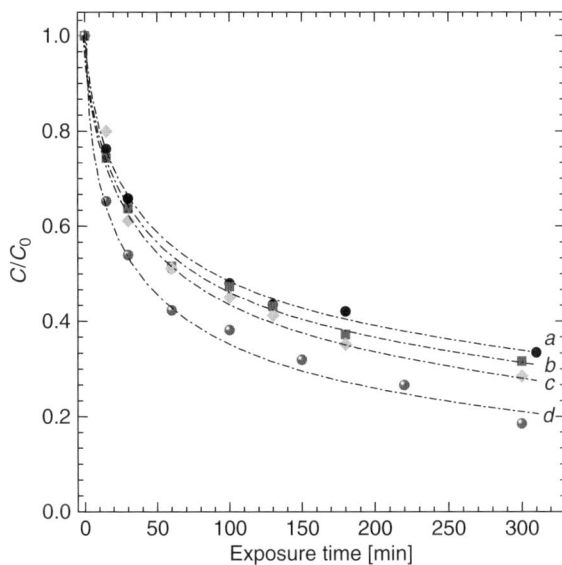

Figure 14 Evolution of carbaryl concentration as a function of time in four CPC reactors with 1 (*a*), 1.5 (*b*), 1.75 (*c*), and 2 (*d*) suns concentration ratio, respectively.

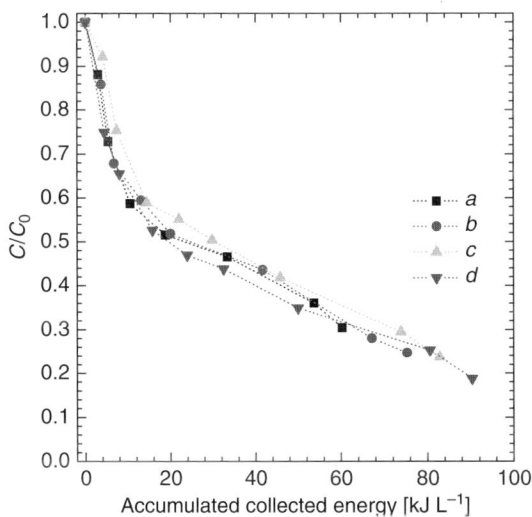

Figure 15 Evolution of carbaryl concentration as a function accumulated collected energy in four CPC reactors with 1 (*a*), 1.5 (*b*), 1.75 (*c*), and 2 (*d*) suns concentration ratio, respectively.

In addition to the above, the reduction in treatment time with concentration is also an attractive feature of the 2 suns collector, in spite of the fact that it misses around 25% of the available radiation. Due to the abundance of solar irradiation in the test site (Temixco, Morelos, Mexico), it does not appear that sacrificing a part of the UV radiation is such a great concern. However, in depth techno-economic analyses are required to support this assertion.

3. RADIATION TRANSFER IN PHOTOCATALYTIC REACTORS

As discussed previously, the radiative quantity of interest in photocatalysis is the local spectral power absorbed by the catalyst per unit volume $Q^m_{abs\text{-}cat,\lambda}(\mathbf{r})$ (W m^{-3}). This parameter is directly related to the number of photons available, per unit volume, and wavelength to carry out the excitation of electron–hole pairs.

Slurry photocatalytic reactors are characterized by the presence of catalyst particles with sizes in the micrometer and submicrometer range. Such small particles produce in general a fair amount of radiation scattering, as will be discussed later. Radiative transfer theory (Mahan, 2002; Modest, 2003) provides the tools to analyze and evaluate the absorption of radiation inside a scattering/absorbing medium.

Radiative transfer models can be very complicated and their solution cumbersome, so it is always desirable to have simplified tools for this analysis. Very simple and intuitive models can be developed in some cases to study radiation transfer inside photocatalytic reactors. However, it must be pointed out that these models must be well grounded in radiative transfer theory, in order to provide results that are consistent and of lasting engineering interest. Here, we present the basic concepts and discuss some approximations and methods that have been used.

3.1. Optical properties of the medium

Optical coefficients of the water/catalyst suspension are needed for carrying out radiative transfer calculations. When a beam of light propagates in a participating medium it can suffer absorption and scattering. Both processes reduce the intensity of this beam; in particular, scattering removes radiation from the beam and delivers it to beams propagating in other directions. The attenuation follows a Lambert–Beer type law, where the diminution of the intensity in a differential distance is proportional to intensity itself and to the distance traveled. The proportionality constant is the extinction coefficient of the medium β_λ (cm^{-1}), which is the sum of the scattering σ_λ (cm^{-1}) and absorption κ_λ (cm^{-1}) coefficients.

$$\beta_\lambda = \kappa_\lambda + \sigma_\lambda \tag{20}$$

The extinction, absorption, and scattering coefficients depend on catalyst particle concentration C_p. If concentration is not extremely high, this dependence is linear

$$\beta_\lambda = C_{\text{catal}}\beta^*_{\lambda,\text{catal}}, \quad \kappa_\lambda = C_{\text{catal}}\kappa^*_{\lambda,\text{catal}}, \quad \sigma_\lambda = C_{\text{catal}}\sigma^*_{\lambda,\text{catal}} \quad (21)$$

where $\beta^*_{\lambda,\text{catal}}$, $\kappa^*_{\lambda,\text{catal}}$, and $\sigma^*_{\lambda,\text{catal}}$ are specific coefficients, independent of concentration. When radiation impinges with a particle, either absorption or scattering must occur. The fraction of radiation that is scattered is given by the quantity known as scattering albedo

$$\omega_\lambda = \frac{\sigma_\lambda}{\beta_\lambda} \quad (22)$$

The directional distribution of scattering is given by the phase function $\Phi_\lambda(\hat{s} \cdot \hat{s}')$. This function depends on the angle between the incident $\hat{s}$ and scattered directions $\hat{s}'$, $\cos\ \theta_s = \hat{s} \cdot \hat{s}'$. The determination of the phase function requires elaborate experiments. It is common to use approximate phase functions, like the Henyey–Greenstein phase function $\Phi_{\text{HG},\lambda}$ (see, for instance, Modest, 2003), which can be used to approximate the real phase function for many types of particles

$$\Phi_{\text{HG},\lambda}(\hat{s} \cdot \hat{s}') = \frac{1 - g_\lambda^2}{[1 + g_\lambda^2 - 2g_\lambda\cos(\hat{s} \cdot \hat{s}')]^{3/2}} \quad (23)$$

Here g_λ is the wavelength-dependent asymmetry parameter (see Equation (36) below).

Several authors have addressed the determination of the optical properties of aqueous titanium dioxide suspensions in the context of photoreactor modeling (Brandi et al., 1999; Cabrera et al., 1996; Curcó et al., 2002; Salaices et al., 2001, 2002; Satuf et al., 2005; Yokota et al., 1999). Among the determined properties are extinction, scattering, and absorption coefficients, as well as the asymmetry parameter of the scattering phase function. In general the procedures involve fitting of a radiative transfer model to the experimental results for reflectance and transmittance of radiation.

In particular, Cabrera et al. (1996), Brandi et al. (1999), and Satuf et al. (2005) have determined optical parameters for TiO_2 particles of several commercial brands. The determinations were carried out by means of spectrophotometry experiments involving the measurement of specular reflectance and beam transmittance, as well as hemispherical transmittance and reflectance, of catalyst suspensions (Cabrera et al., 1996). By radiative transfer calculations with the discrete ordinates method (DOM), the values of the extinction and absorption coefficient and of the asymmetry parameter that better fitted the results of measurements were found. Actually, the extinction coefficients of Satuf et al. (2005) are the same as those of Brandi et al. (1999), but the absorption coefficients are very different. The reason for this is the use of a different phase function in each case: Brandi et al. (1999) assumed an isotropic phase function for simplicity while Satuf et al. (2005) refined this previous work by taking into account the possibility of anisotropic scattering by means of the Henyey–Greenstein phase function.

3.2. Modeling dye degradation

When the photocatalytic degradation of a dye or of any other nontransparent contaminant is considered, the effect of the dye on the optical properties of the aqueous medium has to be accounted for. The dye absorbs radiation but does not scatter it; therefore, the scattering coefficient for the medium is equal to the scattering coefficient of the catalyst particles

$$\sigma_\lambda = C_{catal}\sigma^*_{\lambda,catal} \tag{24}$$

Here $\sigma^*_{\lambda,catal}$ ($cm^2\ g^{-1}$) is the specific scattering coefficient of the catalyst, which is independent of catalyst concentration.

On the other hand, at each wavelength, the absorption coefficient of the medium is a sum of the absorption coefficients of the dye and the catalyst

$$\kappa_\lambda = C_{dye}\kappa^*_{\lambda,dye} + C_{catal}\kappa^*_{\lambda,catal} \tag{25}$$

The specific absorption coefficient of the dye $\kappa^*_{\lambda,dye}$ ($cm^{-1}ppm^{-1}$) is the easiest to determine of all the optical parameters in this kind of problems. For dye concentrations which are not too high, one can expect that Lambert–Beer law is satisfied; that is, the transmittance of an aqueous solution of the dye should follow an exponential decay with concentration. This means that a linear relationship exists between the absorbance of the samples and dye concentration for each measured wavelength. Spectral measurements of the transmittance of solutions of the dye (in the absence of catalyst) for different concentrations can be made on a spectrophotometer. The specific absorption coefficient for different wavelengths is the slope of a straight line fit to these data, divided by the length of the spectrophotometer test cell.

For the optical properties of the catalyst the situation is somewhat more complicated. These optical properties have been determined for suspension of TiO_2 in water; that is, for catalyst particles surrounded by a nonabsorbing medium. In the present case the particles are surrounded by the absorbing aqueous dye solution and there is a possibility that their effective optical properties may be modified by this fact.

There has been some conceptual disagreement in the definition and interpretation of optical cross sections of small particles when the surrounding medium absorbs radiation (Bohren and Gilra, 1979; Lebedev et al., 1999; Mundy et al. 1974). Recently, an operational point of view has been adopted, which allows to carry out calculations for composite media consisting of small particles embedded in absorbing hosts (Lebe-dev et al. 1999; Sudiarta and Chylek, 2001). In the case of commercial TiO$_2$ particles it is difficult to apply scattering theories to determine the optical properties of the catalyst particles due to lack of precise informa-tion about their microstructure. Nevertheless, some conclusions can be drawn from examination of the results reported in the literature: as is found in the cases studied by Mundy et al. (1974), Sudiarta and Chylek (2001), and Bruscaglioni et al. (1993) to observe appreciable effects in the extinction and scattering coefficients of the particles due to the presence of the dye, the imaginary part of the refractive index of the surrounding medium n_{imag} should be at least of the order 10^{-3}. This is not a definitive rule, but one based on the limited information available in the literature. The imaginary part of the index of refraction of the aqueous dye solution is related to the specific absorption coefficient by

$$n_{imag} = \frac{\kappa^*_{\lambda,dye} \, C_{dye} \, \lambda}{4\pi} \qquad (26)$$

For instance, at $\lambda = 355$ nm, for a concentration $C_{dye} = 120$ ppm of Acid Orange 24, it was obtained that $n_{imag} = 3.27 \times 10^{-7}$ (Villafán-Vidales et al., 2007). In that case it was concluded that it was not very likely that there was an important effect of the absorption of the dissolved dye in the optical properties of catalyst particles. However, this is not necessarily so in a general case, where results depend very much on the absorption coefficient of the dye and its concentration. Moreover, special experiments would have to be devised to measure the optical coefficients of TiO$_2$ suspended in strongly colored water.

In the simulation of the evolution of the dye concentration the radiative and kinetic problems cannot be decoupled, as was the case for nonabsorbing contaminants. Therefore, some kind of numerical procedure based on time discretization has to be applied for the solution of Equation (10). One possibility is to apply an Euler-type method that consists in the following steps (Villafán-Vidales et al., 2007):

1. Starting off from the initial dye (variable) and catalyst (constant) concen-trations the absorption and scattering coefficients for these concentra-tions are evaluated. Then, these coefficients are used in a radiative transfer model to obtain e_{L} for a large number of points within the reactor volume.

2. Once the distribution of e_{L} has been evaluated, the integral of Equation (10) is calculated numerically over the illuminated part of the reactor, and from this equation the instantaneous degradation rate is obtained. Multi-plying this rate by the time step of the method, which must be small compared to the total time required for the degradation process, one gets the decrease in dye concentration. That means that concentration

at step $(n+1)$ is obtained from the concentration at step (n) by an equation of the type

$$C_{dye}^{(n+1)} = C_{dye}^{(n)} + \left(\frac{dC_{dye}}{dt}\right)^{(n)} \Delta t \tag{27}$$

3. Once this new dye concentration is obtained, the process is repeated from step (1). This process is carried out for different TiO_2 concentrations.

3.3. The radiation transfer equation

The quantity that fully characterizes radiation in a medium is the spectral radiative intensity $I_\lambda(\mathbf{r}, \hat{s})$ $(W\,m^{-2}\,\mu m^{-1}\,sr^{-1})$. This quantity is the formalization of the intuitive concept of a ray of light. It characterizes the local amount of power traveling along a given direction, per unit wavelength, per unit area normal to this direction, and per unit solid angle. It depends on five variables: three spatial variables for the position vector $\mathbf{r}$ and two angular variables for the direction unit vector $\hat{s}$.

The local spectral power absorbed by the catalyst can be expressed in terms of intensity and of the absorption coefficient of the catalyst by the following equation:

$$Q_{abs\text{-}cat,\lambda}^m = \kappa_{catal,\lambda} G_\lambda(\mathbf{r}) \tag{28}$$

where $G_\lambda(\mathbf{r})$ is called the local incident radiation, which is the total intensity impinging on a point from all directions

$$G_\lambda(\mathbf{r}) = \int_{4\pi} I_\lambda(\mathbf{r}, \hat{s}')d\Omega' \tag{29}$$

Another quantity of interest in the discussions to follow is the radiative flux vector $\mathbf{q}_\lambda(\mathbf{r})$ $(W\,m^{-2}\,sr^{-1})$, which gives the net spectral flux of radiation along the preferential propagation direction in a given point. It is, in simple terms, intensity summed as a vector over all propagation directions

$$\mathbf{q}_\lambda(\mathbf{r}) = \int_{4\pi} \hat{s}'\, I_\lambda(\mathbf{r}, \hat{s}')d\Omega' \tag{30}$$

The distribution of intensity of radiation in a scattering/absorbing medium can be obtained by solving the RTE (Mahan, 2002; Modest, 2003).

$$\frac{dI_\lambda(\mathbf{r}, \hat{s})}{ds} = -\beta_\lambda\, I_\lambda(\mathbf{r}, \hat{s}) + \frac{\sigma_\lambda}{4\pi}\int_{4\pi} I_\lambda(\mathbf{r}, \hat{s}')\,\Phi_\lambda(\hat{s}\cdot\hat{s}')d\Omega' \tag{31}$$

This equation gives the variation of the intensity of radiation, passing through point $\mathbf{r}$ in the medium, along the propagation direction $\hat{\mathbf{s}}$, when it travels a distance ds in that direction. Actually, this can be visualized not as a single equation but as an infinite set of coupled equations, one for each possible propagation direction. As expressed by the first term in the right-hand side of this equation, intensity is diminished due to both scattering and absorption (recall Equation (20)). The second term takes multiple scattering into account and couples all different propagation directions; that is, scattering takes some energy out from propagation direction $\hat{\mathbf{s}}'$ and redirects it into the direction under consideration $\hat{\mathbf{s}}$.

Basically, Equation (31) is a generalization of Lambert–Beer law to include multiple scattering effects; that is, the possibility not only that the intensity of a light beam entering a medium is decreased due to scattering, but also the possibility that some of this radiation may return eventually to the original propagation direction after several scattering events. In the case when the medium does not scatter (as is the case, for instance, in the photo-Fenton processes), this reduces to the ordinary Lambert–Beer law of propagation

$$\frac{dI_\lambda(\mathbf{r}, \hat{\mathbf{s}})}{ds} = -\kappa_\lambda\, I_\lambda(\mathbf{r}, \hat{\mathbf{s}}) \tag{32}$$

3.4. Methods of solution of the RTE

Due to the integro-differential nature of the RTE, it is a complex equation and analytical solutions are not available, except for the simplest problems. Several methods have been devised to overcome this limitation, either numerical solutions of the exact RTE or simplified models that do have analytical solutions. It is also possible that we need to solve numerically such a simplified model, in which case the advantages of the simplification must be carefully considered against the greater exactness of solving the RTE numerically.

Among the numerical models to solve the exact RTE we can mention the DOM (Modest, 2003) and the Monte Carlo method (MCM) (Mahan, 2002). The greatest complication in solving the RTE is how to deal with the integral term in the right-hand side of the equation. The DOM discretizes the infinite number of directions involved in Equation (31) to a finite number of directions $\hat{\mathbf{s}}_i$, optimally chosen according to the geometry of the problem. In such a way, the integral term is reduced to a sum over the chosen propagation directions. The problem then consists in the solution of a coupled system of linear differential equations, one for each propagation direction, at every point inside the reactor volume. The differential term in the left-hand side must also be discretized, for which the left-hand side of the RTE is rewritten in an equivalent form where differentiation is more explicitly expressed

$$\hat{\mathbf{s}} \cdot \nabla I_\lambda(\mathbf{r}, \hat{\mathbf{s}}) = -\beta_\lambda \, I_\lambda(\mathbf{r}, \hat{\mathbf{s}}) + \frac{\sigma_\lambda}{4\pi} \int_{4\pi} I_\lambda(\mathbf{r}, \hat{\mathbf{s}}') \ \Phi_\lambda(\hat{\mathbf{s}} \cdot \hat{\mathbf{s}}') d\Omega' \qquad (33)$$

The present description of the method is somewhat oversimplified, because boundary conditions, directions, and spatial discretization must be carefully established in every case for the geometry at hand.

The DOM has been extensively and successfully applied to photocatalytic reactors by the Santa Fe (Argentina) group (see Cassano and Alfano 2000, and references therein) and verified against experimental results (Brandi et al., 1999; Romero et al., 2003). Also Trujillo et al. (2007) have recently used a variant of the DOM, called finite volume scheme to model the effect of air bubbles injected in a fixed catalyst reactor.

The MCM consist in tracking numerically the propagation of a large sample of individual photons inside the system and evaluating the desired parameters as statistical averages over these samples (Mahan, 2002). Each step in the propagation is considered as a random event and simulated by generating random numbers according to prescribed probability distributions (Arancibia-Bulnes and Ruiz-Suárez, 1999; Yokota et. al., 1999).

For instance, at a wall or at an interface between two media (as may be the water surface), the reflectance and transmittance are interpreted as probabilities of reflection and transmission, respectively. The outcome of the interaction of a photon with the wall is decided by generating a random number uniformly distributed between zero and one, and comparing this number with the respective wall reflectance. The propagation distance before colliding with a catalyst particle follows an exponential probability distribution with β_λ^{-1} as the attenuation coefficient. Then, to simulate propagation a random number is generated according to this distribution to decide where the next collision will happen. At a collision, the scattering albedo ω_λ is interpreted as the probability of scattering and $(1-\omega_\lambda)$ as the probability of absorption. When a photon is absorbed, its history is terminated, but when it is scattered a new propagation direction must be chosen and the process continued. The phase function is interpreted then as a probability distribution for scattering angles.

The MCM has been used to simulate tubular solar photocatalytic reactors, like parabolic troughs (Arancibia-Bulnes et al., 2002a), CPC (Arancibia-Bulnes et al., 2002b), and also of flat plate geometry (Cuevas et al., 2004). Also it has been used to simulate flat lamp reactors (Brucato et al., 2006) or to obtain optical coefficients by comparison with transmission results from an experimental cell (Yokota et al., 1999).

Among the approximate analytical models there are the many-flux models. These models are similar to the DOM, in the sense that the integral term of the RTE is approximated as a sum by taking into account only a limited number of propagation directions. The simplest of these are two flux models, like the classical Schuster–Schwarzchild and Kubelka–Munk (Ishimaru, 1997) models, which are applicable to plane parallel media where transversal directions are much larger than propagation depth. Actually these are the simplest possible models able to take multiple scattering into account. In these models only the forward and backward propagation directions are considered, which correspond to two diffuse radiation fluxes, propagating to the inside and to the outside of the medium, respectively. Two flux models have been used by Brucato and Rizzuti (1997) and Brucato et al. (2006)

Other authors have also used approximate methods to solve the radiation problem. Li Puma and Yue (2003) used a thin film slurry model which does not include scattering effects. More recently, Li Puma et al. (2004), Brucato et al. (2006), and Li Puma and Brucato (2007) have used six flux models for different geometries. Salaices et al. (2001, 2002) used a model which allows for an adequate evaluation of the absorbed radiation in terms of macroscopic balances, based on radiometric measurements. They measured separately total transmitted radiation and nonscattered transmitted radiation, modeling the decay of both radiative fluxes with concentration by exponential functions.

Of a different nature is the P1 approximation (Modest, 2003), also known as the diffusion approximation (Ishimaru, 1997). Here the number of propagation directions is not restricted, but instead it is assumed that energy distributes quite uniformly over all these directions, as will be described in the next section. This approximation is the lowest order of the spherical harmonics method (also known as the Pn approximation). It is more versatile than two- and four-flux models, because it lends itself more easily to different geometries.

4. THE P1 APPROXIMATION

The P1 approximation assumes that the angular distribution of radiation intensity is almost isotropic within the medium; that is, the radiative energy that reaches any given point inside the reactor comes almost equally from all possible directions. This requires a great deal of scattering, in order to erase from the radiation field the directional behavior originated from its sources. Nevertheless, the radiation intensity I_λ cannot be completely uniform for all directions $\hat{s}$, because there would be no net energy propagation in any direction in such a case. Therefore, the P1 approximation assumes that radiation intensity is slightly higher in the direction of the net flux. Mathematically, intensity is expressed as a linear function of the components of the propagation vector $\hat{s} = (s_x, s_y, s_z)$

$$I_\lambda(\mathbf{r}, \hat{s}) = \frac{1}{4\pi}[G_\lambda(\mathbf{r}) + 3\mathbf{q}_\lambda(\mathbf{r}) \cdot \hat{s}] \tag{34}$$

where the second term in the right-hand side should be much smaller than the first, and subsequent terms with higher order dependence on direction are neglected.

Inserting Equation (34) into the RTE, it is possible to carry out analytically the integral in the right-hand side of this equation. After some algebra (Modest, 2003, pp. 511–514) it is obtained that

$$\mathbf{q}_\lambda(\mathbf{r}) = -[3\beta_\lambda(1 - g_\lambda\omega_\lambda)]^{-1} \nabla G_\lambda(\mathbf{r}) \qquad (35)$$

where g_λ is known as asymmetry parameter and is given by

$$g_\lambda = \frac{1}{4\pi} \int_{4\pi} \Phi_\lambda(\hat{\mathbf{s}} \cdot \hat{\mathbf{s}}') (\hat{\mathbf{s}} \cdot \hat{\mathbf{s}}') \, d\Omega' \qquad (36)$$

This parameter describes the scattering directional behavior of the particles; for instance, it equals 1 for purely forward scattering, -1 for purely backwards scattering, and 0 for isotropic scattering.

A physical interpretation of Equation (35) is possible if one notes that it is mathematically analogous to Fourier law of heat conduction. The constant factor in the right-hand side plays the role of thermal conductivity, and the local incident radiation $G_\lambda(\mathbf{r})$ plays the role of temperature. In that sense, differences in the latter variable among neighboring regions in the medium drive the "diffusion" of radiation toward the less radiated zone. Note that the more positive the asymmetry parameter, the higher the "conductivity"; that is, forward scattering accelerates radiation diffusion while backscattering retards it.

Besides the approximate Equation (35), another more fundamental equation from radiative transfer theory also relates G_λ with $\mathbf{q}_\lambda$ (see, for instance, Modest, 2003, pp. 312–314)

$$\nabla \cdot \mathbf{q}_\lambda(\mathbf{r}) = -\kappa_\lambda G_\lambda(\mathbf{r}) \qquad (37)$$

Taking the divergence of Equation (35) and substituting it into Equation (37), a second-order partial differential equation of the Helmholtz type is finally obtained for the P1 approximation

$$\nabla^2 G_\lambda = k_{d,\lambda}^2 \, G_\lambda \qquad (38)$$

where k_d is a wavelength-dependent constant (sometimes called diffusion constant) given by

$$k_{d,\lambda} = \sqrt{3(1 - \omega_\lambda g_\lambda)\kappa_\lambda\beta_\lambda} \qquad (39)$$

To continue the analogy with heat conduction, it can be seen that this equation is analogous with the steady-state equation of heat conduction in the presence of heat sources.

Note that, within the assumptions of the P1 approximation, solving Equation (38) is entirely equivalent to solving the RTE, because once G_λ is known, I_λ can be evaluated by using Equations (34) and (35).

To solve Equation (38) boundary conditions which describe the reflection and transmission of radiation at the boundaries are required. In principle, boundary conditions can only be established in a rigorous manner for the radiative intensity, not for G_λ, because the optical properties of the interfaces depend on the direction of incidence of radiation. Because the P1 approximation solves for an integrated quantity like G_λ instead, approximate boundary conditions must be established (Modest, 2003). One possibility is the Marshak boundary condition (Marshak, 1947), which comes from considering the continuity of the radiative flux through the interface. If this continuity is considered together with the assumption (34) of the P1 approximation and Equation (37), the following equation is obtained (Spott and Svaasand, 2000)

$$(1 - 2\rho_1)\, G_\lambda - (1 + 3\rho_2)\frac{2\hat{n} \cdot \nabla G_\lambda}{3\beta_\lambda(1 - \omega_\lambda g_\lambda)} = 4q_{e,\lambda} \tag{40}$$

Here $q_{e,\lambda}$ accounts for any external flux coming through the surface, $\hat{n}$ the surface normal, and ρ_i is the ith moment of the surface reflectance function $\rho(\hat{n} \cdot \hat{s})$ given by

$$\rho_i = \int_{2\pi} \rho(\hat{n} \cdot \hat{s}')\, (\hat{n} \cdot \hat{s}')^i \, d\Omega' \tag{41}$$

4.1. General solution of the P1 approximation for solar tubular reactors

As discussed previously, several solar photoreactor geometries can be reduced to cylindrical glass tubes externally illuminated by different types of reflectors, like parabolic troughs, CPC, V-grooves, or without reflector, directly illuminated by the sun. In this section the general solution of the P1 approximation for this type of photoreactors is reported. This general solution is applicable to any particular reactor if the flux distribution impinging on the wall of the tubular reaction space is known.

To model the radiation field for a tubular reactor, Equation (38) is written in cylindrical coordinates. If the tube is slender we can neglect end effects and the radiative flux is independent of the longitudinal variable z, then

$$\frac{1}{r}\frac{\partial}{\partial r}\left(r\frac{\partial G_\lambda}{\partial r}\right) + \frac{1}{r^2}\frac{\partial^2 G_\lambda}{\partial \theta^2} = k_{d,\lambda}^2 G_\lambda \tag{42}$$

The general solution of this equation can be obtained by the standard method of separation of variables

$$G_\lambda(r,\ \theta) = \sum_{n=0}^{\infty} [A_n \cos(\delta_n\theta) + B_n \sin(\delta_n\theta)][C_n I_n(k_{d,\lambda}r) + D_n K_n(k_{d,\lambda}r)] \quad (43)$$

In this equation, I_n and K_n are nth order modified Bessel functions of the first and second kind (Olver, 1972), respectively. These functions behave somewhat like increasing and decreasing exponentials, respectively. The eigenvalues δ_n are obtained easily in this case by the consideration that physically local incident radiation must be periodical when the angle θ completes a full turn around the tube

$$G(r,0) = G(r,2\pi) \quad (44)$$

therefore

$$\delta_n = n \quad (45)$$

Another consideration that can be applied is observing that the K_n functions diverge at the center of the reactor ($r=0$) and therefore cannot be present in the solution. Taking this into account, we obtain as general solution for any tubular solar reactor

$$G_\lambda(r,\ \theta) = \sum_{n=0}^{\infty} [E_n \cos(n\theta) + F_n \sin(n\theta)] I_n(k_{d,\lambda}r) \quad (46)$$

In terms of the angular variable, this expression is a Fourier series. To obtain the constants E_n and F_n that appear on this equation one must apply the boundary condition (40) at the tube wall ($r=r_0$), which in this case reduces to

$$(1 - 2\rho_1)\, G_\lambda(r_0) + (1 + 3\rho_2)\, \frac{2}{3\beta_\lambda(1 - \omega_\lambda g_\lambda)} \left.\frac{\partial G_\lambda}{\partial r}\right|_{r=r_0} = 4q_{e,\lambda} \quad (47)$$

For applying this condition, in turn it is necessary to expand the total solar radiative flux entering trough the tube wall (formed by the directly incident part plus that reflected by the mirrors) into a Fourier series

$$q_{e,\lambda}(\theta) = \sum_{n=0}^{\infty} [Q_n \cos(n\theta) + R_n \sin(n\theta)] \quad (48)$$

With this, the unknown coefficients are obtained.

$$E_n = 4Q_n \left\{ (1 - 2\rho_1)\, I_n(k_{d,\lambda}r_0) + 2k_{d,\lambda} \left[\frac{(1 + 3\rho_2)}{3\beta_\lambda(1 - \omega_\lambda g_\lambda)}\right] I'_n(k_{d,\lambda}r_0) \right\}^{-1} \quad (49)$$

$$F_n = 4R_n \left\{ (1 - 2\rho_1) \; I_n(k_{d,\lambda} r_0) + 2k_{d,\lambda} \left[\frac{(1 + 3\rho_2)}{3\beta_\lambda (1 - \omega_\lambda g_\lambda)} \right] I'_n(k_{d,\lambda} r_0) \right\}^{-1} \quad (50)$$

To use this general solution, it is only necessary to obtain the distribution of the radiative flux on the reactor wall, say by applying a ray tracing method to the solar collector, and to expand the resulting distribution of radiative flux in a Fourier expansion like (48). Then the solution inside the reactor is obtained.

Let us consider the particular case of a tube directly illuminated only by beam solar radiation. Graphically this looks like in Figure 16.

The radiative flux on the boundary can be expressed by the simple equation

$$q_{e,\lambda}(\theta) = \tau \; G_{UV,beam} \; \cos(\theta) \quad (51)$$

where τ is the transmittance of the glass tube (assumed constant for simplicity), and $G_{UV,beam}$ is the measured UV irradiance of solar beam radiation. In this case the coefficients of the Fourier expansion are all zero except by

$$Q_1 = \tau \; G_{UV,beam} \quad (52)$$

A slightly different calculation must be done to take also into account the diffuse component of UV solar radiation in the solution.

4.2. Solution for a parabolic trough reactor

The analytical expression for the radiation flux distribution produced on the wall of a tubular receiver by a PTC (Figure 17) has been calculated by Jetter (1986, 1987). Jetter's method is applied here to a particular PTC collector that has been used in our research group to study the degradation of several

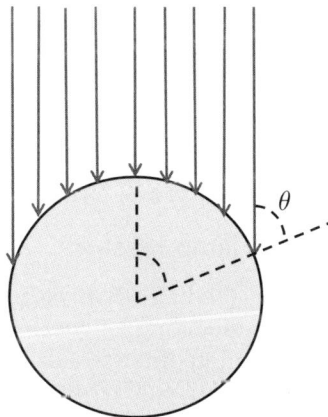

Figure 16 Tubular reactor illuminated by beam solar radiation.

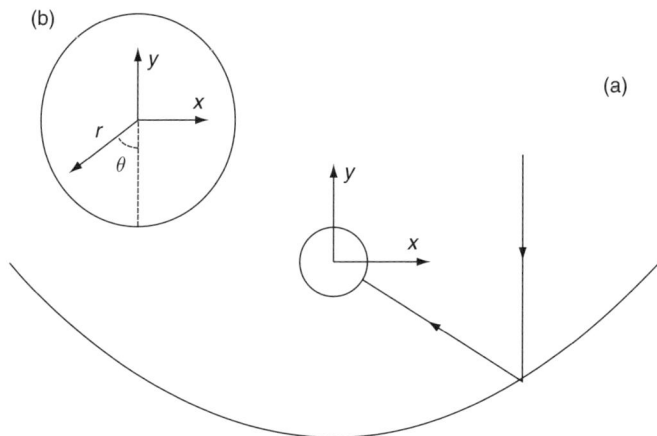

Figure 17 Geometry of the parabolic trough and reactor tube (a), and coordinate systems inside the tube (b) (reprinted from Arancibia-Bulnes and Cuevas, 2004, with permission from Elsevier).

pollutants (Arancibia-Bulnes et al., 2002a,b; Bandala et al., 2002; Gelover et al., 2000; Jiménez et al., 2000). The characteristics of this concentrator are tube diameter 2.45 cm, length 172 cm, concentrator lateral aperture 106 cm, focal length 27 cm, rim angle 90°, average UV reflectance of mirror 75%, and average UV transmittance of glass reactor 85%. The surface reflection error of the mirror has been determined as 7 mrad. The resulting distribution is presented in Figure 18. The distribution is presented as flux concentration, which is the flux distribution divided by the beam solar radiation incident in the collector.

Ten terms were needed in the Fourier series to approximate adequately the distribution in Figure 18 and to reach convergence in the solution for the local incident radiation (Arancibia-Bulnes and Cuevas, 2004). An example of the obtained distribution of volumetric absorbed power is presented in Figure 19, for the cross section of the reactor, at a wavelength of 325 nm, and considering a concentration of $0.15\,\text{g L}^{-1}$ of Aldrich TiO_2 catalyst. The peaks observed on the graph correspond with the peaks of the entering flux of Figure 18.

4.3. Solution for an annular lamp reactor

Even though this chapter is devoted mostly to solar photocatalytic reactors, we would like to discuss the modeling of an annular lamp reactor, as a different example of the application of the P1 approximation. This problem was studied (Cuevas et al., 2007) with reference to a particular reactor known as photo CREC-water II (Salaices et al., 2001, 2002). Equation (38) is again written in cylindrical coordinates. Nevertheless in this case the

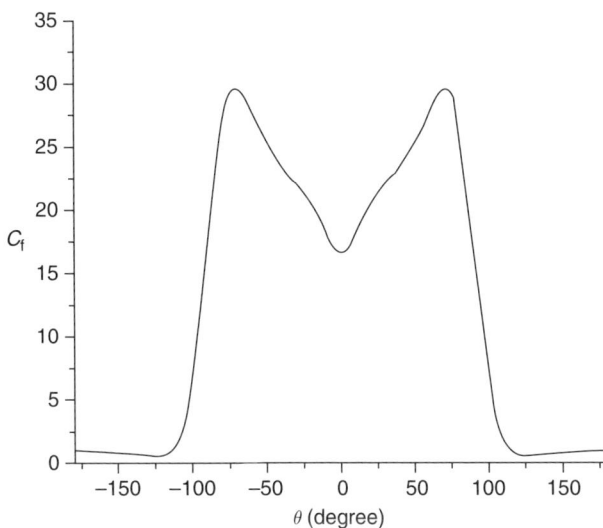

Figure 18 Flux concentration distribution of a 90° rim angle parabolic trough solar concentrator. Adapted from Arancibia-Bulnes and Cuevas (2004), with permission from Elsevier.

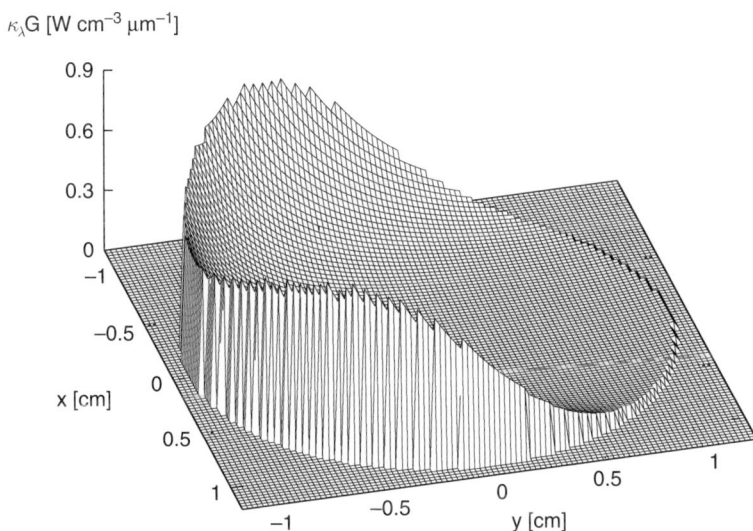

Figure 19 Distribution of spectral volumetric absorbed power inside a tubular reactor illuminated by a 90° rim angle parabolic trough solar concentrator. Wavelength 325 nm and catalyst concentration 0.15 g L^{-1} (Reprinted from Arancibia-Bulnes and Cuevas, 2004, with permission from Elsevier).

incident flux is the same for the whole reactor circumference, but it may depend on z if the flux from the lamp does so.

$$\frac{1}{r}\frac{\partial}{\partial r}\left(r\frac{\partial G_\lambda}{\partial r}\right) + \frac{\partial^2 G_\lambda}{\partial z^2} = k_{d,\lambda}^2 G_\lambda \tag{53}$$

The solution is symmetrical with respect to the center of the lamp ($z = L/2$) in the z coordinate due to the symmetry of the entering flux (Salaices et al., 2002). Therefore the problem is solved only in the range $[0; L/2]$. After applying suitable Marshak boundary conditions in the longitudinal variable to the general solution of this equation it is obtained that

$$G_\lambda(r, z) = \sum_{n=0}^{\infty} f_n(z)[E_n \, I_0(x_n) + F_n \, K_0(x_n)] \tag{54}$$

where

$$x_n = \sqrt{k_{d,\lambda}^2 + \delta_n^2} \, r \tag{55}$$

The eigenfunctions f_n are given by

$$f_n(z) = \left[\cot\left(\frac{\delta_n L}{2}\right)\cos(\delta_n z) + \sin(\delta_n z)\right] C_n^{-1/2} \tag{56}$$

with

$$C_n = \frac{1}{2}\csc^2\left(\frac{\delta_n L}{2}\right)\left[1 - \left(\frac{2}{\delta_n L}\right)\cos\left(\frac{\delta_n L}{2}\right)\sin\left(\frac{\delta_n L}{2}\right)\right] \tag{57}$$

And the eigenvalues δ_n must be obtained this time by the solution of the following transcendental equation

$$\cot\left(\frac{\delta_n L}{2}\right) = \left(\frac{\delta_n L}{2}\right)\frac{4(1 + \rho_p)}{3(1 - \rho_p)(1 - g_\lambda \omega_\lambda)\beta_\lambda L} \tag{58}$$

Boundary conditions are also applied for the variable $\mathbf{r}$, at the two cylindrical walls, from which explicit expressions are obtained for the constants in the solution. Figure 20 presents an example of a distribution of local incident radiation inside the reactor, as a function of the longitudinal and radial coordinates

The theoretical model developed here was compared (Cuevas et al., 2007) with measurements of radiative flux transmission through a series of observation windows in the photo CREC-Water II reactor (Salaices et al., 2001).

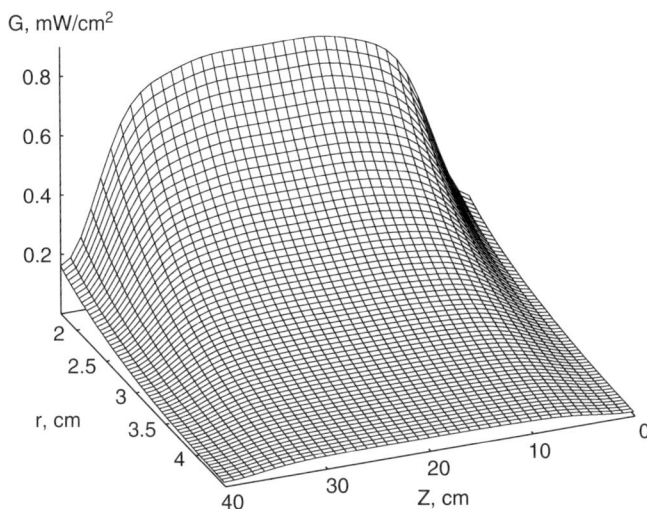

Figure 20 Distribution of the local incident radiation inside the reactor volume, for a catalyst concentration of $0.05\,\mathrm{g\,L^{-1}}$ (Adapted from Cuevas et al., 2007, with permission from Elsevier).

4.4. Applicability of the P1 approximation

The main limitation of the P1 approximation seems to be its applicability mainly to participating media with high optical depths (Cuevas et al., 2007). Optical depth is defined as

$$d_{\mathrm{opt}} = \beta_\lambda D = (\beta^*_{\lambda,\mathrm{catal}} C_{\mathrm{catal}} + \beta^*_{\lambda,\mathrm{dye}} C_{\mathrm{dye}})D \qquad (59)$$

where D is a characteristic spatial distance that radiation crosses inside the participating medium (reaction space); for example, the tube radius in a parabolic trough, the depth of a flat plate photoreactor, or the radial distance between the concentric tubes in an annular photoreactor. Optical depth equals the number of photon mean free paths that this characteristic distance spans. A photon traversing the medium would experiment on the average at least a number of collisions equal to this optical depth before reaching a wall. So, high optical depths imply a great deal of scattering, which is the necessary requirement to have almost isotropic intensity distribution, as required by the P1 approximation.

It would appear that the above requirement imposes strong limitations to the use of the P1 approximation, but this is really not so; in practice, the prevailing situation is operation of the reactors at high optical depths, based on the characteristic dimensions of reacting spaces and the typical catalyst concentrations used (Cuevas et al., 2007). This seems contradictory at first, because from the point of view of radiative transfer the optimal catalyst concentrations correspond to optical depths of order unity (Pasqualli et al., 1996). For smaller optical depths there is the risk that much of the radiation go out from the reactor without being absorbed, while for optical depths of order 10, radiation is absorbed almost immediately as it enters the reactor, leaving a large portion of the volume almost in the dark (Arancibia-Bulnes and Cuevas, 2004). In spite of this, much larger catalyst concentrations may be seen to give the maximum reaction rates, due to the need for more adsorption sites for the pollutants. A trade-off occurs: catalyst concentration should decrease in inverse proportion to the increase in the size of the reaction space, to maintain good illumination of the whole volume, while at the same time it should stay proportional to the pollutant concentration, to ensure availability of adsorption sites.

5. CONCLUSIONS AND PERSPECTIVES

This chapter reviews some of the main topics involved in the design and modeling of solar photocatalytic reactors, with particular emphasis on the authors' research experience. Solar photons are source of energy that initiates photocatalytic degradation. Thus, proper consideration of radiative processes is key to address this subject. The determination of the directional and spectral characteristics of solar UV radiation, the interaction of the catalyst with radiation inside reaction spaces, the optical design of solar collectors, and the optical properties of the materials involved are all subjects where these concepts are necessary. Therefore, developments in this area should be solidly grounded on the fields of solar collector optics and radiative transfer, besides the more traditional chemical engineering aspects involved. This requires a multidisciplinary approach.

Although different solar photoreactors have been developed in the last 20 years, each one with its own advantages and limitations, there is still room for new designs and innovative ideas. Nowadays, CPC appear as one of the most promising alternatives among solar photocatalytic reactors. The concepts of nonimaging optics, with its emphasis in efficient energy collection can also be a very useful in future developments.

The supporting of the TiO_2 in extended surfaces, glass beads, or other types of supports offers many possibilities for new designs and concepts. Modeling of radiative transfer in both slurry and fixed catalyst reactors is a necessary tool for the full understanding of the effect of the different design parameters in the performance of photoreactors. This should be accompanied with advances in material science, in order to develop improved catalysts that help reducing costs, making these processes economically viable.

ACKNOWLEDGMENTS

This work has been partially supported by CONACYT grants 56918, 49895-Y, UNAM grant 372311721 and PUNTA program. J. J. Quiñones Aguilar, R. Morán Elvira, M. L. Ramón García, I. Salgado, and F. Payán are acknowledged for technical support.

LIST OF SYMBOLS

A_a	aperture area of solar collector, m^2
A_n	constant
A_r	receiver area of solar collector, m^2
B_n	constant
c	speed of light in vacuum, $2.998 \times 10^8 \, m \, s^{-1}$
C_{catal}	catalyst concentration, $g \, L^{-1}$
C_{dye}	dye concentration, $g \, L^{-1}$
$C_{dye}^{(n)}$	dye concentration at time step n, $g \, L^{-1}$
C_n	normalization constant for the nth eigenfunction
C_n	constant
C_p	local pollutant concentration, mM
$C_{p,av}$	average pollutant concentration, mM
C_R	concentration ratio, suns
$C_{R,max}$	maximum solar concentration ratio of collector, suns
d_{opt}	optical depth
D	characteristic reactor spatial dimension, cm
e_L	local volumetric rate of photon absorption, $E \, m^{-3} \, s^{-1}$
$E_{available}$	accumulated available energy, $J \, m^{-3}$
$E_{collect}$	accumulated collected energy, $J \, m^{-3}$
E_n	constant
E_λ	photon energy, J
F_n	constant
f_n	normalized nth eigenfunction of an ordinary differential equation
F_1	function dependent on component concentration, $mM \, s^{-1}$
F_2	function depending on the LVREA
g_λ	asymmetry parameter
$G_{UV,global}$	global UV solar irradiance, $W \, m^{-2}$
$G_{UV,diffuse}$	diffuse UV solar irradiance, $W \, m^{-2}$
$G_{UV,beam}$	beam UV solar irradiance, $W \, m^{-2}$
$G_{UV,collect}$	collected UV solar irradiance, $W \, m^{-2}$
G_λ	local incident radiation, $W \, m^{-2} \, \mu m^{-1}$
h	Planck's constant, $6.626 \times 10^{-34} \, J \, s$
I_λ	radiation intensity, $W \, m^{-2} \, sr^{-1} \, \mu m^{-1}$
I_0, K_0	modified Bessel functions of order zero

I_n, K_n	modified Bessel functions of nth order
$k_{d,\lambda}$	radiation diffusion coefficient, cm^{-1}
$k_{O,i}$	kinetic parameter of the ith intermediate oxidation, mM^{-1}s^{-1}
k_e	electron–hole generation rate constant, s^{-1}
k_E	constant for electron capture by oxidant agent, s^{-1}
k_F	constant for hole capture by reducing agent, s^{-1}
k_R	electron–hole recombination rate constant, mM^{-1}s^{-1}
k_α	kinetic constant for pollutant oxidation, mM^{-1}s^{-1}
L	reactor tube length, cm
$\hat{n}$	unit vector normal to a surface
n_{imag}	imaginary part of the index of refraction of the aqueous dye solution
N_A	Avogadro's number, 6.022×10^{23} mol^{-1}
$Q^m_{abs-cat,\lambda}$	volumetric power absorbed by the catalyst, W m^{-2}μm^{-1}
Q_n	constant
$\mathbf{q}_\lambda$	radiative flux vector, W m^{-2}μm^{-1}
$q_{e,\lambda}$	external flux entering through the reactor wall, W m^{-2}μm^{-1}
$\mathbf{r}$	position vector, m
r	radial coordinate, m
r_0	radius of tubular receiver, m
$[R_i]$	concentration of the ith degradation intermediate, mM
R_n	constant
R_{opt}	reaction rate optical factor, cm^3
s	linear coordinate along propagation direction $\hat{s}$, m
$\hat{s}$	unit propagation direction vector
s_x, s_y, s_z	Cartesian components of direction vector
t	time, s
V_T	treated volume, m^3
V_R	illuminated (reactor) volume, m^3
x, y, z	Cartesian coordinates, m
x_n	scaled radial coordinate

Greek letters

β_λ	extinction coefficient of the medium, cm^{-1}
$\beta^*_{\lambda,catal}$	specific absorption coefficient of the catalyst, cm^{2}g^{-1}
$\beta^*_{\lambda,dye}$	specific absorption coefficient of the dye, cm^{2}g^{-1}
γ	lumped kinetic parameter, E$^{-1/2}$cm$^{3/2}$s$^{1/2}$
δ_n	nth eigenvalue of an ordinary differential equation
κ_λ	absorption coefficient of the medium, cm^{-1}
$\kappa^*_{\lambda,catal}$	specific absorption coefficient of the catalyst, cm^{2}g^{-1}
$\kappa^*_{\lambda,dye}$	specific absorption coefficient of the dye, cm^{2}g^{-1}
λ	wavelength, μm
θ	polar angle inside tubular reaction volume, rad

θ_s scattering angle, rad
ρ reflectance function of a surface
ρ_i ith moment of the reflectance function
ρ_P reflectance of the polyethylene wall
ρ_g reflectance of the glass wall
σ_λ scattering coefficient of the medium, cm^{-1}
$\sigma^*_{\lambda,catal}$ specific scattering coefficient of the catalyst, $cm^2\ g^{-1}$
τ surface transmittance function
φ polar angle around receiver tube, m
Φ_λ scattering phase function
$\Phi_{HG,\lambda}$ Henyey–Greenstein scattering phase function
ω_λ scattering albedo
Ω solid angle, sr

ABBREVIATIONS

CPC   compound parabolic collector
DOM   discrete ordinates method
LVRPA   local volumetric rate of photon absorption
MCM   Monte Carlo method
RTE   radiative transfer equation
TC   tubular collector
UV   ultraviolet
VTC   V-trough collector

REFERENCES

Alfano, O.M., Bahnemann, D., Cassano, A.E., Dillert, R., and Goslich, R. *Catal. Today* **58**, 199 (2000).
Alfano, O.M., Cabrera, M.I., and Cassano, A.E. *J. Catal.* **172**, 370 (1997).
Alpert, D.J., Sprung, J.L., Pacheco, J.E., Prairie, M.R., Reilly, H.E., Milne, T.A., and Nimlos, M.R. *Sol. Energy Mater.* **24**, 594 (1991).
Arancibia-Bulnes, C.A., Bandala, E.R., and Estrada, C.A. *Catal. Today* **76**, 149 (2002a).
Arancibia-Bulnes, C.A., Bandala, E.R., and Estrada, C.A. Radiation absorption in parabolic trough and CPC solar photocatalytic reactors. *in* A. Steinfeld (Ed.), "Proceedings of the 11th Solar PACES Symposium on Concentrated Solar Power and Chemical Energy Technologies", Sept. 4–6, 2001, Zurich, Paul Scherrer Institut, Zurich (2002b), p. 445.
Arancibia-Bulnes, C.A., and Cuevas, S.A. *Sol. Energy* **76**, 615 (2004).
Arancibia-Bulnes, C.A., and Ruiz-Suárez, J.C. *Appl. Opt.* **38**, 1877 (1999).
Bahnemann, D. *Sol. Energy* **77**, 445 (2004).
Bahnemann, D.W., Bockelmann, D., and Goslich, R. *Sol. Energy Mater.* **24**, 175 (1991).
Ballari, M.M., Brandi, R., Alfano, O., and Cassano, A. *Chem. Eng. J.* **136**, 50 (2008).
Bandala, E.R., Gelover, S., Leal, M.T., Arancibia-Bulnes, C., Jiménez, A., and Estrada, C.A. *Catal. Today* **76**, 189 (2002).
Bandala, E.R., Arancibia-Bulnes, C.A., Orozco, S.L., and Estrada, C.A. *Sol. Energy* **77**, 503 (2004).
Bandala, E.R., and Estrada, C. *J. Sol. Energy Eng.* **129**, 22 (2007).

Bedford, J., Klausner, J.F., Goswami, D.Y., and Schanze, K.S. *J. Sol. Energy Eng.* **116**, 8 (1994).

Blanco, J., Malato, S., Fernandez, P., Vidal, A., Morales, A., Trincado, P., Oliveira, J.C., Minero, C., Musci, M., Casalle, C., Brunotte, M., Tratzky, S., Dischinger, N., Funken, K.-H., Sattler, C., Vincent, M., Collares-Pereira, M., Mendes, J.F., and Rangel, C.M. *Sol. Energy* **67**, 317 (1999).

Blanco-Galvez, J., Fernández-Ibáñez, P., and Malato-Rodríguez, S. *J. Sol. Energy Eng.* **129**, 4 (2007).

Bockelmann, D., Weichgrebe, D., Goslich, R., and Bahnemann, D. *Sol. Energy Mater. Sol. Cells* **38**, 441. (1995).

Bohren, C.F., and Gilra, D.P. *J. Colloid Interface Sci.* **72**, 215 (1979).

Brandi, R.J., Alfano, O.M., and Cassano, A.E. *Chem. Eng. Sci.* **54**, 2817 (1999).

Brinker, J.C., and Scherer, G.W., "Sol-gel Science, The Physics and Chemistry of Sol-Gel Processing". Academic Press, New York (1990).

Brucato, A., Cassano, A.E., Grisafi, F., Montante, G., Rizzuti, L., and Vella, G. *AIChE J.* **52**, 3882 (2006).

Brucato, A., and Rizzuti, L. *Ind. Eng. Chem. Res.* **36**, 4748 (1997).

Bruscaglioni, P., Ismaelli, A., and Zaccanti, G. *Waves Random Media* **3**, 147 (1993).

Cabrera, M.I., Alfano, O.M., and Cassano, A.E. *J. Phys. Chem.* **100**, 20043 (1996).

Carvalho, M.J., Collares-Pereira, M., Gordon, J.M., and Rabl, A., *Sol. Energy* **35**, 393 (1985).

Cassano, A.E., and Alfano, O.M. *Catal. Today* **58**, 167 (2000).

Chaves, J., and Collares Pereira, M. *J. Sol. Energy Eng.* **129**, 16 (2007).

Cuevas, S.A., Arancibia-Bulnes, C., and Serrano, B. *Int. J. Chem. Reactor Eng.* **5**, A58 (2007).

Cuevas, S.A., Villafán, H.I., and Arancibia-Bulnes, C.A. in "Proceedings of the XIII International Materials Research Congress". Academia Mexicana de Ciencia de Materiales, México (2004).

Curcó, D., Giménez, J., Addardak, A., Cervera-March, S., and Esplugas, S. *Catal. Today* **76**, 177 (2002).

Curcó, D., Malato, S., Blanco, J., and Giménez, J. *Sol. Energy Mater. Sol. Cells* **44**, 199 (1996a).

Curcó, D., Malato, S., Blanco, J., Giménez, J., and Marco, P. *Sol. Energy* **56**, 387 (1996b).

Diaz, J., Rodríguez, J., Ponce, S., Solís, J., and Estrada, W. *J. Sol. Energy Eng.* **129**, 94 (2007).

Gelover, S., Leal, T., Bandala, E.R., Román, A., Jimenez, A., and Estrada C. *Water Sci. Technol.* **42**, 101 (2000).

Gelover, S., Mondragón, P., and Jiménez, A. *J. Photochem. Photobiol. A: Chem.* **165**, 241 (2004).

Gernjak, W., Maldonado, M.I., Malato, S., Cáceres, J., Krutzler, T., Glaser, A., and Bauer R. *Sol. Energy* **77**, 567 (2004).

Giménez, J., Curcó, D., Queral, M.A. *Catal. Today* **54**, 229 (1999).

Goslich, R., Dillert, R., and Bahnemann, D. *Water Sci. Technol.* **35**, 137 (1997).

Goswami, D.Y. Engineering of the solar photocatalytic detoxification and disinfection processes. in K.W. Boer (Ed.), "Advances in Solar Energy, Vol. 10". American Solar Energy Society, Boulder (1995), p 165.

Goswami, D.Y. *J. Sol. Energy Eng.* **119**, 101 (1997).

Goswami, D.Y., Kreith, F., and Kreider, J.F., "Principles of Solar engineering, 2nd Edn". Taylor & Francis, Philadelphia (2000), Chap. 10.

Gryglik, D., Miller, J.S., and Ledakowickz, S. *Sol. Energy* **77**, 615 (2004).

Guillard, C., Disdier, J., Monnet, C., Dussaud, J., Malato, S., Blanco, J., Maldonado, M.I., and Herrmann, J.-M. *Appl. Catal. B: Environ.* **46**, 319 (2003).

Hulstrom, R., Bird, R., and Riordan, C. *Sol. Cells* **15**, 365 (1985).

Ishimaru, A. "Wave Propagation and Scattering in Random Media". Oxford University Press, Oxford (1997).

Jetter, S.M. *Sol. Energy* **37**, 335 (1986).

Jetter, S.M. *Sol. Energy* **39**, 11(1987).

Jiménez, A.E., Estrada, C.A., Cota, A.D., and Román, A. *Sol. Energy Mater. Sol. Cells* **60**, 85 (2000).

Jiménez, A.E., and Salgado, I. in preparation (2008).

Jiménez González, A.E., and Gelover Santiago, S. *Semicond. Sci. Technol.* **22**, 709 (2007).

Jiménez-González, A.E., Salgado-Tránsito, I., Payán-Martínez, L.F., and Ramón-García, M.L., submitted to *J. Photochem. Photobiol. A: Chem.* (2009).

Lebedev, A.N., Gartz, M., Kreibig, U., and Stenzel, O. *Euro. Phys. J. D* **6**, 365 (1999).

Li Puma, G., and Brucato, A. *Catal. Today* **122**, 78 (2007).

Li Puma, G., Khor, J.N., and Brucato, A. *Environ. Sci. Technol.* **38**, 3737 (2004).

Li Puma, G., and Yue, P.L. *Chem. Eng. Sci.* **58**, 2269 (2003).

Mahan, J.R. "Radiation Heat Transfer: A Statistical Approach". Wiley, New York (2002).

Malato, S., Blanco, J., Richter, C., Curco, D., and Giménez, J. *Water Sci. Technol.* **35**, 157 (1997).

Malato, S., Blanco, J., Alarcón, D.C., Maldonado, M.I., Fernández-Ibáñez, P., and Gernjak, W. *Catal. Today* **122**, 137 (2007).

Malato Rodríguez, S., Blanco Gálvez, J., Maldonado Rubio, M.I., Fernández Ibañez, P., Alarcón Padilla, D., Collares Pereira, M., Farinha Mendes, J., Correia de Oliveira, J. *Sol. Energy* **77**, 513 (2004).

Marshak, R.E. *Phys. Rev.* **71**, 443 (1947).

McLoughlin, O.A., Fernández Ibañez, P., Gernjak, W., Malato Rodríguez, S., and Gill, L.W. *Sol. Energy* **77**, 625 (2004a).

McLoughlin, O.A., Kehoe, S.C., McGuigan, K.G., Duffy, E.F., Al Touati, F., Gernjak, W., Oller Abedrola, I., Malato Roidríguez, S., Gill, L.W. *Sol. Energy* **77**, 657 (2004b).

Meichtry, J.M., Lin, H.J., de la Fuente, L., Levy, I.K., Gautier, E.A., Blesa, M.A., Litter, M.I. *J. Sol. Energy Eng.* **129**, 119 (2007).

Minero, C., Pelizzetti, E., Malato, S., and Blanco, J. *Chemosphere* **26**, 2103 (1993).

Modest, M.F. "Radiative Heat Transfer", 2nd Edn. Academic Press, New York (2003).

Mundy, W.C., Roux, J.A., and Smith, A.M. *J. Opt. Soc. Am.* **64**, 1593 (1974).

Navntoft, C., Araujo, P., Litter, M.I., Apella, M.C., Fernández, D., Puchulu, M.E., Hidalgo, M.V., and Blesa, M.A. *J. Solar Energy Eng.* **129**, 127 (2007).

Okamoto, K., Yamamoto, Y., Tanaka, H., and Itaja, A. *Bull. Chem. Soc. Jpn* **58**, 2023 (1985).

Olver, F.W.J. *in* M. Abramowitz, I.A. Stegun (Eds.), "Handbook of Mathematical Functions". Dover, New York (1972).

Oyama, T., Aoshima, A., Horikoshi, S., Hidaka, H., Zhao, J., and Serpone, N. *Sol. Energy* **77**, 525 (2004).

Pasquali, M., Santarelli, F., Porter, J.F., and Yue, P.-L., *AIChE J.* **42**, 532 (1996).

Pichat, P., Vannier, S., Dussaud, J., and Rubis, J.-P. *Sol. Energy* **77**, 533 (2004).

Rabl, A. *Sol. Energy* **18**, 112 (1976).

Rabl, A. "Active Solar Collectors and their Applications". Oxford University Press, New York (1985).

Román Rodríguez, A. Master's degree Thesis, UNAM. México (in Spanish) (2001).

Romero, R.L., Alfano, O.M., and Cassano, A.E. *Ind. Eng. Chem. Res.* **42**, 2479 (2003).

Salaices, M., Serrano, B., and de Lasa, H.I. *Ind. Eng. Chem. Res.* **40**, 5455 (2001).

Salaices, M., Serrano, B., and de Lasa, H.I. *Chem. Eng. J.* **40**, 219 (2002).

Satuf, M.L., Brandi, R.J., Cassano, A.E., and Alfano, O.M. *Ind. Eng. Chem. Res.* **44**, 6643 (2005).

Spott, T., and Svaasand, L.O. *Appl. Opt.* **39**, 6453 (2000).

Sudiarta, I.W., and Chylek, P. *J. Quant. Spectrosc. Radiat. Transfer* **18**, 709 (2001).

Trujillo, F.J., Safinski, T., and Adesina, A.A. *J. Sol. Energy Eng.* **129**, 27 (2007).

Turchi, C.S., and Ollis, D.F. *J. Catal.* **122**, 178 (1990).

van Well, M., Dillert, R.H.G., Bahnemann, D.W., Benz, V.W., and Mueller, M.A. *J. Sol. Energy Eng.* **119**, 114 (1997).

Villafán-Vidales, H.I., Cuevas, S.A., and Arancibia-Bulnes, C.A. *J. Sol. Energy Eng.* **129**, 87(2007).

Winston, R., Miñano, J.C., and Benítez, P. "Nonimaging Optics". Elsevier Academic Press, Amsterdam (2005).

Yokota, T., Cesur, S., Suzuki, H., Baba, H., and Takahata, Y. *J. Chem. Eng. Jpn.* **32**, 314 (1999).

CHAPTER 7

Scaling-Up of Photoreactors: Applications to Advanced Oxidation Processes

Orlando M. Alfano[*] and **Alberto E. Cassano**

Contents			
	1.	Introduction	230
	2.	Scaling-up of a Photocatalytic Wall Reactor with Radiation Absorption and Reflection	234
		2.1 Reaction scheme and kinetic model	235
		2.2 Laboratory reactor: description and modeling	237
		2.3 Kinetics results	243
		2.4 Pilot scale reactor	243
		2.5 Reactor model	243
		2.6 The radiation field	246
		2.7 Validation	249
	3.	Scaling-Up of a Homogeneous Photochemical Reactor with Radiation Absorption	250
		3.1 Reaction scheme and kinetic model	250
		3.2 Laboratory reactor: description and modeling	251
		3.3 Kinetic results	256
		3.4 Pilot scale reactor	257
		3.5 Reactor model	258
		3.6 Radiation model	259
		3.7 Validation	262
	4.	Scaling-Up of a Heterogeneous Photocatalytic Reactor with Radiation Absorption and Scattering	263
		4.1 Reaction scheme and kinetic model	263
		4.2 Laboratory reactor: description and modeling	266
		4.3 Kinetic results	273
		4.4 Pilot scale reactor	277

Sections 2, 3, and 4 of this Chapter correspond to contributions that have been written with the following coworkers: Horacio A. Irazoqui, Carlos A. Martín, Rodolfo J. Brandi, Cristina S. Zalazar, Marisol D. Labas, Gustavo E. Imoberdorf, and María L. Satuf.

INTEC (Universidad Nacional del Litoral and CONICET), 3000 Santa Fe, Argentina

[*] Corresponding author.
E-mail address: alfano@santafe-conicet.gov.ar

Advances in Chemical Engineering, Volume 36
ISSN 0065-2377, DOI: 10.1016/S0065-2377(09)00407-4

4.5   Reactor model   277
4.6   Radiation model   279
4.7   Validation   280
5. Conclusions   282
Acknowledgments   283
Notation   283
References   286

1. INTRODUCTION

The analysis and design of homogeneous or heterogeneous photochemical reactors is a particular case of a chemical reactor in which case the radiation contribution of the conservation energy principle has two special characteristics: (i) it will always be present regardless of the operating temperature and (ii) since the majority of these reactions are carried out at a fixed, close to room temperature, and, usually, maintained under isothermal conditions, the radiation contribution can be uncoupled from the thermal energy equation. With this consideration, the modeling of the reactor based on first principles requires as usual, the solution of the momentum, thermal energy, and multicomponent mass conservation equations adding, in a photoreactor, the radiation energy contribution, in terms of a photon balance. However, it must be taken into account along the design, that the main objectives of its inclusion are the resulting kinetics effects derived by its dominant participation in initiating the reaction.

It is rather atypical that a photochemical reaction will proceed in a single molecular pathway. Thus, several elementary steps are involved. Normally, the majority of them are dark (thermal) reactions while, ordinarily, one activation step is produced by radiation absorption by a reactant molecule or a catalyst. From the kinetics point of view, dark reactions do not require a different methodological approach than conventional thermal or thermal-catalytic reactions. Conversely, the activation step constitutes the main distinctive aspect between thermal and radiation activated reactions. The rate of the radiation activated step is proportional to the absorbed, useful energy through a property that has been defined as the local volumetric rate of photon absorption, LVRPA (Cassano et al., 1995; Irazoqui et al., 1976) or the local superficial rate of photon absorption, LSRPA (Imoberdorf et al., 2005). The LVRPA represents the amount of photons that are absorbed per unit time and unit reaction volume and the LSRPA the amount of photons that are absorbed per unit time and unit reaction surface. The LVRPA is a property that must be used when radiation absorption strictly occurs in a well-defined three-dimensional (volumetrical) space. On the other hand, to treat wall-catalyzed photochemical reactions of superficial nature, the LSRPA is the most apt property to treat this special physical configuration.

The concept of absorbed useful energy is very important to avoid the frequent confusion of considering the radiation source output energy as equivalent to the useful absorbed photons. The second comprises the consideration of all the physical processes that are involved from the moment and place that a photon leaves the lamp boundaries to the time and position in which the photon is finally taken up by the radiation absorbing species.

The methodology to treat the activation step is much better understood if one resort to the typical approach used in chemical reaction engineering. Figure 1 illustrates the concept. Starting from the necessary first step, one writes a mass balance. Either in the differential equation or in its boundary conditions the reaction rate for each participating species must be formulated. When the expression for the reaction rate of the species under consideration involves the activation step, it necessarily involves the photon absorption rate (either volumetric or superficial). Its evaluation is performed stating first the radiative transfer equation (RTE) that requires the appropriate constitutive equations for radiation absorption, emission, scattering and seldom internal spontaneous or induced emission (Ozisik, 1973). In the most general case, the RTE is an integro-differential equation that involves all the constitutive equations for the comprised physical phenomena previously mentioned as well as the appropriate boundary conditions. For participating media, the constitutive equations invariably involve physical properties that depend of each particular case: absorption and scattering coefficients as well as the scattering phase function. Differently, absorption and

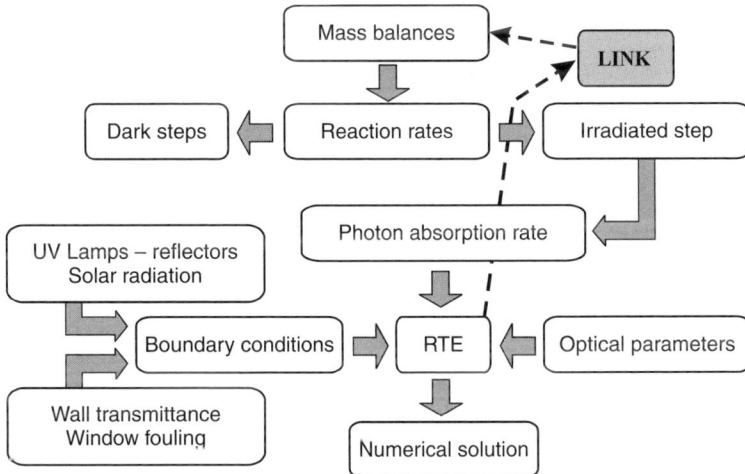

Figure 1 Evaluation of the photon absorption rate.

refection coefficients are needed for nonparticipating media. In its turn, the boundary conditions for the RTE must account for the existence of UV lamps and reflectors (or direct and diffuse radiation for solar reactors) as well as the effects of the reactor walls (for example, wall transmittance, including reflections and refractions, and sometimes, fouling of the radiation entrance window). The result is a completely defined RTE that will be applied to each reactor configuration. In the detailed description of the method in this chapter, the resulting RTE will be successively applied to a reaction space for a nonparticipating medium (Section 2), a homogeneous medium where there is only radiation absorption (Section 3), and a heterogeneous medium where radiation absorption and scattering are present (Section 4).

A very important point to emphasize is that the RTE is written considering two important aspects of the radiation propagation: (i) the direction of flight (movement in space) of the radiation beam and (ii) the energy transported by the photon traveling in such direction. Thus, the RTE is derived for a single propagating direction and a single wavelength. This is better realized resorting to the graphical illustration presented by Alfano and Cassano (2008). Figure 2a shows a photon distribution in directions and wavelengths (or frequencies) in an elementary volume V in space, bounded by a surface A having an outwardly directed normal $\underline{n}$. This figure illustrates the case of photons with indiscriminate directions and different wavelengths in the mentioned elementary volume. For example, white dots can represent a wavelength λ_n and black dots a wavelength λ_m. Then, in Figure 2b, photons with any direction and having a single wavelength (white dots or λ_n) are exemplified. Finally, Figure 2c illustrates photons having a single direction and a single wavelength (λ_n).

This visual conception helps to define the fundamental property for characterizing the radiation field in a photochemical reactor: the spectral

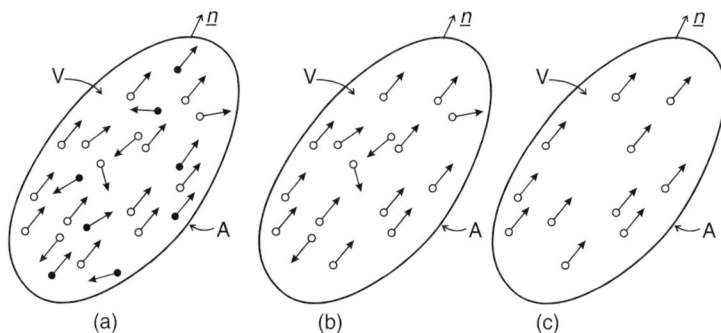

Figure 2 Characterization of the photon distribution in directions and wavelengths.

specific intensity. This radiation property is defined as the amount of energy per unit time, per unit normal area, per unit solid angle, and per unit wavelength interval:

$$I_{\underline{\Omega}\lambda}(\underline{x},t) = \lim_{(\Delta S,\Delta\Omega,\Delta t,\Delta\lambda)\to 0} \left(\frac{\Delta E_\lambda}{\Delta A \cos\theta \, \Delta\Omega \, \Delta t \, \Delta\lambda} \right) \qquad (1)$$

having units of Einstein s^{-1} m^{-2} sr^{-1} m^{-1} (or nm^{-1}). Note that the monochromatic (λ) radiation intensity is a function of position ($\underline{x}$), direction ($\underline{\Omega}$), and time (t). This means that its complete definition involves three spatial coordinates, two directional coordinates (usually in a spherical coordinate system), and a temporal coordinate. The property that is used to formulate the RTE is the starting point to calculate the Incident Radiation at any material point inside a reaction space or the Radiation Flux Density for wall-catalyzed photochemical reactions.

Employing the spectral specific intensity, it is then possible to write a photon balance in a bounded volume as follows (Cassano et al., 1995; Ozisik, 1973):

Time rate of change of ($\lambda,\underline{\Omega}$) photons in the volume V		Net flux of ($\lambda,\underline{\Omega}$) photons leaving the volume V across the surface A		Net gain of ($\lambda,\underline{\Omega}$) photons owing to emission, absorption, in- and out-scattering in the volume V
	$+$		$=$	

Considering the two source terms (emission and in-scattering) and the two sink terms (absorption and out-scattering) defined on the right-hand side of the previous expression, one can transform the preceding fundamental law in a more formal mathematical definition:

$$\frac{1}{c}\frac{\partial I_{\lambda,\underline{\Omega}}}{\partial t} + \underline{\nabla}\cdot(I_{\lambda,\underline{\Omega}}) = W^{em.}_{\lambda,\underline{\Omega}} - W^{abs.}_{\lambda,\underline{\Omega}} - W^{out\text{-}scatt.}_{\lambda,\underline{\Omega}} + W^{in\text{-}scatt.}_{\lambda,\underline{\Omega}} \qquad (2)$$

It is now possible to introduce two simplifications rationalized as follows: (i) since the selected applications will be usually operated at room temperatures, internal emission can be safely neglected ($W^{em.}_{\lambda,\underline{\Omega}} = 0$) and (ii) considering the magnitude of the light speed c, the first term is unequivocally neglected in all practical cases ($(1/c)(\partial I_{\lambda,\underline{\Omega}}/\partial t) \cong 0$.

As written, Equation (2) is the general formulation of a radiation balance for a single direction and a given amount of transported energy, but it is not very useful because we do not have sufficient information to evaluate the terms in its right-hand side. With the same approximation rigorously employed in continuous mechanics, the problem is solved with the introduction of a constitutive equation for each term, resorting to the measurement of the discernible behavior of each phenomenon and its relationship with the involved observable physical properties affecting each one of the listed events. With this consideration, the RTE takes the following form:

$$
\frac{dI_{\underline{\Omega},\lambda}}{ds}(s,t) + \underbrace{\kappa_\lambda(s,t)I_{\underline{\Omega},\lambda}(s,t)}_{\text{Absorption}} + \underbrace{\sigma_\lambda(s,t)\,I_{\underline{\Omega},\lambda}(s,t)}_{\text{Out-scattering}}
$$

$$
= \underbrace{\frac{\sigma_\lambda(s,t)}{4\pi} \int\limits_{\Omega'=4\pi} p_\lambda(\underline{\Omega}' \to \underline{\Omega})\, I_{\underline{\Omega},\lambda}(s,t)\, d\Omega'}_{\text{In-scattering}}
\tag{3}
$$

A significant part of the potential applications of photoreactors are related with their capacity to be used in different processes of air, water, and soil remediation and are recognized with the generic name of advanced oxidation technologies. They are characterized in the majority of the cases by resorting to reactions that produce hydroxyl radicals as the main oxidant agent. These practices will be used to illustrate the proposed scaling-up procedures.

The general methodology for all applications is the same. One must model precisely the laboratory reactor even if it is very small with almost the same tools that are used in the larger scale. Figure 3, depicts the different steps of the method. Advancing to the change of scale, the kinetic model for the reaction obtained in the laboratory experiments (and the same catalyst when corresponds) must be used. This introduces the strict requirement that the obtained kinetics must be a point value function, valid for any reactor shape and/or arrangement. It can be seen that in the above assertion there is an obvious exception: the photon absorption rate in the kinetic expression must be calculated from a different radiation balance derived for the particular geometry of the larger reactor. For this reason, it should not be unexpected that in changing scales the mass and radiation balances will almost always be very different, because they are very dependent on the new proposed, reactor size, shape, configuration, and operation. In what follows, it will be also shown that both the laboratory reactor and the large-scale reactor must be strictly using the same wavelength spectral distribution of the employed input power.

2. SCALING-UP OF A PHOTOCATALYTIC WALL REACTOR WITH RADIATION ABSORPTION AND REFLECTION

The degradation of PCE in an air stream with different degrees of relative humidity was used in this first example. It was performed in a multitubular reactor of annular cross section (Imoberdorf et al., 2007). However, the design of this reactor is the last operation in the procedure illustrated by Figure 3.

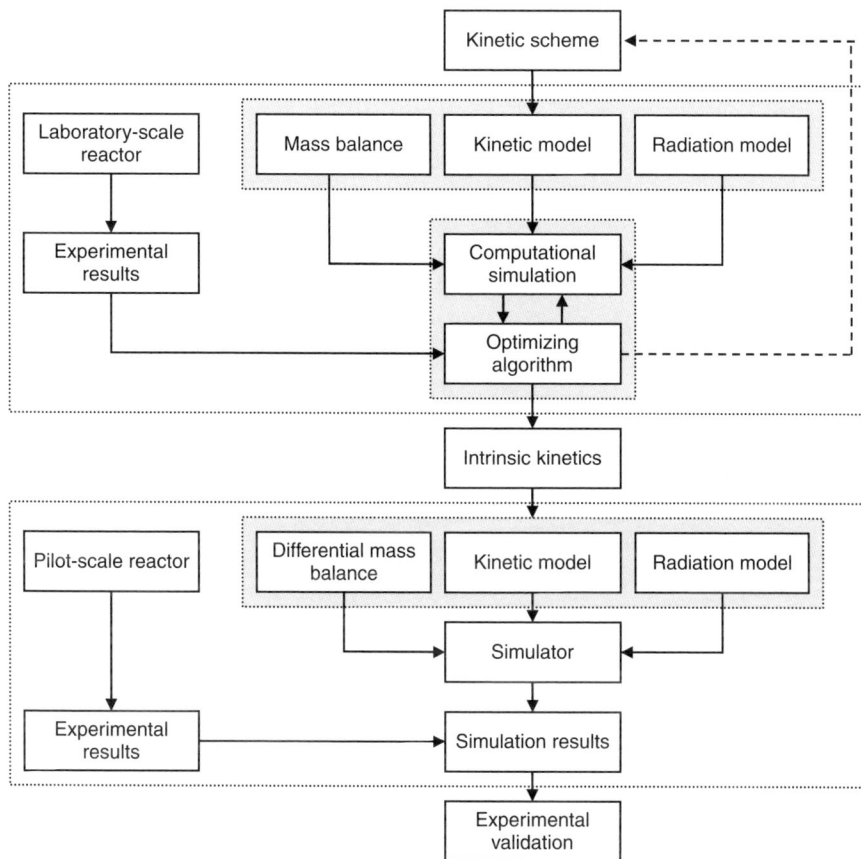

Figure 3 Scaling-up methodology.

2.1. Reaction scheme and kinetic model

The design of any form of photoreactor is greatly facilitated if a complete reaction sequence (even better if it is a true reaction mechanism) is known. On the basis of previous work, particularly the one reported by Yamazaqui and Araki (2002), the kinetic mechanism described in Table 1, was adopted. However, a complete reaction model and its kinetic parameters are needed. This is the first important step in the method.

To derive the kinetic model from the reaction scheme, the following assumptions were made: (i) the formation of OH• in steps 2 and 3 can be grouped in a single kinetic constant, (ii) the PCE degradation occurs through an elementary reaction involving the PCE attack by OH• radicals followed by a sequence of steps that leads to the generation of a chlorine atom and a chain reaction initiated by the attack of the Cl• on the PCE,

Table 1 Proposed kinetic scheme for the PCE degradation

Radiation activation	$TiO_2 \rightarrow TiO_2 + e^- + h^+$	R_g
OH• generation	$h^+ + H_2O_{ads} \rightarrow OH\cdot + H^+$	$\left\{ k_1 \right.$
	$h^+ + HO^-_{ads} \rightarrow OH\cdot$	
	$e^- + O_2 \rightarrow O_2^-\cdot$	k_2
Cl• generation	$C_2Cl_{4\ ads} + OH\cdot \rightarrow C_2Cl_4OH\cdot$	k_3
	$C_2Cl_4OH\cdot + O_2 \rightarrow C_2Cl_4OHOO\cdot$	k_4
	$2C_2Cl_4OHOO \rightarrow 2C_2Cl_4OHO\cdot + O_2$	k_5
	$C_2Cl_4OHO\cdot \rightarrow C_2Cl_3OHO + Cl\cdot$	k_6
Chain propagation	$C_2Cl_{4ads} + Cl\cdot \rightarrow C_2Cl_5\cdot$	k_7
	$C_2Cl_5\cdot + O_2 \rightarrow C_2Cl_5OO\cdot$	k_8
	$2C_2Cl_5OO\cdot \rightarrow 2C_2Cl_5O\cdot + O_2$	k_9
	$C_2Cl_5O\cdot \rightarrow CCl_2O + CCl_3\cdot$	k_{10}
	$C_2Cl_5O\cdot \rightarrow C_2Cl_4O + Cl\cdot$	k_{11}
	$CCl_3\cdot + O_2 \rightarrow CCl_3OO\cdot$	k_{12}
	$2CCl_3OO\cdot \rightarrow 2CCl_3O\cdot + O_2$	k_{13}
	$CCl_3O\cdot \rightarrow CCl_2O + Cl\cdot$	k_{14}
Termination reactions	$e^- + h^+ \rightarrow heat$	k_{15}
	$Cl\cdot + M \rightarrow products$	k_{16}
Phosgene hydrolysis (fast)	$CCl_2O + H_2O \rightarrow CO_2 + 2HCl$	k_{17}
Adsorption	$C_2Cl_4 + Sites \leftrightarrow C_2Cl_{4ads}$	K_{PCE}
	$H_2O + Sites \leftrightarrow H_2O_{ads}$	K_W

Adapted from Imoberdorf et al. (2005).

(iii) the PCE decomposition in the first step of the chain propagation is much faster than that involving the direct attack by OH• and this result is translated into the degradation rate, (iv) the net generation rate of free radicals, atomic species, free electrons, and holes is null, (v) the termination reactions of atomic Cl• can be grouped in a single reaction between chlorine atoms and water vapor, other radical and atomic species, reactor walls and other surfaces in the system, and (vi) surface concentrations of adsorbed PCE and water can be related to the gas phase concentrations through adsorption equilibrium constants. All these assumptions have shown to be good approximations in the abundant literature concerning TCE degradation mentioned by Imoberdorf et al. (2005). Assuming that the kinetic information will be obtained in the absence of mass transfer limitations, with these approximations, the following expression was obtained:

$$R_{PCE} = -\frac{\alpha_1 [PCE]_{gas} [H_2O]_{gas}}{(1 + K_{PCE}[PCE]_{gas} + K_W[H_2O]_{gas})^2}$$

$$\times \left(-1 + \sqrt{\frac{1 + K_{PCE}[PCE]_{gas} + K_W[H_2O]_{gas}}{[H_2O]_{gas}} \alpha_2 R_g(\underline{x}) + 1} \right) \tag{4}$$

where the kinetic parameters α_1 and α_2 were defined by

$$\alpha_1 = \frac{k_1 k_2 k_7 K_{PCE} K_W [Sites]^2 [O_2]}{2 k_{15} k_{16} [M]} \text{ and } \alpha_2 = \frac{4 k_{15}}{k_1 k_2 K_W [Sites][O_2]} \tag{5}$$

The individual specific rate and adsorption equilibrium constants are defined in Table 1. In Equations (4) and (5) [Sites]refer to the available concentration of sites for adsorption on the TiO_2 film, $[O_2]$ to the liquid phase oxygen concentration, [M] to the concentration of water, atomic or free radical species, reactor walls or other surfaces trapping atomic chlorine, and R_g to the superficial rate of electrons and holes generation.

The local superficial rate of electron–hole pair generation can be computed considering a wavelength averaged primary quantum yield for the generation of charge carriers on the catalytic surface $\bar{\Phi}$:

$$R_g(\underline{x}) = \bar{\Phi} \sum_\lambda e_\lambda^{a,s}(\underline{x}) = \bar{\Phi} e^{a,s}(\underline{x}) \tag{6}$$

The adopted average is needed because it is very difficult to obtain monochromatic primary quantum values. $e_\lambda^{a,s}(\underline{x})$ is the spectral LSRPA on the surface of the catalytic wall. Substituting Equation (6) into Equation (4) gives

$$R_{PCE}(\underline{x}, t) = -\frac{\alpha_1 [PCE]_{gas} [H_2O]_{gas}}{(1 + K_{PCE}[PCE]_{gas} + K_W[H_2O]_{gas})^2}$$

$$+ \left(-1 + \sqrt{\frac{1 + K_{PCE}[PCE]_{gas} + K_W[H_2O]_{gas}}{[H_2O]_{gas}} \alpha_2 \bar{\Phi} e^{a,s}(\underline{x}) + 1} \right) \tag{7}$$

From a plausible reaction sequence in Table 1 and reliable approximations, a local expression for the reaction kinetics in terms of observable and independent variables has been obtained. α_1 and α_2 are lumped kinetic parameters.

2.2. Laboratory reactor: description and modeling

The kinetics must be obtained in a reactor that should be as simple as possible with a well-characterized geometry in such a way that the obtained parameters are point value results of position and time. The laboratory device shown in Figures 4 and 5 is a continuous flow, well-mixed, recirculating reactor. The details are described in Table 2. The reader interested in

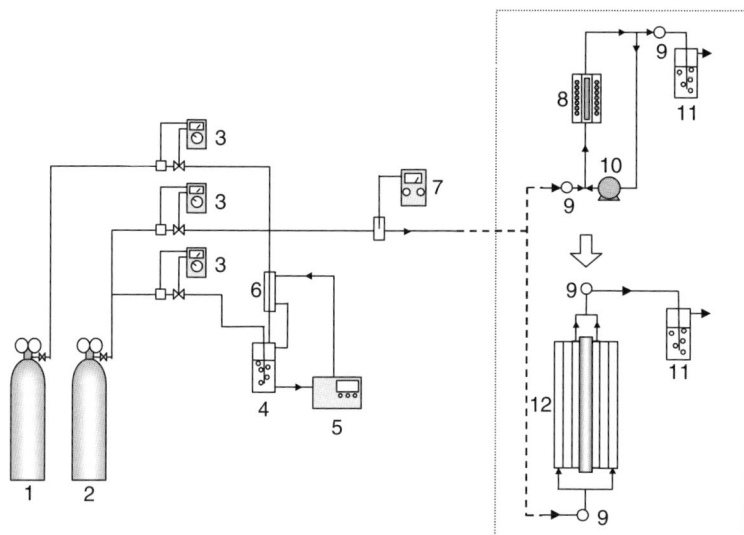

Figure 4 Laboratory and pilot scale photocatalytic reactors. Keys: (1) PCE + air, (2) air, (3) mass flowmeter, (4) air humidifier, (5) thermostatic bath, (6) heat exchanger, (7) thermohygrometer, (8) flat plate photoreactor, (9) sampling device, (10) recycle pump, (11) gas scrubber, (12) multiannular photocatalytic reactor.

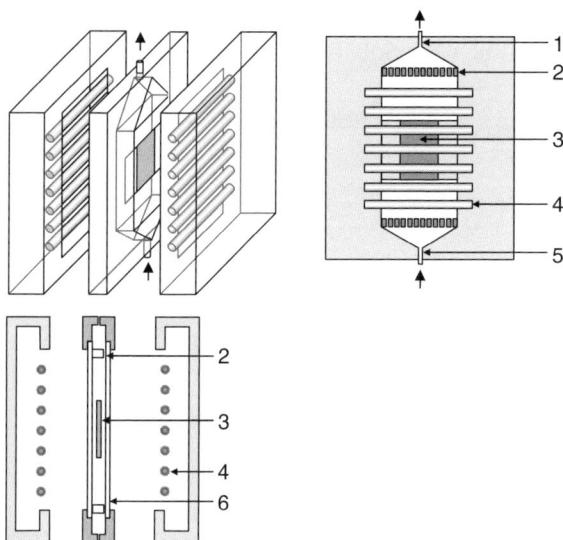

Figure 5 Schematic representation of the laboratory photoreactor. Keys: (1) gas outlet, (2) flow homogenizer, (3) photocatalytic plate, (4) UV lamps, (5) gas inlet, (6) acrylic windows.

Table 2 Laboratory and pilot scale reactor description and operating conditions

Item	Specification/Dimensions	
	Laboratory reactor	Pilot scale reactor
Type	Catalytic wall flat plate	Catalytic wall. Annular. Three concentric cylinders
Total surface	$81\ cm^2$	$5{,}209\ cm^2$
Dimensions	Two sides. Each: $x = 4.5\ cm^2 \times z = 9\ cm$	$Z_R = 48\ cm$, $R_{R,int} = 1.48\ cm$, $R_{R,ext} = 4.26\ cm$
Annulus I		$\chi_1 R_1 = 1.69\ cm$, $R_1 = 2.31\ cm$
Annulus II		$\chi_2 R_2 = 2.51\ cm$, $R_2 = 3.30\ cm$
Annulus III		$\chi_3 R_3 = 3.53\ cm$, $R_3 = 3.94\ cm$
Lamps	Seven in each side	One
	Philips TL 4W/08 F4T5/BLB	Philips TL 18W/08 F4TS/BLB
	Output power: 0.5 W each	Out put power: 3.5 W
	$\Delta\lambda = 300\text{–}420\ nm$	$\Delta\lambda = 300\text{–}420\ nm$
	$Z_L = 13.6\ cm$, $R_L = 0.8\ cm$	$Z_L = 59\ cm$, $R_L = 1.4\ cm$
Catalyst TiO_2	Sol-gel deposition	Sol-gel deposition
Operation Continuous		
Feed flow rate	With recirculation: $2{,}000\ cm^3\ min^{-1}$ $20\text{–}200\ cm^3\ min^{-1}$	$120\text{–}1{,}800\ cm^3\ min^{-1}$

Table 2 (*Continued*)

Specification/Dimensions	Laboratory reactor	Pilot scale reactor
Temperature	20°C	20°C
Pressure	1 atm.	1 atm.
Inlet PCE concentration	10–30 mg m^{-3}	50 mg m^{-3}
Relative humidity	10–100%	10–90%
Irradiation level	Local net radiation flux on the reacting surface: $q = 4 \times 10^{-3}$ W cm^{-2}(100%)	LSRPA ($e^{a,s}$): From 1.6×10^{-9} to 1×10^{-11} Einstein cm^{-2} s^{-1}
Variation	(24–100%)	

Adapted from Imoberdorf et al. (2005, 2006, 2007).

more experimental information such as catalyst preparation, analytical procedures, start-up method, reaction times, conditions to exclude mass transfer limitations, and so on, can resort to the original work (Imoberdorf et al., 2005).

The mass conservation equation is very simple:

$$\langle R_{PCE} \rangle_{A_R} = \frac{Q^{in}\left(\langle C_{PCE} \rangle^{out} - \langle C_{PCE} \rangle^{in} \right)}{A_R} \tag{8}$$

Considering the possibility of nonuniform coating (not occurring here) and nonuniform irradiation, the results are expressed in terms of an average value:

$$\langle R_{PCE} \rangle_{A_R} = \frac{1}{A_R} \int_{x=0}^{x=Xr} \int_{z=0}^{z=Zr} R_{PCE}(x,z) dz dx \tag{9}$$

It is clear that a model for the radiation field is required. It should be taken into account that PCE does not absorb radiation in the wavelength range of the employed radiation sources. Thus, the space surrounding the catalytic walls is not participative. According to Figure 6, the absorbed radiation must be calculated according to

$$e_\lambda^{a,s}(x,z) = (q_{dir,\lambda}^i - q_{dir,\lambda}^t) + (q_{ind,\lambda}^i - q_{ind,\lambda}^t) \tag{10}$$

where

$$q_\lambda(x,z) = \underline{n}_g \cdot \underline{q}_\lambda = \int_{\Omega_L} I_\lambda(x,y,\underline{\Omega})\underline{\Omega} \cdot \underline{n}_g d\Omega \tag{11}$$

To solve Equation (11) one can resort to the three-dimensional source with superficial emission model (Cassano et al., 1995) and the ray tracing technique (Siegel and Howell, 2002). The integration limits depend on the geometry and dimensions of the reacting system and the set of the 14 employed lamps (Figures 5 and 6).

$$q_\lambda(r,z) = \sum_{i=1}^{n} \int_{\phi_{min,L_i}}^{\phi_{max,L_i}} \int_{\theta_{min,L_i}(\phi)}^{\theta_{max,L_i}(\phi)} I_\lambda(x,y,\phi,\theta)\sin^2\theta \sin\phi\, d\theta d\phi \tag{12}$$

The boundary conditions result

$$I_\lambda(r,z,\phi,\theta) - \begin{cases} 0 & (\phi,\theta) < (\phi_{min,i}, \theta_{min,i}) \\ I_{\lambda,L} = \dfrac{P_{\lambda,L}}{2\pi^2 R_L Z_L}(\phi_{min,i}, \theta_{min,i}) < (\phi,\theta) < (\phi_{max,i}, \theta_{max,i}) \\ 0 & (\phi_{max,i}, \theta_{max,i}) < (\phi,\theta) \end{cases} \tag{13}$$

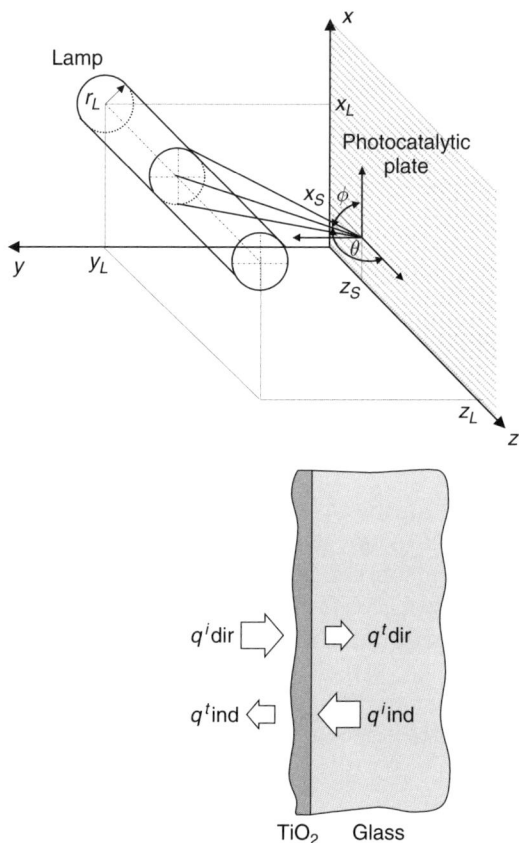

Figure 6 Coordinate system for the radiation model of the laboratory photoreactor.

The limiting values for θ and ϕ can be calculated from geometrical considerations of the lamps and reactor dimensions and their relative positions. The final result is

$$
e^{a,s}(x, z) = \frac{P_{\lambda,L}}{2\pi^2 R_L Z_L} \sum_\lambda \sum_{\ell=1}^{7} \int_\theta \int_\phi \exp\left(-\frac{\kappa_{\lambda,a}\ell_a}{\cos\theta_n}\right) \left[1 - \exp\left(-\frac{\kappa_{\lambda,f}\ell_f}{\cos\theta_n}\right)\right]
$$

$$
\times \sin^2\phi \sin\theta \, d\phi \, d\theta
$$

$$
+ \frac{P_{\lambda,L}}{2\pi^2 R_L Z_L} \sum_\lambda \sum_{\ell=8}^{14} \int_\theta \int_\phi \exp\left(-\frac{\kappa_{\lambda,a}\ell_a}{\cos\theta_n} - \frac{\kappa_{\lambda,g}\ell_g}{\cos\theta_n} - \frac{\kappa_{\lambda,f}\ell_f}{\cos\theta_n}\right)
$$

$$
\times \left[1 - \exp\left(-\frac{\kappa_{\lambda,f}\,\ell_f}{\cos\theta_n}\right)\right] \sin^2\phi \sin\theta \, d\phi \, d\theta
$$

(14)

2.3. Kinetics results

The photocatalytic degradation of PCE was studied in the laboratory photo-reactor depicted in Figure 5, for different values of PCE inlet concentrations, relative humidities, and irradiation levels (Imoberdorf et al., 2005). It was experimentally found that the PCE reaction rate shows (i) first-order kinetics with respect to the PCE concentration in the gas phase, (ii) linear dependence with respect to the irradiation level, and (iii) site-competitive kinetics for the dependence with the relative humidity. Thus, Equation (7) is reduced to a rather simple analytic expression:

$$R_{PCE}(\underline{x}, t) = -\alpha \frac{[PCE(x,t)]_{gas}}{1 + K_W[H_2O(x,t)]_{gas}} e^{a,s}(\underline{x}, t) \tag{15}$$

with

$$\alpha = \frac{\alpha_1 \alpha_2 \bar{\Phi}}{2} = \frac{k_7 K_{PCE}[Sites]\bar{\Phi}}{k_{16}[M]} \tag{16}$$

Comparing the experimental data with the kinetic model (reaction scheme, mass balance, and radiation model) and resorting to a nonlinear regression procedure, the kinetic parameters can be obtained. The results are given in Table 3. Figure 7 shows the quality of the results.

2.4. Pilot scale reactor

The reactor is sketched in Figure 8 where three concentric annular spaces with the six catalytic walls are shown. The reactor is fed through the outer annular space (maximum PCE concentration) where the radiation field has its minimum value. Exit of reactants and products occurs from the inner annular space. All details of the reactor assembly and operating conditions are described in Table 2. For more details the reader can resort to references (Imoberdorf et al., 2006, 2007).

2.5. Reactor model

In this case the reactor is very different and it operates in a continuous fashion. Under the employed experimental conditions the maximum

Table 3 Reaction kinetic parameters for PCE decomposition

Parameter	Value	95% Confidence interval	Units
α	1.54×10^8	0.19×10^8	cm^3 Einstein^{-1}
K_w	3.21×10^{-4}	0.48×10^{-4}	m^3 mg^{-1}

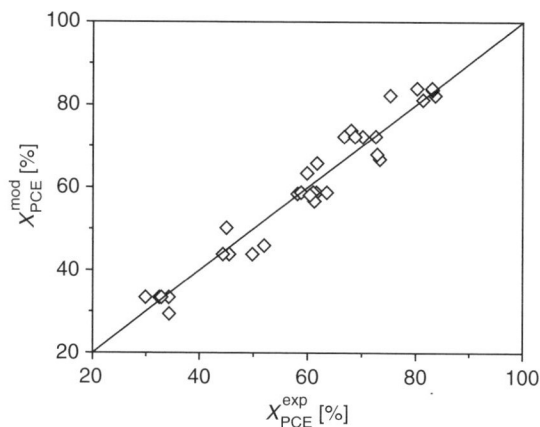

Figure 7 Experimental and predicted outlet conversions for the laboratory photoreactor.

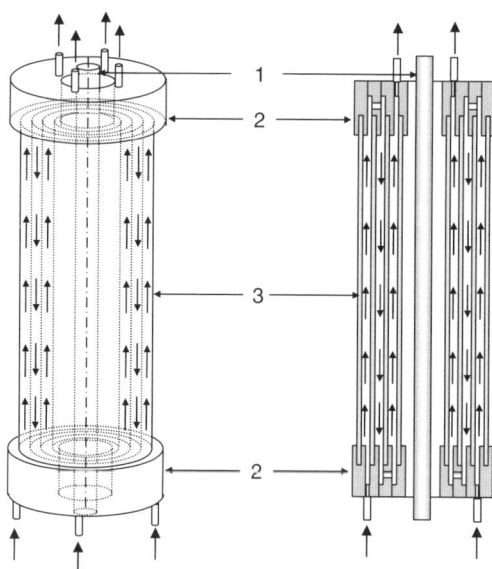

Figure 8 Schematic representation of the pilot scale photocatalytic reactor. Keys: (1) UV lamp, (2) distribution heads, (3) borosilicate glass tubes.

Reynolds number was 25. Therefore, the reactor is operating under well-defined laminar flow conditions. It is also clear that diffusive fluxes of PCE (the only species that experiments significant changes in concentration) can not affect the velocity profile because of the very low contaminant concentrations. The velocity profile for the j annular space results

$$
v_{z,j}(r) = (-1)^{j+1} \frac{2Q}{\pi R_j^2} \frac{\ln \chi_j}{\left[\left(1 - \chi_j^4\right)\ln \chi_j + \left(1 - \chi_j^2\right)^2\right]}
$$
$$
\times \left[1 - \left(\frac{r}{R_j}\right)^2 - \frac{(1 - \chi_j^2)}{\ln \chi_j}\ln\left(\frac{r}{R_j}\right)\right] \quad j = 1, 2, 3 \tag{17}
$$

It can be noticed that the momentum balance was not required in the laboratory reactor due to its particular operating conditions.

The mass transfer equation is written in terms of the usual assumptions. However, it must be considered that because the concentration of the more abundant species in the flowing gas mixture (air), as well as its temperature, are constant, all the physical properties may be considered constant. The only species that changes its concentration along the reactor in measurable values is PCE. Therefore, the radial diffusion can be calculated as that of PCE in a more concentrated component, the air. This will be the governing mass transfer mechanism of PCE from the bulk of the gas stream to the catalytic boundaries and of the reaction products in the opposite direction. Since the concentrations of nitrogen and oxygen are in large excess they will not be subjected to mass transfer limitations. The reaction is assumed to occur at the catalytic wall with no contributions from the bulk of the system. Then the mass balance at any point of the reactor is

$$
\frac{\partial C_{PCE}(r,z)}{\partial z}v_{z,j}(r) = \frac{D^0_{PCE-Air}}{r}\frac{\partial}{\partial r}\left(r\frac{\partial C_{PCE}(r,z)}{\partial r}\right)
$$
$$
(0 < z < Z_R; \quad \chi_j R_j < r < R_j; \quad j = 1,2,3) \tag{18}
$$

with the boundary conditions:

$$
D^0_{PCE-Air}\frac{\partial C_{PCE}(r,z)}{\partial r}\bigg|_{r=R_j} = R_{PCE}[C_{PCE}(R_j,z), C_{H_2O}, e^{a,s}(R_j,z)]
$$
$$
(0 < z < Z_R; \quad j = 1,2,3) \tag{19}
$$
$$
D^0_{PCE-Air}\frac{\partial C_{PCE}(r,z)}{\partial r}\bigg|_{r=\chi_j R_j} = -R_{PCE}[C_{PCE}(\chi_j R_j,z), C_{H_2O}, e^{a,s}(\chi_j R_j,z)]
$$
$$
(0 < z < Z_R; \quad j = 1,2,3)
$$

$$C_{PCE}(r,z)\big|_{z=0} = C_{PCE}^0; \qquad\qquad (\chi_3 R_3 < r < R_3)$$

$$C_{PCE}(r,z)\big|_{z=Z_R} = \frac{\displaystyle\int_{\chi_3 R_3}^{R_3} C_{PCE}(r, Z_R) v_{z,3}(r) r dr}{\displaystyle\int_{\chi_3 R_3}^{R_3} v_{z,3}(r) r dr}; \qquad (\chi_2 R_2 < r < R_2) \qquad (20)$$

$$C_{PCE}(r,z)\big|_{z=0} = \frac{\displaystyle\int_{\chi_2 R_2}^{R_2} C_{PCE}(r, 0) v_{z,2}(r) r dr}{\displaystyle\int_{\chi_2 R_2}^{R_2} v_{z,2}(r) r dr}; \qquad (\chi_1 R_1 < r < R_1)$$

It can be observed that the differences between the set of Equations (18)–(20) with respect to Equation (8) are evident. It is also important to recognize that in this mathematical modeling it can be observed the correct formulation of the nature of the catalytic wall reaction and its spatial dependence with respect to the position variables.

The PCE conversion at the reactor outlet is calculated with the mixing cup average concept:

$$X_{PCE}[\%] = \left(1 - \frac{\displaystyle\int_{\chi_1 R_1}^{R_1} C_{PCE}(r, Z_R) v_{z,1}(r) r dr}{C_{PCE}^0 \displaystyle\int_{\chi_1 R_1}^{R_1} v_{z,1}(r) r dr}\right) \times 100 \qquad (21)$$

At this point a third important difference must be noticed. The value of R_{PCE} in Equation (19) is not exactly equal to Equation (15) because the value of $e^{a,s}$ will be utterly different from one reactor and the other. This aspect will be considered in the next section.

2.6. The radiation field

The concepts employed in developing Equations (11)–(14) can be described in a simpler way here, because this reactor has a single lamp (Figure 9). The calculus of the LSRPA is made as follows:

$$e_\lambda^{a,s}(r,z) = q_\lambda^i(r,z) - q_\lambda^t(r,z) \qquad (22)$$

with

$$q_\lambda(r,z) = \int_{\phi_{min}(r)}^{\phi_{max}(r)} \int_{\theta_{min}(r,z,\phi)}^{\theta_{max}(r,z,\phi)} I_{\lambda,L} \cos\phi \, \sin^2\theta \, d\theta d\phi \qquad (23)$$

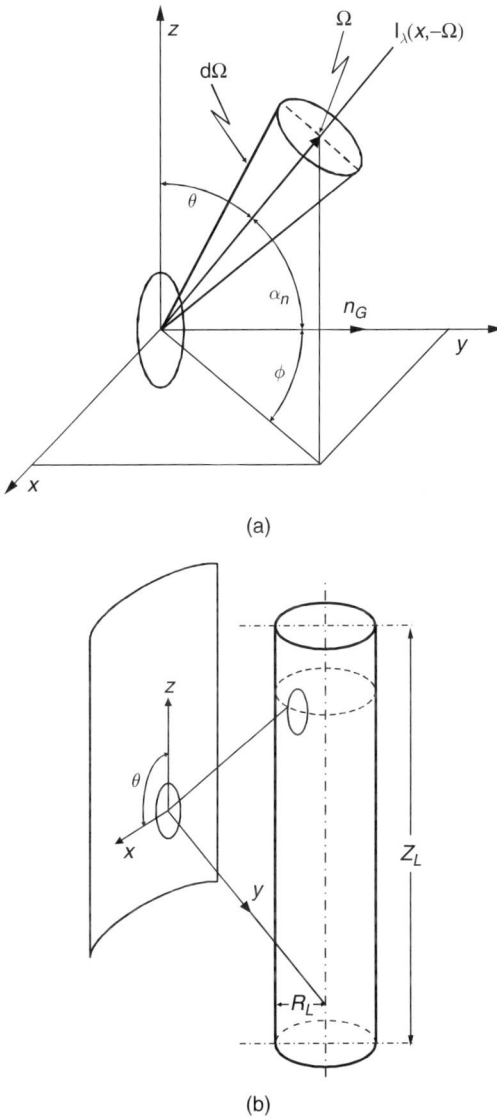

Figure 9   Coordinate system for the radiation model of the pilot scale photoreactor. Adapted from Imoberdorf et al. (2006).

where the limits are defined by the dimensions of the lamp contour as seen from the point of incidence according to the relative distances between the lamp and the reactor (Figure 9):

$$\theta_{min}(r, z, \phi) = \tan^{-1}\left\{\frac{r\cos\phi - [R_L^2 - r^2\sin^2\phi]^{1/2}}{(Z_L - z)}\right\} \qquad (24)$$

$$\theta_{max}(r, z, \phi) = \tan^{-1}\left\{\frac{r\cos\phi - [R_L^2 - r^2\sin^2\phi]^{\frac{1}{2}}}{z}\right\} \tag{25}$$

$$\phi_{min} = -\cos^{-1}\left[\frac{(r^2 - R_L^2)^{\frac{1}{2}}}{r}\right] \qquad \phi_{max} = \cos^{-1}\left[\frac{(r^2 - R_L^2)^{\frac{1}{2}}}{r}\right] \tag{26}$$

And the boundary conditions result

$$I_\lambda(r, z, \phi, \theta) = \begin{cases} 0 & (\phi, \theta) < (\phi_{min}) \\ I_{\lambda,L} = \dfrac{P_{\lambda,L}}{2\pi^2 R_L Z_L} & (\phi_{min}) < (\phi, \theta) < (\phi_{max}) \\ 0 & (\phi_{max}) < (\phi, \theta) \end{cases} \tag{27}$$

The value of the LSRPA in the thin catalytic film is finally obtained as follows (Table 4):

$$
\begin{aligned}
e^{a,s}(r, z) &= \frac{P_{\lambda,L}}{2\pi^2 R_L Z_L} \sum_{\lambda=300\,nm}^{420\,nm} \int_{\phi_{min}(r)}^{\phi_{max}(r)} \int_{\theta_{min}(r,z,\phi)}^{\theta_{max}(r,z,\phi)} \\
&\times \exp\left(-n_g(r)\frac{\kappa_{\lambda,g}e_g}{\cos\alpha_n} - n_f(r)\frac{\kappa_{\lambda,f}e_f}{\cos\alpha_n}\right) \\
&\times \left[1 - \exp\left(-\frac{\kappa_{\lambda,f}e_f}{\cos\alpha_n}\right)\right] \cos\phi \sin^2\theta\,d\theta\,d\phi
\end{aligned}
\tag{28}
$$

Variables are defined in the nomenclature section. Differing from the case of the flat plate reactor where the radiation field in the employed part of the surface was almost uniform (differences were never larger than 9%), in this reactor the whole system exhibit only azimuthal symmetry and significant differences were observed in both directions (r, z). However, even more important is the contrast of the LSRPA between the value in the inner annulus and the outer one. This result must be used in Equation (19).

Table 4   Keys to interpret the values of the parameters employed in Equation (28)

Geometry of the multiannular reactor	Space #	$n_g(r)$	$n_f(r)$
First annular space. Inner radius	1	1	0
First annular space. Outer radius	1	1	1
Second annular space. Inner radius	2	2	2
Second annular space. Outer radius	2	2	3
Third annular space. Inner radius	3	3	4
Third annular space. Outer radius	3	3	5

2.7. Validation

The experimental conversion is calculated according to

$$X_{PCE}\% = \frac{C_{PCE}^{in} - C_{PCE}^{out}}{C_{PCE}^{in}} \times 100 \tag{29}$$

Experimental values obtained in this way must be compared with those calculated according to Equation (21). Predicted versus experimental values are compared in Figure 10. The root mean square error (RMSE) is less than 5.6%. It should be noted that no adjustable parameters have been employed and the method can be used for any reactor size without limitations.

As mentioned in Imoberdorf et al. (2007) it is very difficult that a practical application will use annular spaces thicker than the ones employed in this reactor without producing mass transfer limitations. Similar effects should be expected if lamps of larger output power were employed. It is also apparent that no additional annular spaces are needed because almost no UV radiation is transmitted by the outer reactor tube. The only variables that are available to the design engineers are (i) the reactor length Z_R, (ii) the flow rate, taken into account by v_z, and (iii) larger lamps (defined by $P_{\lambda,L}$, R_L, and Z_L). All these variables are included in the design equations previously presented in Sections 2.5 and 2.6 of this work.

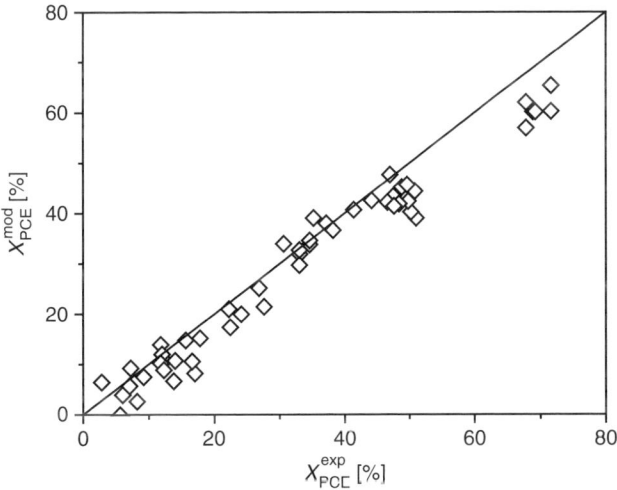

Figure 10 Experimental and predicted outlet conversions for the pilot scale photoreactor.

3. SCALING-UP OF A HOMOGENEOUS PHOTOCHEMICAL REACTOR WITH RADIATION ABSORPTION

This case will be illustrated with a rather simple reaction in liquid phase: the decomposition of low concentrations of formic acid in aqueous solutions employing hydrogen peroxide and UVC radiation (253.7 nm). Formic acid is a byproduct of the degradation of many organic compounds (Labas et al., 2002).

3.1. Reaction scheme and kinetic model

The same considerations made before are valid for this case and it is very important to have an available validated reaction mechanism. It can be obtained from three main sources (Blelski et al., 1985; Buxton et al., 1988; Stefan and Bolton, 1998) and it is shown in Table 5. With the available information about the constant k_2, k_3, k_5, k_6, and k_7, it could be possible to solve a system of four differential equations and extract from the experimental data, the missing constants ϕ and k_4 (that in real terms is $k_4^\star/C_{O_2}$). This method would provide good information about the kinetic constants, but it is not the best result for studying temperature effects if the same information is not available for the pre-exponential factors and the activation energies. Then, it is better to look for an analytical expression even if it is necessary to make some approximations. This is particularly true in this case, where the direct application of the micro steady-state approximation (MSSA) is more difficult due to the existence of a recombination step that includes the two free radicals formed in the reaction. From the available information, it is possible to know that to calculate the pseudo-steady-state

Table 5 Proposed kinetic scheme for formic acid mineralization

Activation	
$H_2O_2 + h\nu \rightarrow 2OH\bullet$	ϕ
Main reaction	
$H_2O_2 + OH\bullet \rightarrow HO_2\bullet + H_2O$	k_2
$HO_2\bullet + H_2O_2 \rightarrow H_2O + O_2 + OH\bullet$	k_3
$OH\bullet + HCOOH \xrightarrow{C_{O_2}} CO_2 + H_2O + HO_2\bullet$	k_4
Termination steps	
$HO_2\bullet + OH\bullet \rightarrow H_2O + O_2$	k_5
$HO_2\bullet + HO_2\bullet \rightarrow H_2O_2 + O_2$	k_6
$OH\bullet + OH\bullet \rightarrow H_2O_2$	k_7

$k_4 = k_4^\star/C_{O_2}$; $C_{O_2} =$ constant $\cong$ 7.5–8.0 ppm.
Adapted from Labas et al. (2002).

concentration of the OH· radicals, a good presumption is to consider that step 5 can be neglected (because k_5 and HO_2· are relatively small), but the same cannot be done to calculate the concentration of HO_2·. Also, it is widely recognized (Buxon et al., 1988; Stefan and Bolton, 1998) that the main attack to formic acid is produced by the OH· radical as compared with the one produced by the HO_2· radical. Applying the MSSA the following equations are obtained:

$$R_F = - \frac{\phi e_\lambda^a}{1/2 + r(\alpha_1)} \tag{30}$$

$$R_P = - \phi e_\lambda^a - \frac{\phi e_\lambda^a}{1 + (1/\alpha_1)^{\frac{1}{r}}} \tag{31}$$

where $r = C_P/C_F$ and $\alpha_1 = k_4/k_2$. Here $e_\lambda^a(\underline{x}, t)$ is the LVRPA a function of position and time because (i) the radiation field is intrinsically not uniform and the radiation absorbing species concentration changes with time. In this case, since there is no direct photolysis of formic acid (negligible radiation absorption at 253.7 nm) the chemical species that is responsible for the initiation reaction is hydrogen peroxide (see Table 5).

3.2. Laboratory reactor: description and modeling

It is always important to design a laboratory reactor that has the simplest mathematical representation for both the mass and the radiation balance. In this case the work was carried out in a flat plate reactor with circular windows made of quartz. A removable shutter permits to obtain steady-state operation of the whole system (including lamps) before the run commences. The start of the reaction ($t = 0$) occurs when it is taken off. Other features are described in Figures 11 and 12 and Table 6. Details on all the experimental procedure can be found in Labas et al. (2002).

A simple mass balance can be obtained for the recycling system when the following conditions are fulfilled: (i) the whole system operates under well-stirred conditions, (ii) the ratio V_R/V_T is $<< 1$, and (iii) the recirculating flow rate is high such as to have differential conversion per pass in the photoreactor and, at the same time, improve mixing. Then, it can be shown (Cassano and Alfano, 2000) that the changes in concentration in the tank are related to the reaction rates according to Equations (30) and (31)

$$\left. \frac{dC_j(t)}{dt} \right|_{\text{Tank}} = \frac{V_R}{V_T} \langle R_{\text{Hom.},j,\lambda} (\underline{x}, t) \rangle_{V_R} \tag{32}$$

With the initial condition that $C_j(t = 0) = C_j^0$

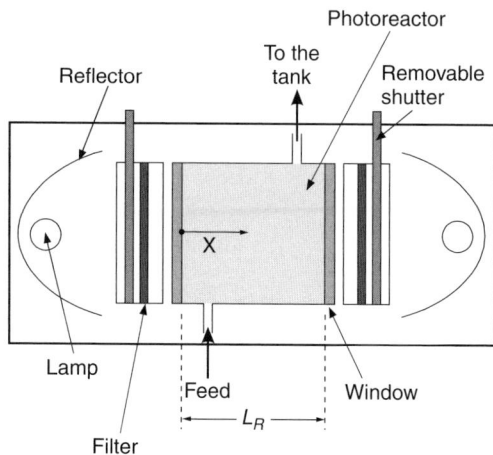

Figure 11   Laboratory reactor.

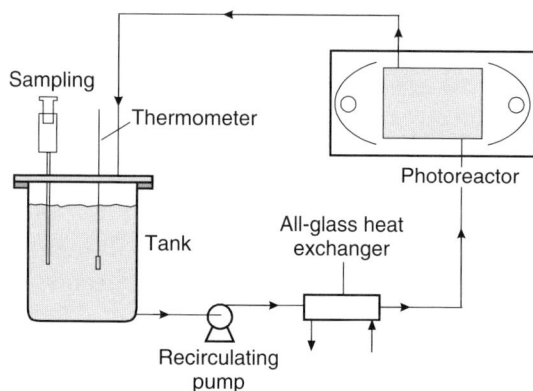

Figure 12   Schematic representation of laboratory reactor set-up.

In spite of the prevailing well-stirred conditions, since the radiation field is not uniform, the volume average of the homogeneous rate must be calculated. It will be seen in what follows, that the significant variable is x and the averaging integral is

$$\langle R_{\text{Hom}.j,\lambda}\,(\underline{x},t)\rangle_{V_R} = \langle R_{\text{Hom}.j,\lambda}(x,t)\rangle_{L_R} = \frac{1}{L_R}\int_{L_R} R_{\text{Hom}.j,\lambda}(x,t)\mathrm{d}x \qquad (33)$$

The reaction rate requires the evaluation of the LVRPA. In the absence of emission and scattering, the transport of photons in the

Table 6 Laboratory and pilot scale reactor description and operating conditions

Item	Specification/Dimensions	
	Laboratory reactor	Pilot scale reactor
Type	Flat plate. Batch, with a recycle. Two circular windows made of quartz, Suprasil quality	Annular shape. Continuous flow, tubular reactor in two sections of one meter each
Total reactor volume	69.9 cm³	13,000 cm³
Dimensions	Two sides. Each: $D = 4.26$ cm $L_R = 4.9$ cm	$Z_R = 200$ cm, $R_{int} = 2.4$ cm, $R_{ext} = 5.35$ cm $R_{R,int} = 2.5$ cm, quartz $R_{R,ent} = 5.2$ cm
Lamps	One in each side	Two. One in each reactor section Philips TUV 40W each.
Germicidal	Philips TUV 15 W Heraeus UV-C 40 W	Output power: 0.105 W cm⁻¹ Total output power: 25.2 W Used output power: 21 W $Z_{L,tot} = 2 \times 120$ cm, $Z_{L,used}: 2 \times 100$ cm $R_L = 1.3$ cm
Almost monochromatic	Philips TUV 15 W with filter	
Wavelength	(90%) $\lambda = 253.7$ nm	(90%) $\lambda = 253.7$ nm
Reflectors	Parabolic cylinder. Made of Aluminum, specularly finished	None
Incident radiation at the wall	Measured by actinometry (potassium ferrioxalate) $(13.9{-}5.45{-}2.33) \times 10^{-9}$ Einstein cm⁻² s⁻¹	Calculated with a volumetric emission model

Table 6 (*Continued*)

Item	Specification/Dimensions	
	Laboratory reactor	Pilot scale reactor
Operation		
Flow rate	Batch, with recirculation: 2,000 cm^3 min^{-1}	Continuous: 1,500–5,000 cm^3 min^{-1}
Temperature	20°C	20°C
Pressure	1 atm	1 atm
Inlet COOH concentration	40–140 ppm	46–110 ppm
Inlet H$_2$O$_2$ concentration	Molar ratio: 1–32	Molar ratio: 2–7.5

reaction space is given by Equation (34) and the appropriate boundary conditions.

$$\underbrace{\frac{dI_{\lambda,\underline{\Omega}}(\underline{x},t)}{ds}}_{\substack{\text{Changes in } I_{\lambda,\underline{\Omega}} \text{ along} \\ \text{the distance } ds \text{ in a} \\ \text{three-dimensional space}}} + \underbrace{\kappa_\lambda(\underline{x},t)I_{\lambda,\underline{\Omega}}(x,t) = 0}_{\substack{\text{Radiation absorption} \\ \text{along } ds}} \tag{34}$$

with the boundary condition that at $s = 0 \rightarrow I_{\lambda,\underline{\Omega}}(\underline{x},t) = I^0_{\lambda,\underline{\Omega}}(\underline{x},t)$.

It has been shown that under some geometric restrictions that involve conditions in distances and dimensions of the complete experimental device; that is, lamps, reflectors, and reactors, the radiation field produced by the tubular lamp, and the parabolic reflector can be modeled by a one-dimensional representation (Alfano et al., 1986). These limitations were imposed on the equipment design of this work. Since κ_λ is a function of the radiation-absorption species concentration, in this case, Equation (34) is coupled with Equation (32).

When Equation (34) is solved, monochromatic specific intensities are obtained. From these values, the monochromatic incident radiation G_λ (photons of a given energy per unit normal area of incidence, unit time, and unit wavelength interval) and the monochromatic LVRPA $e^a_{j,\lambda}$ (absorbed photons of a given energy by the intervening radiation absorption species, per unit reaction volume, unit time and unit wavelength interval) can be readily calculated according to Equations (35) and (36), respectively. No direct photolysis of formic acid (F) was observed; then, in this case, j is just H_2O_2 (P).

$$G_\lambda(x,t) = \int_\Omega I_{\lambda,\underline{\Omega}}(x,t)d\Omega \tag{35}$$

Note that in calculating G_λ all possible radiation absorbing species must be counted; for example, if inner filtering effects are present they must be included.

$$e^a_\lambda(x,t) = \kappa_{\lambda,P}(x,t)\,G_\lambda(x,t) \tag{36}$$

In Equation (36) the subscript P was included to indicate that in calculating the LVRPA from the value of $G(x,t)$ only the absorption coefficient of the reactant radiation absorbing species must be included. When Equation (36) is averaged over the reactor volume, it renders the required expression for Equation (32). This direct substitution is possible only because the reactor is assumed to be well mixed and concentrations are uniform, taking on a single spatial value (resulting from a hydrodynamic average). Since mono-chromatic radiation is used, no integration over wavelengths is necessary. The final local result for $e_\lambda^a(x,t)$ is

$$
\begin{aligned}
e_\lambda^a(x,t) = \kappa_{\lambda,P} G_{w,\lambda} \Big\{ &\exp\Big[-\Big(\sum_j \kappa_{\lambda,j}(t)\Big)x\Big] \\
&+ \exp\Big[-\Big(\sum_j \kappa_{\lambda,j}(t)\Big)(L_R - x)\Big]\Big\}
\end{aligned}
\tag{37}
$$

In this particular case, j is also equal to P. As indicated in Equation (34), the solution of the problem requires the evaluation of the boundary condition. In the one-dimensional model, this requisite is translated into the evaluation of $G_{w,\lambda}(x, t)$ as shown in Equation (37). Thus G_w at $x=0$ and $x=L_R$ must be known.

For a laboratory reactor like the one used in this work, these values can be obtained resorting to conventional actinometry employing the well-known potassium ferrioxalate reaction. The details of the method can be found in Zalazar et al. (2005). The obtained results for the boundary conditions are indicated in Table 6.

3.3. Kinetic results

The experimental data were used to extract from the modeling Equations (30) and (31) the values of the parameters. For this purpose, a multipara-meter, nonlinear estimator was used. It turns out that resulting from these estimations, the equations can be further simplified for $r < 50$ (a conservative assumption), rendering:

$$
R_F(x,t) = -2\,\phi\,e_\lambda^a(x,t)
\tag{38}
$$

$$
R_P(x,t) = -\alpha_1\,\phi\,e_\lambda^a(x,t)r(t)
\tag{39}
$$

The final estimation gives the following results with a 95% confidence interval:

$\phi = 0.372 \pm 0.0114$ Einstein mol^{-1}
$\alpha_1 = 6.67 \times 10^{-3} \pm 1.86 \times 10^{-3}$ (dimensionless).

Figure 13a, b shows two typical outcomes where the solid lines are the theoretical simulations from the simplified model. These parameters can be now applied to the large-scale reactor and, as before, a new value of the LVRPA will be necessary according to its particular configuration.

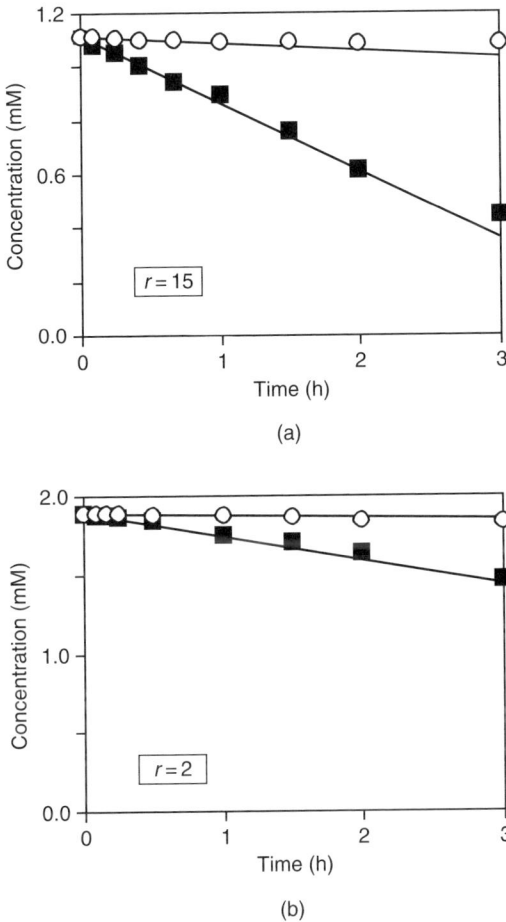

Figure 13   Typical results of laboratory experiments and results from the model (solid lines). (a) $r = 25$, (b) $r = 2$. ($\bigcirc$) Hydrogen peroxide, ($\blacksquare$) formic acid.

3.4. Pilot scale reactor

The larger reactor operates under a steady state, continuous flow conditions and was made of two 1 m cylindrical reactors of annular shape in order to use conventional Germicidal lamps (Figure 14). The system of tanks shown in the flow sheet was used to (i) feed the reactor with a constant flow rate and (ii) wash the system after each experimental run. The actual operating length (Z_L) of each lamp (1.2 m long) was 1 m. Operation could be made with just one reactor or the two in series.

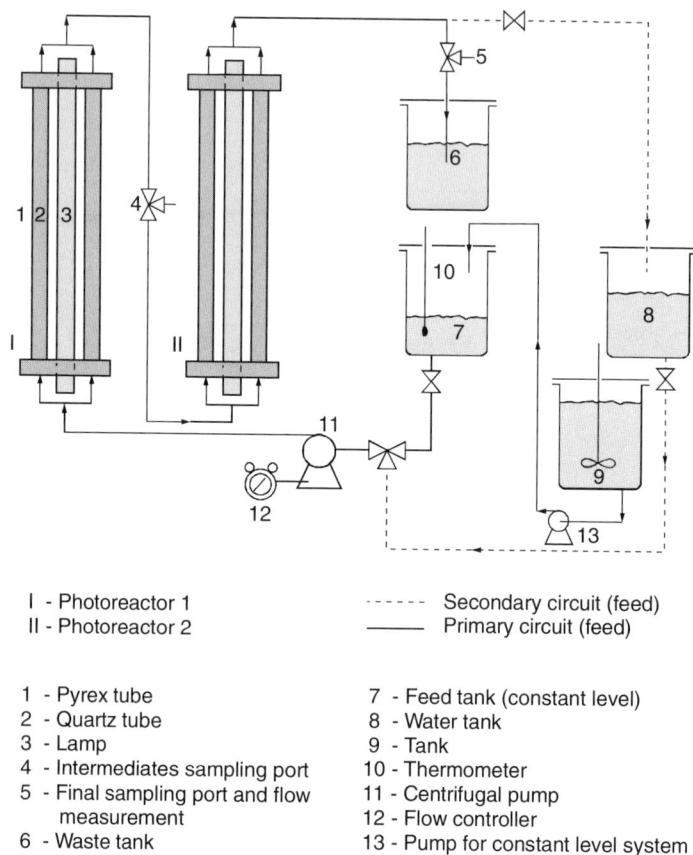

Figure 14   Schematic representation of pilot size setup for a continuous flow reactor.

I - Photoreactor 1
II - Photoreactor 2

----- Secondary circuit (feed)
——— Primary circuit (feed)

1 - Pyrex tube
2 - Quartz tube
3 - Lamp
4 - Intermediates sampling port
5 - Final sampling port and flow measurement
6 - Waste tank
7 - Feed tank (constant level)
8 - Water tank
9 - Tank
10 - Thermometer
11 - Centrifugal pump
12 - Flow controller
13 - Pump for constant level system

3.5. Reactor model

The momentum balance is similar to the one used in Section 2.5 but for a single annular space:

$$v_z(r) = \frac{\Delta P R_{R,\text{ext}}^2}{4\mu Z_R}\left[1 - \left(\frac{r}{R_{R,\text{ext}}}\right)^2 + \left(\frac{1-\chi^2}{\ln(1/\chi)}\right)\ln\left(\frac{r}{R_{R,\text{ext}}}\right)\right] \qquad (40)$$

In this equation $\chi = R_{R,\text{int}}/R_{R,\text{ext}}$.

For the mass balance, the following assumptions and operating conditions are considered: (i) steady state, (ii) unidirectional, incompressible, continuous flow of a Newtonian fluid under laminar flow regime, (iii) only ordinary diffusion is significant for a mixture where the main component is water, (iv) azimuthal symmetry, (v) axial diffusion neglected as compared to the convective flow, and (vi) constant physical and transport properties. The following model equation in cylindrical coordinates (r, z) holds:

$$v_z(r)\frac{\partial C_i(z,r)}{\partial z} - D_{i,\text{water}}\left[\frac{1}{r}\frac{\partial}{\partial r}\left(r\frac{\partial C_i(z,r)}{\partial r}\right)\right] = R_{\text{Hom},i}\ (z,r), \quad i = P, F \quad (41)$$

Note the difference between this equation and Equation (18). In this case, the homogeneous reaction is a part of the differential equation. Equation (41) must be integrated with the following initial and boundary conditions:

$$z = 0 \quad R_{R,\text{int}} \leq r \leq R_{R,\text{ext}} \quad C_i = C_{i,0} \tag{42}$$

$$r = R_{R,\text{int}} \quad 0 \leq z \leq Z_R \quad \frac{\partial C_i}{\partial r} = 0 \tag{43}$$

$$r = R_{R,\text{ext}} \quad 0 \leq z \leq Z_R \quad \frac{\partial C_i}{\partial r} = 0 \tag{44}$$

In the right-hand side of Equation (41), we must insert the results of the kinetic model, that is, Equations (38) and (39) with the kinetic parameters obtained in the laboratory reactor. The solution of the partial differential equation provides formic acid and hydrogen peroxide exit concentrations as a function of the radial position.

3.6. Radiation model

As in Section 2.6, in order to use Equations (38) and (39) in Equation (41) a model for the photon distribution in the annular space is needed. In this case, no actinometric methods can be applied because, in the general case, the reactor cannot be built before the design is completed; that is, there is no equipment to make the measurements and, consequently, no experimental data concerning the incoming radiation are available. Thus, an emission model must be used. It can be obtained with the three-dimensional source with volumetric emission model, the TDSVE Model and the ray tracing technique (Cassano et al., 1995; Irazoqui et al., 1976; Siegel and Howell, 2002). It is necessary to know the optical characteristics of the reactor wall, the optical characteristics of the reaction space and the geometrical dimensions of the lamp–reactor system.

The emission model is based on the following assumptions: (1) the lamp has an extension given by its used length (Z_L) and its radius (R_L); in this extension, emitters are uniformly distributed. (2) Each elementary volume of the lamp is an emitter. The specific intensity associated with each bundle of radiation coming from each emitter, at each wavelength, is spherical, isotropic, and proportional to its extension. (3) Each elementary differential volume of the lamp is transparent to the energy emerging by each emitter located in its surroundings. (4) The lamp is a perfect cylinder whose boundaries are mathematical surfaces without thickness. (5) In this case, end effects in the lamp electrodes are avoided; that is, the used length is shorter that the lamp length. (6) A spherical coordinate system located at each point of radiation reception (I_n)inside the reactor can characterize the arriving specific intensity. It is necessary to know the distance from such point to the centerline of the lamp and two pairs of angular coordinates $[(\theta_{max}, \theta_{min}); \ (\phi_{max}, \phi_{min})]$ that define the extension of the useful volume of the lamp. (7) The radiant power of the lamp (P, in units of Einstein s^{-1}) can be calculated knowing the dimensions of the lamp and the isotropic characteristics of its emission.

Figure 15a shows the emission along the direction (θ, ϕ) produced by a small volume element of the lamp with volumetric emission that reaches the reactor at $s = s_R$. Looking at Figure 15a, Equation (34) can be integrated along this given direction of propagation (defined by the θ and ϕ coordinates) from $s = s_R$ (at an arbitrary point on the surface of radiation entrance to the reactor) to a point of incidence I_n inside the reactor

$$I_\lambda\,(\underline{x},\,\theta,\,\phi,t)\,=\,I_\lambda^0\,(\theta,\,\phi,t)\,\exp\left[-\int_{\bar{s}=s_R(\underline{x}_0,\theta,\phi)}^{\bar{s}=s_I(\underline{x},\theta,\phi)}\kappa_{\lambda,p}(\bar{s},t)\mathrm{d}\bar{s}\right] \qquad (45)$$

where $I_\lambda^0\,(\theta,\phi,t) = I_\lambda\,(s_R,\underline{\Omega},t)$ is the boundary condition for I_λ at the point of entrance and for an arbitrary direction $\underline{\Omega}$. This boundary condition is provided by the lamp emission model. For steady irradiation and the listed assumptions, the result is

$$I_\lambda^0(\theta,\phi) = \frac{P_{\lambda,L}}{4\pi^2 R_L^2 Z_L}\frac{(R_L^2 - r^2\,\sin^2\phi)^{1/2}}{\sin\theta}Y_{\lambda,R}(\theta,\phi) \qquad (46)$$

In Equation (46)$Y_{\lambda,R}$ is a compounded transmission coefficient of the reactor wall (considering absorption and reflections). The value of $P_{\lambda,L}$ was verified with radiometer measurements. Equations (45) and (46) give the radiation contribution of an arbitrary direction (θ,ϕ) to the point $I_n(\underline{x},\theta,\phi)$ inside the reactor. The next step is to integrate all possible directions of irradiation from the lamp volume of emission to the point I_n (Figure 15b).

According to Cassano et al. (1995), knowing the value of the specific intensity at each point, the value of the Incident Radiation $\left[G_\lambda(\underline{x}) = \int_\Omega I_\lambda(\underline{x},\underline{\Omega})\mathrm{d}\Omega\right]$ at a point $\underline{x}$ inside the reactor, is obtained by integration over the solid angle of incidence ($\mathrm{d}\Omega = \sin\theta\mathrm{d}\theta\mathrm{d}\phi$). The limits

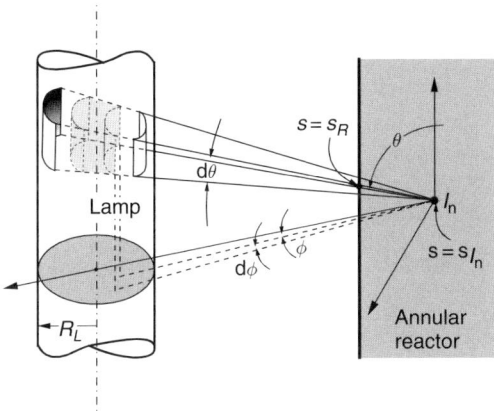

(a)

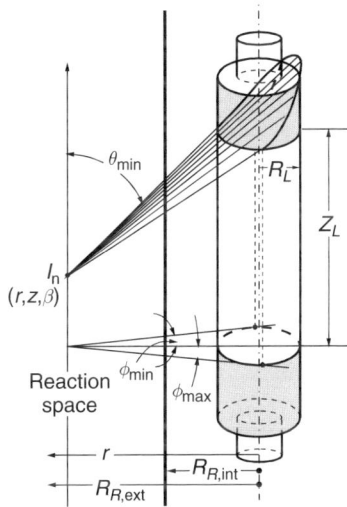

(b)

Figure 15 (a) Emission model for the volumetric lamp. (b) Limits of integration for the lamp contour.

for the integration are obtained from the lamp model, according to the dimensions of the radiation source, defined by the spherical coordinates $[(\theta_{max}, \theta_{min}); \quad (\phi_{max}, \phi_{min})]$:

$$G_\lambda(\underline{x}, t) = \int_{\phi_{min}}^{\phi_{max}} d\phi \int_{\theta_{min}(\phi)}^{\theta_{max}(\phi)} d\theta \sin\theta\, I_{\lambda,\underline{\Omega}}^0 (\theta, \phi, t) \exp\left[-\int_{\bar{s}=s_R(\underline{x}_0,\theta,\phi)}^{\bar{s}=s_I(\underline{x},\theta,\phi)} \kappa_\lambda(\bar{s}, t) d\bar{s} \right]$$

(47)

In the double integral θ accounts for the lamp length and ϕ for the lamp diameter. The integration limits are given by Equations (24)–(26).

Finally, at any point inside the annular reaction space, the LVRPA is

$$
e_\lambda^a(\underline{x}, t) = \kappa_\lambda(\underline{x}, t) \int_{\phi_{\min}}^{\phi_{\max}} d\phi \int_{\theta_{\min}(\phi)}^{\theta_{\max}(\phi)} d\theta \, \sin\theta \, I_{\lambda,\underline{\Omega}}^0(\theta, \phi, t)
$$

$$
\times \exp\left[-\int_{\bar{s}=s_R(\underline{x}_0,\theta,\phi)}^{\bar{s}=s_I(\underline{x},\theta,\phi)} \kappa_\lambda(\bar{s}, t) d\bar{s} \right]
$$

(48)

Equation (48) has to be inserted into Equations (38) and (39) to provide the reaction rates for Equation (41).

3.7. Validation

In order to compare these results with data from the actual reactor and considering the velocity distribution of the outgoing flow, the bulk or flow-average concentration of the reactor outgoing stream must be calculated:

$$
C_{\text{exit},i} = \frac{\displaystyle\int_{R_{R,\text{int}}}^{R_{R,\text{ext}}} v_z(r) C_i\,(r, z = Z_R) r\, dr}{\displaystyle\int_{R_{R,\text{int}}}^{R_{R,\text{ext}}} v_z(r) r\, dr}
$$

(49)

These values must be compared with the experimental measurements. The results are shown in Table 7. There are some differences and two main reasons to explain some of the discrepancies: (i) the fully developed laminar flow assumption of Equation (40) is not fulfilled and (ii) in the connection between both reactor lengths there is some mixing that has not been taken into account by the model. The problems occurring with the velocity field can certainly be solved without difficulties, substituting Equation (40) by a more realistic velocity distribution inside the reaction space resorting to available CFD programs.

Table 7 Scale-up: predictions versus experiments

Run	C_F (ppm)	r	$X_{\text{Pred.}}$	$X_{\text{Exp.}}$	% error
1	46	2	19.0	17	11.8
2	92	2	19.2	23.5	18.3
3	110	7.5	32.7	35	6.6

$X_F = \dfrac{C_{F,0} - C_{F,\text{exit}}}{C_{F,0}}$.

In this reactor some additional information should be added to the last paragraph of Section 2.7. In this case, (i) longer (up to 220 cm) and better lamps (having a significant output power at lower wavelengths, where radiation absorption by hydrogen peroxide is stronger) can be used and (ii) wider annular spaces can also be adopted because radiation absorption by H_2O_2 is rather weak. All other observations have the same validity.

4. SCALING-UP OF A HETEROGENEOUS PHOTOCATALYTIC REACTOR WITH RADIATION ABSORPTION AND SCATTERING

This case shows the degradation of 4-chlorophenol (4-CP) employing UVA radiation and Aldrich titanium dioxide catalyst ($S_g = 9.6 \, cm^2 \, g^{-1}$, nominal diameter of the elementary particle: 200 nm) in a slurry reactor at pH 2.5, which was found to provide the most efficient reaction condition (Satuf et al., 2007a, b, 2008).

4.1. Reaction scheme and kinetic model

The reaction scheme is shown in Table 8 and was constructed on the basis of information existing in the literature, mainly in the following contributions: Turchi and Ollis (1990), Terzian et al. (1990), Minero et al. (1992), Mills et al. (1993), Mills and Morris (1993), Theurich et al. (1996), Alfano et al. (1997), Almquist and Biswas (2001), Dijkstra et al. (2002), and Palmisano et al. (2007) among others. The different reaction steps show the typical characteristics of this heterogeneous reaction: (i) UV catalyst activation, (ii) electron–hole recombination, (iii) hole trapping, (iv) electron trapping, (v) hydroxyl radical attack to 4-CP and the main reaction intermediates along two different routes (Figure 16), and (vi) all the involved adsorption steps indicating that 4-CP and the principal byproducts [4-chlorocatechol (4-CC) and hydroquinone (HQ)] absorb in different catalytic sites than oxygen.

The kinetic model is based on the following assumptions: (i) photocatalytic reactions occur among adsorbed species adsorbed on the catalytic surface, (ii) dynamic equilibrium exits between the bulk and the surface concentrations of adsorbed species, (iii) ·OH radical attack is the main route for all degradation reactions of organic compounds, (iv) the MSSA applies for unstable intermediate species, and (v) there are no mass transfer limitations due to the operating mixing conditions. Most of these assumptions were first proposed and discussed by Turchi and Ollis (1990) for the photocatalytic degradation of several organic water pollutants in TiO_2 slurries and then applied by Cabrera et al. (1997) and Brandi et al. (2002) to obtain a kinetic model of the photocatalytic decomposition of trichloroethylene in aqueous suspensions.

Table 8   Reaction mechanism

Activation	$TiO_2 + h\nu \rightarrow TiO_2 + e^- + h^+$	R_g
Recombination	$e^- + h^+ \rightarrow \text{heat}$	k_2
Electron trapping	$e^- + O_{2,ads} \rightarrow \bullet O_2^-$	k_3
Hole trapping	$h^+ + H_2O_{ads} \rightarrow \bullet OH + H^+$	k_4
	$h^+ + OH^-_{\ ads} \rightarrow \bullet OH$	
Hydroxyl attack	$4\text{-}CP_{ads} + \bullet OH \rightarrow 4\text{-}CC$	k_5
	$4\text{-}CP_{ads} + \bullet OH \rightarrow HQ$	k_6
	$4\text{-}CC_{ads} + \bullet OH \rightarrow X_i$	k_7
	$HQ_{ads} + \bullet OH \rightarrow X_j$	k_8
	$Y_{In,ads} + \bullet OH \rightarrow Y_m$	k''_{In}
Adsorption	$site_{O_2} + O_2 \leftrightarrow O_{2,ads}$	K_{O_2,S_1}
	$site_{H_2O} + H_2O \leftrightarrow H_2O_{ads}$	K_{H_2O}
	$site_{H_2O} + H_2O \leftrightarrow OH^-_{\ ads} + H^+$	
	$site_{4\text{-}CP} + 4\text{-}CP \leftrightarrow 4\text{-}CP_{ads}$	$K_{4\text{-}CP,S_2}$
	$site_{4\text{-}CP} + 4\text{-}CC \leftrightarrow 4\text{-}CC_{ads}$	$K_{4\text{-}CC,S_2}$
	$site_{4\text{-}CP} + HQ \leftrightarrow HQ_{ads}$	K_{HQ,S_2}
	$site_{Y_{In}} + Y_{Y_{In}} \leftrightarrow Y_{Y_{In,\ ads}}$	K_{In}

$S_1 = $ Site type 1; $S_2 = $ Site type 2 with competitive adsorption.
$C_{O_2} \cong$ constant. $C_{H_2O_{ads}}$ and $C_{OH^-_{ads}} \cong$ constant.
$X_i = $ organic degradation products of $4\text{-}CC_{ads}$.
$X_j = $ organic degradation products of HQ_{ads}.
$Y_{In} = $ inorganic radicals and species that compete to react with $\cdot OH$.
Adapted from Satuf et al. (2008).

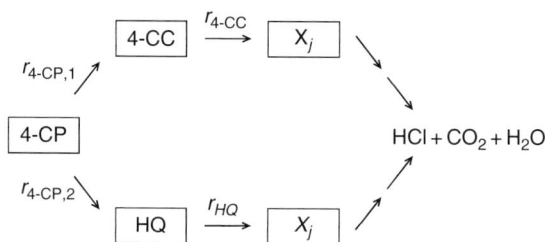

Figure 16   Reaction pathway suggested for the photocatalytic degradation of 4-chlorophenol in acidic medium.

Resorting to a balance of sites for competitive adsorption in sites type 2 and considering that the rate of electrons and holes can be expressed in terms of

$$R_g(x) = \frac{\bar{\Phi}}{a_V} \sum_\lambda e_\lambda^a(\underline{x}) = \frac{\bar{\Phi}}{a_V} e^a(\underline{x}) \tag{50}$$

the following reaction rates can be obtained:

$$R_{\text{4-CP},1}(\underline{x}, t) = \frac{\alpha_{2,1} C_{\text{4-CP}}(t)}{1 + \alpha_3 C_{\text{4-CP}}(t) + \alpha_1' C_{\text{4-CC}}(t) + \alpha_2' C_{\text{HQ}}(t)}$$
$$\times \left(-1 + \sqrt{1 + \frac{\alpha_1}{a_v} e^a(\underline{x})} \right) \tag{51}$$

$$R_{\text{4-CP},2}(\underline{x}, t) = \frac{\alpha_{2,2} C_{\text{4-CP}}(t)}{1 + \alpha_3 C_{\text{4-CP}}(t) + \alpha_1' C_{\text{4-CC}}(t) + \alpha_2' C_{\text{HQ}}(t)}$$
$$\times \left(-1 + \sqrt{1 + \frac{\alpha_1}{a_v} e^a(\underline{x})} \right) \tag{52}$$

$$R_{\text{4-CC}}(\underline{x}, t) = \frac{\alpha_4 C_{\text{4-CC}}(t)}{1 + \alpha_3 C_{\text{4-CP}}(t) + \alpha_1' C_{\text{4-CC}}(t) + \alpha_2' C_{\text{HQ}}(t)}$$
$$\times \left(-1 + \sqrt{1 + \frac{\alpha_1}{a_v} e^a(\underline{x})} \right) \tag{53}$$

$$R_{\text{HQ}}(\underline{x}, t) = \frac{\alpha_5 C_{\text{HQ}}(t)}{1 + \alpha_3 C_{\text{4-CP}}(t) + \alpha_1' C_{\text{4-CC}}(t) + \alpha_2' C_{\text{HQ}}(t)}$$
$$\times \left(-1 + \sqrt{1 + \frac{\alpha_1}{a_v} e^a(\underline{x})} \right) \tag{54}$$

In Equations (51)–(54), it must be noted that (i) $e^a(\underline{x}) = e^a[(\underline{x}), C_{\text{cm}}]$ to recall that the LVRPA is a strong function of the catalyst loading and (ii) there are eight lumped kinetic parameters that depend on intrinsic properties of the system according to

$$\alpha_1 = \frac{\bar{\Phi} 4 k_2}{k_4 k_3 C_{\text{H}_2\text{O,ads}} C_{\text{O}_2,\text{ads}}} \tag{55}$$

$$\alpha_{2,1} = \frac{k_5 k_4 k_3 C_{\text{H}_2\text{O,ads}} C_{\text{O}_2,\text{ads}} C_{\text{site,4-CP}} K_{\text{4-CP}}}{2 k_2 \left(\sum_{\text{In}} k_{\text{In}}'' C_{Y_{\text{In,ads}}} \right)} \tag{56}$$

$$\alpha_{2,2} = \frac{k_6 k_4 k_3 C_{\text{H}_2\text{O,ads}} C_{\text{O}_2,\text{ads}} C_{\text{site,4-CP}} K_{\text{4-CP}}}{2 k_2 \left(\sum_{\text{In}} k_{\text{In}}'' C_{Y_{\text{In,ads}}} \right)} \tag{57}$$

$$\alpha_3 = \frac{C_{site,4\text{-}CP}(k_5 + k_6)K_{4\text{-}CP} + K_{4\text{-}CP}}{\sum_{In} k''_{In} C_{In,ads}} \tag{58}$$

$$\alpha'_1 = \frac{C_{site,4\text{-}CP}k_7 K_{4\text{-}CP} + K_{4\text{-}CC}}{\sum_{In} k''_{In} C_{In,ads}} \tag{59}$$

$$\alpha'_2 = \frac{C_{site,4\text{-}CP}k_8 K_{HQ} + K_{HQ}}{\sum_l k''_l C_{l,ads}} \tag{60}$$

$$\alpha_4 = \frac{k_7 k_4 k_3 C_{H_2O,ads} C_{O_2,ads} C_{site,4\text{-}CP,T} K_{4\text{-}CC}}{2k_2 \left(\sum_{In} k''_{In} C_{Y_{In,ads}} \right)} \tag{61}$$

$$\alpha_5 = \frac{k_8 k_4 k_3 C_{H_2O,ads} C_{O_2,ads} C_{site,4\text{-}CP,T} K_{HQ}}{2k_2 \left(\sum_{In} k''_{In} C_{Y_{In,ads}} \right)} \tag{62}$$

4.2. Laboratory reactor: description and modeling

The laboratory reactor was designed with the same criteria described in Section 3.2. in order to simplify the modeling. The details are given in Table 9 and Figures 17, 18a, b, and 19. There are two important differences with respect to the previous case: (i) both windows are mobile in the direction of the reactor length and the displacement is made with a single consolidate unit of the window and the irradiation system to keep the irradiation rate on each surface constant and (ii) interposed between the reactors windows made of borosilicate glass and the position built to insert the shutters (to start the reaction at a desired time) and the filters (to attenuate the incoming radiation when needed), there are two fixed circular plates made of ground glass to produce diffuse irradiation on each surface of radiation entrance. Diffuse incoming radiation means azimuthal symmetry in both boundary conditions (Figure 18a). With these arrangements, three different reactor lengths were selected (0.5, 1, and 5 cm) to study the effect of the optical thickness (catalyst concentration plus radiation path) allowing for a two orders of magnitude change on the variable. Similarly, with the neutral density filters the incoming radiation was changed in three levels: 100, 67, and 30%. For more details about the experimental setup and the operating conditions the reader is referred to Satuf et al. (2007a, b, 2008).

Table 9 Laboratory and pilot scale reactor description and operating conditions

Item	Specification/Dimensions	
	Laboratory reactor	Pilot scale reactor
Type	Flat plate. Batch, with a recycle. Two circular windows made of borosilicate glass	Flat plate with recycle. Irradiated from one wall with two tubular lamps and parabolic reflectors
Reactor volume	Photoreactor: V_R = from 29.03 to 290.29 cm^3 Total: V_T = 1,000 cm^3	Photoreactor: V_R = 734.4 cm^3 Total system volume: V_T = 5,000 cm^3
Dimensions	Two sides. $D(inner)$ = 8.6 cm L_R = Variable from 0.5 to 5 cm	Length: Z_R = 34.0 cm Width: Y_R = 18.0 cm Thickness: X_R = 1.2 cm
Lamps Pilot reactor: Actinic: without built-in reflectors	Four in each side (total 8) Philips: TL 4W/08 Black Light UVA Output power: 4 W each Arc length: 13.6 cm R_L = 0.8 cm	Two. On the front Philips: TLK40/09 Actinic UVA Output power: 40 W each Z_L = 60 cm R_L = 3.8 cm
Wavelength	300–400 nm (peak: 350 nm)	310–400 nm (peak: 350 nm)
Reflectors	None	Two parabolic cylinders. Made of aluminum, specularly finished

Table 9 (*Continued*)

Item	Specification/Dimensions	
	Laboratory reactor	Pilot scale reactor
Incident radiation at the wall	Measured by actinometry (potassium ferrioxalate) 100% (without filters): 7.55×10^{-9} Einstein cm^{-2} s^{-1} (at each window)	Calculated with a superficial emission model
Operation		
Flow rate	Recirculation: 6,000 cm^3 min^{-1}	Recirculation: 6,000 cm^3 min^{-1}
Temperature	20°C	20°C
Pressure	1 atm.	1 atm.
Catalyst concentration	0.05, 0.1, 0.5, and 1.0×10^{-3} g cm^{-3}	0.05, 0.1, 0.5. and 1.0×10^{-3} g cm^{-3}
Initial 4-CP concentration	1.5×10^{-7} mol cm^{-3}	1.5×10^{-7} mol cm^{-3}
Initial pH	2.5	2.5

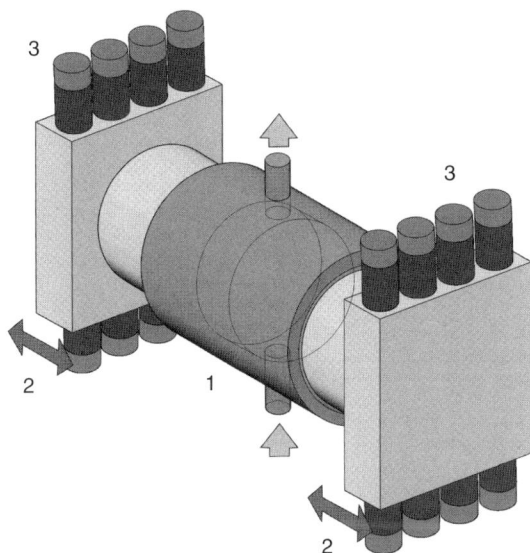

Figure 17 Schematic representation of the photocatalytic reactor. Keys: (1) reactor, (2) mobile windows system, (3) radiation emitting system.

Employing a high recirculating flow rate in this small laboratory reactor, the following assumptions can be used: (i) there is a differential conversion per pass in the reactor, (ii) the system is perfectly stirred, (iii) there are no mass transport limitations. Also, it can be assumed that (iv) the chemical reaction occurs only at the solid–liquid interface (Minero et al., 1992) and (v) direct photolysis is neglected (Satuf et al., 2007a). As a result, the mass balance for the species i in the system takes the following form (Cassano and Alfano, 2000):

$$\varepsilon_L \frac{dC_i(t)}{dt}\bigg|_{Tk} = \frac{V_R}{V_T} a_v \langle v_i \mathsf{R}(\underline{x}, t) \rangle_{A_R} \quad i = 4\text{-CP}, 4\text{-CC}, \text{HQ} \quad (63)$$

where ε_L is the liquid hold-up ($\varepsilon_L \cong 1$), C_i is the molar concentration of the component i and $\langle R(\underline{x}, t) \rangle_{A_R}$ is the superficial reaction rate averaged over the catalytic reaction area.

The primary oxidation products of 4-CP are 4-CC, HQ, and benzoquinone (BQ) (Theurich et al., 1996). Under the experimental conditions employed in our work, at pH 2.5, the oxidation via 4-CC represents the main pathway for the degradation of 4-CP. In second place appears the formation of HQ. Because of the low concentrations of BQ found during the experiments, this intermediate product is not considered in the kinetic model. Then, the

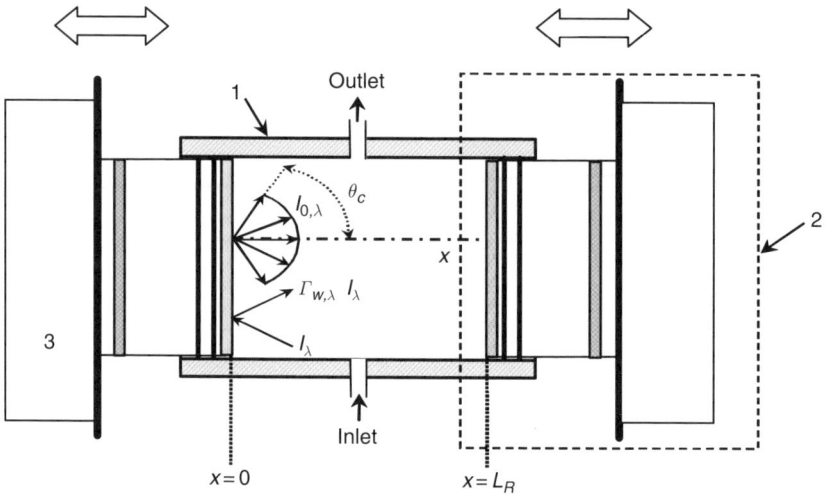

(a)

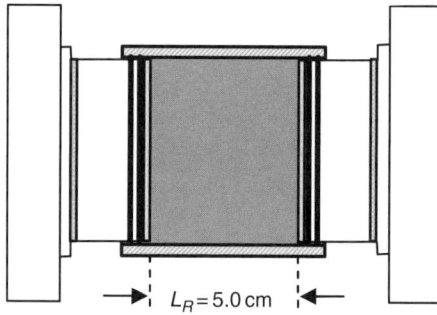

$L_R = 5.0\,\text{cm}$

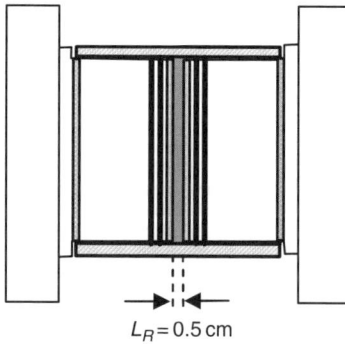

$L_R = 0.5\,\text{cm}$

(b)

Figure 18 (a) Diagram of mobile windows mechanism. (b) Sketch of maximum and minimum reactor length.

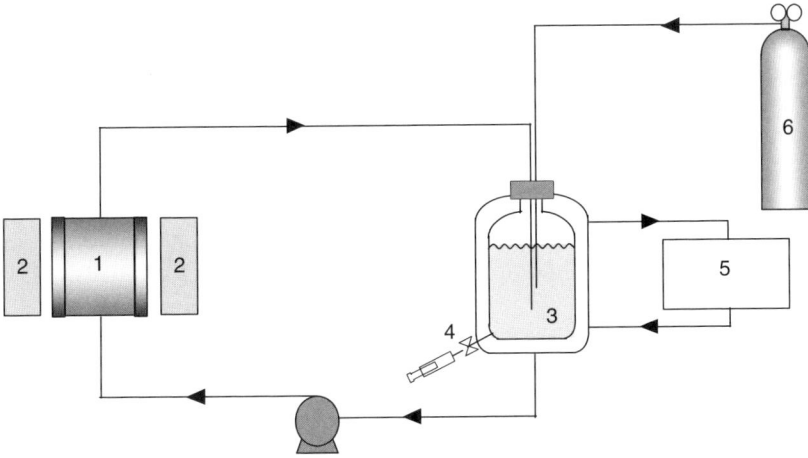

Figure 19 Schematic representation of the experimental setup. Keys: (1) reactor, (2) radiation emitting system, (3) tank, (4) sampling valve, (5) thermostatic bath, (6) oxygen supply, (7) pump.

mass balances for 4-CP, 4-CC, and HQ, with the corresponding initial conditions, are

$$\varepsilon_L \frac{dC_{4\text{-CP}}(t)}{dt} = -\frac{V_R}{V_T} a_v \left\{ \left\langle R_{4\text{-CP},1}(\underline{x},t) \right\rangle_{A_R} + \left\langle R_{4\text{-CP},2}(\underline{x},t) \right\rangle_{A_R} \right\}$$
$$\times C_{4\text{-CP}}(t=0) = C_{4\text{-CP},0} \tag{64}$$

$$\varepsilon_L \frac{dC_{4\text{-CC}}(t)}{dt} = \frac{V_R}{V_T} a_v \left\{ \left\langle R_{4\text{-CP},1}(\underline{x},t) \right\rangle_{A_R} - \left\langle R_{4\text{-CC}}(\underline{x},t) \right\rangle_{A_R} \right\}$$
$$\times C_{4\text{-CC}}(t=0) = 0 \tag{65}$$

$$\varepsilon_L \frac{dC_{\text{HQ}}(t)}{dt} = \frac{V_R}{V_T} a_v \left\{ \left\langle R_{4\text{-CP},2}(\underline{x},t) \right\rangle_{A_R} - \left\langle R_{\text{HQ}}(\underline{x},t) \right\rangle_{A_R} \right\}$$
$$\times C_{\text{HQ}}(t=0) = 0 \tag{66}$$

As shown in Equations (51)–(54) and Figure 16, two parallel reaction pathways are postulated for the degradation of 4-CP: $R_{4\text{-CP},1}$ represents the degradation rate to give 4-CC, whereas $R_{4\text{-CP},2}$ is the rate that leads to the formation of HQ. $R_{4\text{-CC}}$ and R_{HQ} denote the degradation rates of 4-CC and HQ.

Equations (51)–(54) indicate the need for calculating e_λ^a and $\Sigma_\lambda e_\lambda^a$.

The evaluation of the LVRPA inside the reactor was achieved by solving the RTE for the heterogeneous system. The radiation model considers that

(Figure 18a) (i) the main changes in the spatial distribution occur along the x coordinate axis due to the significant extinction produced by the TiO$_2$ catalyst particles; thus, a one-dimensional (x in space) model can be applied and (ii) the arrangement of the four UV lamps and the ground glass plates at the external side of the reactor windows ensure the arrival of diffuse radiation with azimuthal symmetry; consequently, radiation intensity in the medium is independent of the azimuthal angle and a one-directional (θ in direction) radiation model can be assumed. Then, a one-dimensional, one-directional radiation transport model is applied (Alfano et al., 1995; Cassano et al., 1995):

$$\mu \frac{\partial I_\lambda(x, \mu)}{\partial x} + \beta_\lambda I_\lambda(x, \mu) = \frac{\sigma_\lambda}{2} \int_{\mu'=-1}^{1} I_\lambda(x, \mu')\, p(\mu, \mu')\mathrm{d}\mu' \qquad (67)$$

In Equation (67) β_λ is the extinction coefficient, μ the direction cosine of the ray for which the RTE is written, μ' the cosine of an arbitrary ray before scattering, and "p" the phase function for scattering.

The Henyey and Greenstein phase function ($p_{HG,\lambda}$) was adopted to model the radiation scattering of the TiO$_2$ suspensions (Siegel and Howell, 2002):

$$p_{HG,\lambda}(\mu_0) = \frac{1 - g_\lambda^2}{(1 + g_\lambda^2 - 2g_\lambda\mu_0)^{3/2}} \qquad (68)$$

It is worth noting that $p_{HG,\lambda}$ is determined by a single parameter, the dimensionless asymmetry factor (g_λ), that varies from isotropic ($g_\lambda = 0$) to a narrow forward peak ($g_\lambda = 1$) or to a narrow backward peak ($g_\lambda = -1$); this parameter can be estimated employing the method described in Satuf et al. (2005).

Equation (67) must be solved with the appropriate boundary conditions in terms of the radiation intensity. At $x = 0$, radiation intensities are the result of the transmitted portion of the radiation arriving to the reactor glass plate and the reflected portion of the radiation coming from the aqueous suspension (Figure 18a).

It should be also noted that (i) the radiation coming from the lamps arrives in a diffuse way at the external side of the glass window. The angular directions of the intensities entering the suspension ($I_{0,\lambda}$) are comprised between 0 and the critical angle θ_c; this fact can be explained taking into account that radiation must cross two interfaces: air-glass and glass suspension and (ii) the intensities coming from the reacting medium undergo multiple specular reflections at the reactor window. For this case, a global reflection coefficient ($\Gamma_{W,\lambda}$) can be defined; it represents the reflected fraction of the radiation that goes back to the suspension.

A similar analysis can be applied for the boundary condition at $x = L_R$:

$$I_\lambda(0, \mu) = I_{0,\lambda} + \Gamma_{W,\lambda}(-\mu)I_\lambda(0, -\mu) \quad 1 \geq \mu \geq \mu_c \qquad (69a)$$

$$I_\lambda(0,\mu) = \Gamma_{W,\lambda}(-\mu)I_\lambda(0,-\mu) \quad \mu_c \geq \mu \geq 0 \tag{69b}$$

$$I_\lambda(L_R,-\mu) = I_{L_R,\lambda} + \Gamma_{W,\lambda}(\mu)I_\lambda(L_R,\mu) \quad 1 \geq \mu \geq \mu_c \tag{69c}$$

$$I_\lambda(L_R,-\mu) = \Gamma_{W,\lambda}(\mu)I_\lambda(L_R,\mu) \quad \mu_c > \mu \geq 0 \tag{69d}$$

Here μ_c is the cosine of the critical angle θ_c. More details on the one-dimensional, one-directional radiation model can be found in Satuf et al. (2007a).

To solve Equations (67)–(69), the Discrete Ordinate Method was applied (Duderstadt and Martin, 1979). From the solution of the RTE, the monochromatic radiation intensity at each point and each direction inside the reactor can be obtained. Considering constant optical properties of the catalyst and steady radiation supply by the emitting system, the radiation field can be considered independent of time.

The LVRPA for polychromatic radiation is calculated from the values of the monochromatic radiation intensity as

$$e^a(x) = 2\pi \int_\lambda \kappa_\lambda \int_{\mu=-1}^1 I_\lambda(x,\mu)\mathrm{d}\mu\mathrm{d}\lambda \tag{70}$$

Figure 20 shows LVRPA profiles for different TiO$_2$ concentrations and for the three values of the reactor lengths employed in the experimental runs. In Figure 20a, the strongest variations of LVRPA are observed for the highest catalyst loading $(C_m \geq 0.5 \times 10^{-3}$ g cm$^{-3})$ and for the largest reactor $(L_R = 5.0$ cm). Smoother profiles are obtained as long as the catalyst loading and the reactor length decrease.

4.3. Kinetic results

A nonlinear, multiparameter regression procedure (Levenberg–Marquardt method) was applied to estimate the kinetic parameters involved in Equations (51)–(54). The experimental concentrations of the pollutant (4-CP) and of the main intermediate species (4-CC and HQ) at different reaction times were compared with model predictions. Under the operating conditions of the experimental runs, it was found that the terms $\alpha_3 C_{4\text{-CP}}(t)$, $\alpha_1' C_{4\text{-CC}}(t)$, and $\alpha_2' C_{HQ}(t)$ were much lower than 1. As a result, the final expressions employed for the regression of the kinetic parameters are the following:

$$R_{4\text{-CP},1}(x,t) = \alpha_{2,1}C_{4\text{-CP}}(t)\left(-1 + \sqrt{1 + \frac{\alpha_1}{a_v}e^a(x)}\right) \tag{71}$$

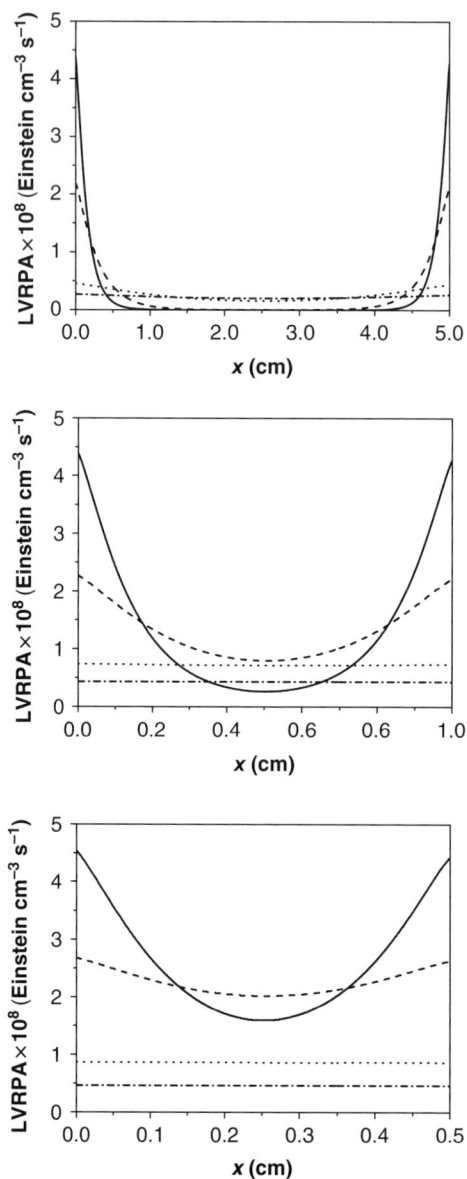

Figure 20 LVRPA profiles for different catalyst concentrations and different reactor lengths. (——): $C_m = 1.0 \times 10^{-3}$ g cm^{-3}; (– – – –): $C_m = 0.5 \times 10^{-3}$ g cm^{-3}; (· · · ·): $C_m = 0.1 \times 10^{-3}$ g cm^{-3}; (– · – ·): $C_m = 0.05 \times 10^{-3}$ g cm^{-3}. (a) $L_R = 5.0$ cm, (b) $L_R = 1.0$ cm, (c) $L_R = 0.5$ cm.

$$R_{4\text{-}CP,2}(x,t) = \alpha_{2,2}C_{4\text{-}CP}(t)\left(-1 + \sqrt{1 + \frac{\alpha_1}{a_v}e^a(x)}\right) \tag{72}$$

$$R_{4\text{-}CC}(x,t) = \alpha_4 C_{4\text{-}CC}(t)\left(-1 + \sqrt{1 + \frac{\alpha_1}{a_v}e^a(x)}\right) \tag{73}$$

$$R_{HQ}(x,t) = \alpha_5 C_{HQ}(t)\left(-1 + \sqrt{1 + \frac{\alpha_1}{a_v}e^a(x)}\right) \tag{74}$$

Table 10 provides the values of the five kinetic parameters, with the corresponding 95% confidence interval. Note that $\alpha_{2,1} \cong 5\alpha_{2,2}$, showing that the 4-CP degradation via 4-CC is favored, being 4-CC the most abundant primary intermediate (Satuf et al., 2008).

Figures 21a, b show the 4-CP, 4-CC, and HQ concentrations derived from inserting the estimated parameters in the kinetic model and a comparison with the experimental data under different operating conditions. Symbols correspond to experimental data and solid lines to model predictions calculated with Equations (64)–(66) and Equations (71)–(74). For these experimental runs, the RMSE was less than 14.4%. These experimental 4-CC and HQ concentrations are in agreement with the proposed kinetic mechanism of parallel formation of the intermediate species (Figure 16), and also with the "series–parallel" kinetic model reported by Salaices et al. (2004) to describe the photocatalytic conversion of phenol in a slurry reactor under various operating conditions.

The term $\psi = \left(-1 + \sqrt{1 + \frac{\alpha_1}{a_v}e^a(x)}\right)$ in Equations (71)–(74) can be used to study the dependence of reaction kinetics on the photon absorption rate. The following limiting cases may be readily obtained: (i) for sufficiently high values of e^a, it holds that $\frac{\alpha_1}{a_v}e^a(x) \gg 1$ and $\sqrt{1 + \frac{\alpha_1}{a_v}e^a(x)} \approx \sqrt{\frac{\alpha_1}{a_v}e^a(x)}$; additionally, if $\sqrt{\frac{\alpha_1}{a_v}e^a(x)} \gg -1$, then $\psi \approx \sqrt{\frac{\alpha_1}{a_v}e^a(x)}$. Therefore, under high

Table 10 Kinetic parameters of 4-CP decomposition

Parameter	Value	Confidence interval (95%)	Units
α_1	1.09×10^{11}	$\pm 0.07 \times 10^{11}$	s cm^2 Einstein^{-1}
$\alpha_{2,1}$	9.43×10^{-6}	$\pm 0.26 \times 10^{-6}$	cm s^{-1}
$\alpha_{2,2}$	2.18×10^{-6}	$\pm 0.46 \times 10^{-6}$	cm s^{-1}
α_4	1.29×10^{-5}	$\pm 0.08 \times 10^{-5}$	cm s^{-1}
α_5	9.21×10^{-6}	$\pm 0.96 \times 10^{-6}$	cm s^{-1}

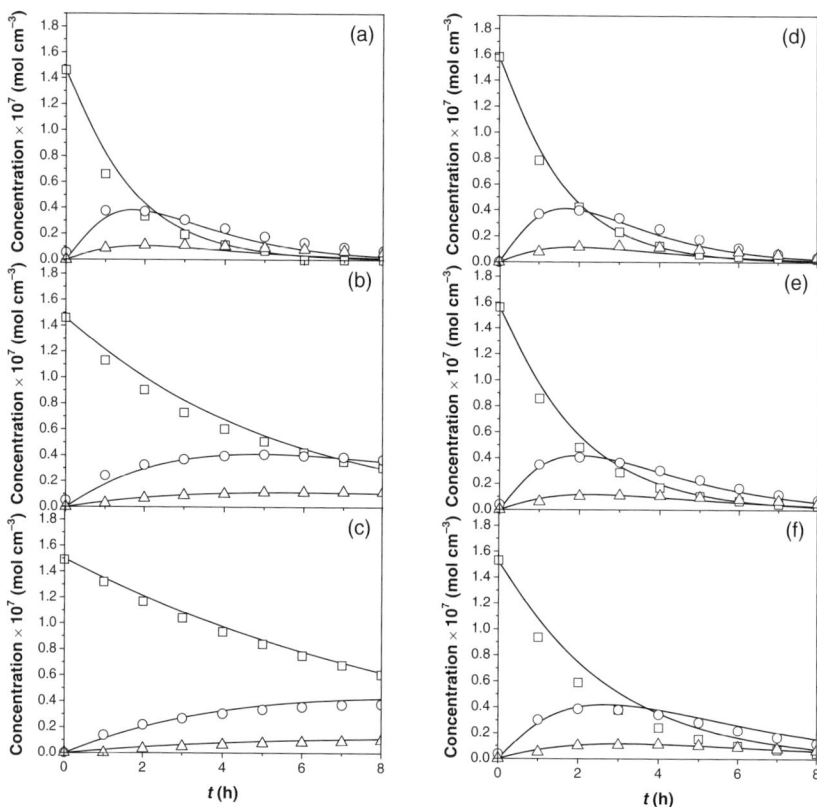

Figure 21 Experimental and predicted concentrations of 4-CP, 4-CC, and HQ versus time. Left: $L_R = 1.0$ cm, $E_L = 100\%$ and different C_m. Experimental data: □: 4-CP; ○: 4-CC; △: HQ. Model results: solid lines. (a) $C_m = 1.0 \times 10^{-3}$ g cm^{-3}; (b) $C_m = 0.1 \times 10^{-3}$ g cm^{-3}; (c) $C_m = 0.05 \times 10^{-3}$ g cm^{-3}. Right: $C_m = 0.5 \times 10^{-3}$ g cm^{-3}, $E_L = 100\%$ and different L_R. Experimental data: □: 4-CP; ○: 4-CC; △: HQ. Model results: solid lines. (d) $L_R = 5.0$ cm; (e) $L_R = 1.0$ cm; (f) $L_R = 0.5$ cm.

values of e^a, the reaction rate is proportional to the square root of the LVRPA. (ii) For rather low values of e^a: $\frac{\alpha_1}{a_v} e^a(x) \ll 1$; thus, from a Taylor series expansion, it can be shown that $\sqrt{1 + \frac{\alpha_1}{a_v} e^a(x)} \approx 1 + \frac{1}{2}\frac{\alpha_1}{a_v} e^a(x)$ (Alfano et al., 1997). Hence, for low values of e^a, $\psi \approx \frac{1}{2}\frac{\alpha_1}{a_v} e^a(x)$ and the reaction rate has a linear dependence with respect to the LVRPA. Table 11 describes the numerical results obtained for two laboratory reactor lengths and two catalyst concentrations: (i) for $L_R = 5.0$ cm, $C_m = 1.0 \times 10^{-3}$ g cm^{-3}, and $x = 0$ (at the reactor window), the dependence of the reaction rate is nearly proportional to the square root of the LVRPA, whereas at $x = L_R/2$ (in the center of the reactor) the same dependence is linear. (ii) For $L_R = 0.5$ cm,

Table 11 Dependence of the reaction rate on the photon absorption rate

L_R (cm)	C_m ($\times 10^3$ g cm^{-3})	X	$\frac{\alpha_1}{a_v} e^a(x)$	$\sqrt{\frac{\alpha_1}{a_v} e^a(x)}$
5	1.0	0	49.7	7.05
5	1.0	$L_R/2$	1.15×10^{-4}	1.07×10^{-2}
0.5	0.05	0	105.6	10.3
0.5	0.05	$L_R/2$	105.6	10.3

$C_m = 0.05 \times 10^{-3}$ g cm^{-3}, the reaction rate is proportional to the square root of the LVRPA throughout the reaction space. Consequently, in a photocatalytic slurry reactor, where low, intermediate, and high photon absorption rates can coexist, the complete reaction rate equation must be used.

4.4. Pilot scale reactor

Figures 22a, b provides a schematic representation of the pilot scale reactor. Essentially it is a rectangular parallelepiped limited by two parallel windows made of borosilicate glass and operated as a slurry reactor inside the loop of a batch recycling system. Irradiation of one of the reactor faces is obtained using two tubular lamps that were placed along the focal axis of two parabolic reflectors made of specularly finished aluminum (Brandi et al., 1996, 1999, 2002). The specific information concerning the experimental device is presented in Table 9, and more details can be found in Satuf et al. (2007b).

4.5. Reactor model

The mass balance in the reactor is derived under the following assumptions: (i) unsteady state operation, (ii) convective laminar Newtonian flow in the axial direction z (the Reynolds number is below the transition regime), (iii) diffusion in the z direction is negligible with respect to convection, (iv) symmetry in the y direction (the lamp length is much larger than the reactor width), and (v) constant physical properties. The local mass balance for a species i in the reactor and the corresponding initial and boundary conditions are

$$\frac{\partial C_{i,R}}{\partial t} + v_z \frac{\partial C_{i,R}}{\partial z} - D_i \frac{\partial^2 C_{i,R}}{\partial x^2} - R_i = 0; \quad i = \text{4-CP, 4-CC, HQ} \tag{75}$$

$$C_{i,R}(x, z, t = 0) = C_{i,0} \tag{76}$$

$$C_{i,R}(x, z = 0, t) = C^0_{i,\text{Tk}}(t) \tag{77}$$

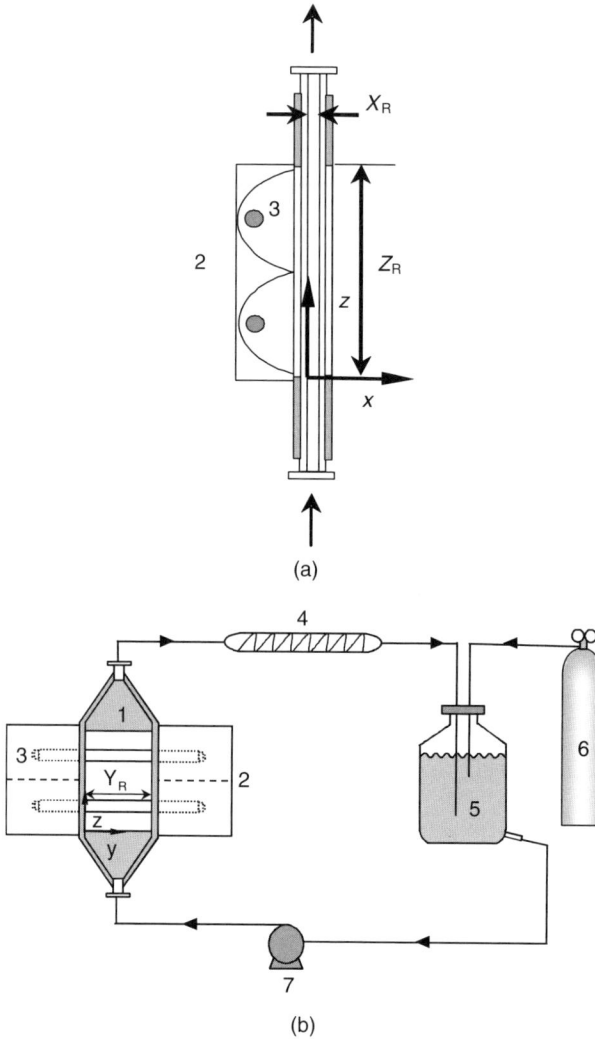

Figure 22 Schematic representation of the pilot scale experimental device. (a) Lateral view of the reactor. (b) Setup. Keys: (1) front view of the reactor, (2) reflectors, (3) lamps, (4) heat exchanger, (5) tank, (6) oxygen supply, (7) pump.

$$\frac{\partial C_{i,R}}{\partial x}\bigg|_{x=0} = 0 \qquad \frac{\partial C_{i,R}}{\partial x}\bigg|_{x=X_R} = 0 \qquad (78)$$

$C_{i,0}$ is the initial concentration and $C_{i,\mathrm{Tk}}^{o}$ is the exit concentration from the tank.

The reaction rate R_i in Equation (75) for the species 4-CP, 4-CC, and HQ was replaced by the Equations (71)–(74) with the kinetic parameters reported in Table 10.

The concentrations of 4-CP, 4-CC, and HQ as a function of x, z, and t can be obtained solving Equations (75)–(78). The average exit concentration from the reactor at a given time t is given by

$$\langle C_{i,R}(x, z = Z_R, t) \rangle_{A_R} = \frac{\int_{x=0}^{x=X_R} C_{i,R}(x, z = Z_R, t) v_z(x) dx}{\int_{x=0}^{x=X_R} v_z(x) dx} = C_{i,Tk}^i(t) \qquad (79)$$

Equation (79) gives the inlet condition for the mass balance in the tank, $C_{i,Tk}^i(t)$.

Considering that the tank operates under unsteady state and well-stirred conditions, the mass balance and the initial condition for a species i yields

$$\varepsilon_L \frac{dC_i(t)}{dt}\bigg|_{Tk} = \frac{1}{\tau_{Tk}} \left[C_{i,Tk}^i(t) - C_{i,Tk}^o(t) \right]; \quad i = \text{4-CP, 4-CC, HQ} \qquad (80)$$

$$C_{i,Tk}^i(t = 0) = C_{i,0} \qquad (81)$$

where τ_{Tk} is the mean residence time in the tank.

4.6. Radiation model

The existence of the LVRPA in the reaction rate expression (Equation 75) makes necessary the solution of the RTE in the pilot scale reactor. As explained before, the lamps length is significantly larger than the reactor width Y_R, thus uniformity of radiation is considered along the y direction. Therefore, a two-dimensional (x, z) model for the spatial variations of the radiation field was adopted. The angular distribution of radiation was modeled with the spherical coordinates (θ, ϕ). The RTE for a two-dimensional, two-directional model is (Brandi et al., 1996, 1999),

$$\mu \frac{\partial I_\lambda(x, z, \underline{\Omega})}{\partial x} + \eta \frac{\partial I_\lambda(x, z, \underline{\Omega})}{\partial z} + \beta_\lambda I_\lambda(x, z, \underline{\Omega})$$
$$= \frac{\sigma_\lambda}{4\pi} \int_{\Omega'=4\pi} I_\lambda(x, z, \underline{\Omega}') p(\underline{\Omega}' \rightarrow \underline{\Omega}) d\underline{\Omega}' \qquad (82)$$

where $\mu = \cos \phi \sin \theta$ and $\eta = \sin \phi \sin \theta$. The boundary conditions are

$$I_\lambda(x=0,z,\underline{\Omega}=\underline{\Omega}^i)=\chi(\text{properties corresponding to the emitting}$$
$$\text{system and the reactor wall})$$

$$I_\lambda(x=X_R,z,\underline{\Omega}=\underline{\Omega}^i)=\chi(\text{properties corresponding to the reactor wall})$$
$$I_\lambda(x,z=0,\underline{\Omega}=\underline{\Omega}^i)=0; I_\lambda(x,z=Z_R,\underline{\Omega}=\underline{\Omega}^i)=0 \qquad (83)$$

where $\underline{\Omega}^i$ represents the directions of radiation intensity entering the reactor.

The radiation flux at the wall of radiation entrance (Figure 22) was determined by actinometric measurements (Zalazar et al., 2005). Additionally, the boundary condition for this irradiated wall ($x=0$) was obtained using a lamp model with superficial, diffuse emission (Cassano et al., 1995) considering: (i) direct radiation from the two lamps and (ii) specularly reflected radiation from the reflectors (Brandi et al., 1996). Note that the boundary conditions at the irradiated and opposite walls consider the effect of reflection and refraction at the air–glass and glass–liquid interfaces, as well as the radiation absorption by the glass window at low wavelengths (the details were shown for the laboratory reactor). The radiation model also assumes that no radiation arrives from the top and bottom reactor walls (x–y plane at $z=0$ and $z=Z_R$).

The discrete ordinate method (Duderstadt and Martin, 1979) was employed to solve the RTE (Equations 82 and 83). Afterward, the LVRPA was obtained according to

$$e^a(x,z)=\int_\lambda \kappa_\lambda \int_\Omega I_\lambda(x,z,\underline{\Omega})\mathrm{d}\Omega\mathrm{d}\lambda \qquad (84)$$

where the limit Ω indicates all radiation arrival directions at a given point (x,z).

4.7. Validation

Figure 23a shows the predicted and experimental concentrations versus time of 4-CP, 4-CC, and HQ for a catalyst mass concentration of 0.5×10^{-3} g cm^{-3}. As can be observed, the 4-CP concentration decreases throughout the experimental run following a first-order kinetics and the pollutant is completely degraded after 6 h of irradiation. This figure also shows the formation and destruction of 4-CC and HQ, with a maximum at approximately 1 h. Then, these two main intermediate species decrease gradually until they almost disappear at the end of the run. The changes in the 4-CC and HQ concentrations are consistent with the proposed kinetic mechanism reported in Section 4.3, and with the "series–parallel" kinetic model

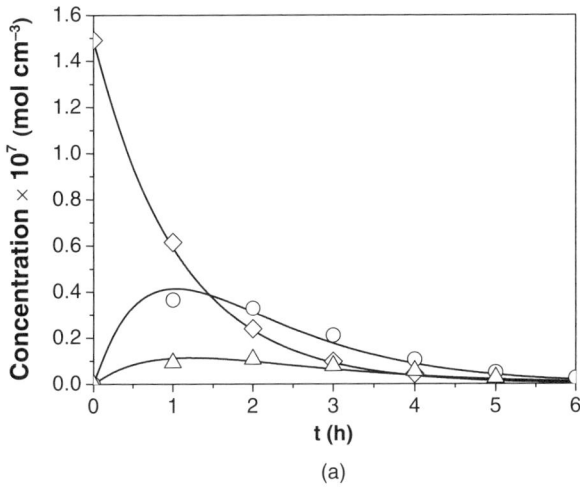

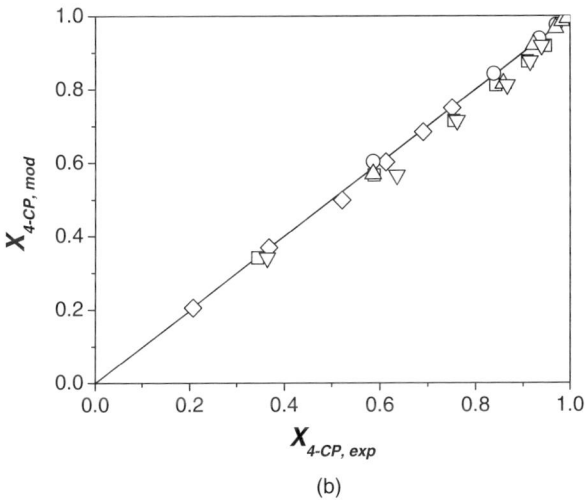

Figure 23 (a) Experimental and predicted concentrations versus time for $C_m = 0.5 \times 10^{-3}$ g cm^{-3}. Solid line: model predictions. Symbols: experimental data. $\Diamond$: 4-CP, $\bigcirc$: 4-CC, $\triangle$: HQ. (b) Experimental versus predicted conversions of 4-CP. Symbols correspond to different runs: $\Diamond$: $C_{4\text{-}CP,0} = 1.4 \times 10^{-7}$ mol cm^{-3} and $C_m = 0.05 \times 10^{-3}$ g cm^{-3}, $\square$: $C_{4\text{-}CP,0} = 1.4 \times 10^{-7}$ mol cm^{-3} and $C_m = 0.1 \times 10^{-3}$ g cm^{-3}, $\bigcirc$: $C_{4\text{-}CP,0} = 1.4 \times 10^{-7}$ mol cm^{-3} and $C_m = 0.5 \times 10^{-3}$ g cm^{-3}, $\triangle$: $C_{4\text{-}CP,0} = 1.4 \times 10^{-7}$ mol cm^{-3} and $C_m = 1.0 \times 10^{-3}$ g cm^{-3}, ∇: $C_{4\text{-}CP,0} = 0.7 \times 10^{-7}$ mol cm^{-3} and $C_m = 0.1 \times 10^{-3}$ g cm^{-3}.

proposed by Salaices et al. (2004) for the photocatalytic degradation of phenolic compounds in an annular slurry recycling reactor. The total organic carbon (TOC) evolution has been also reported by Satuf et al. (2008) for a typical experimental run (results not shown here). The TOC concentration decreases slowly, remaining a significant amount of organic carbon at the end of the run. This fact can be explained by the generation of aliphatic compounds, mainly aldehydes and carboxylic acids, not detected by high performance liquid chromatography analysis.

The quality of the scaling-up procedure can be seen when experimental and predicted 4-CP conversions for the pilot scale reactor are compared in Figure 23b. The symbols correspond to the 4-CP conversions obtained at different reaction times for all the experimental runs performed in this reactor. Predicted concentrations of 4-CP and 4-CC compared with experimental results show an RMSE lower than 9.91%. HQ was not included in the computation of the RMSE due to the low concentrations obtained during the experiments.

5. CONCLUSIONS

To some extent, the conclusions are very simple. These three examples, with three very different applications, corroborate the statement made in Section 1.

We have been able to scale-up results obtained in small-scale laboratory reactors results to reactors of completely different sizes, shapes, configurations, and operating conditions. The method has been illustrated with three typical and distinct applications: (i) a wall reactor, with an immobilized catalyst in a nonparticipating gaseous medium, (ii) a homogeneous photochemical process in a continuous flow, annular reactor operated in an aqueous medium, and (iii) a slurry reactor, working in a heterogeneous, participating medium, employing a photocatalyst with strong absorption and scattering properties. In all cases, the results have shown to be very satisfactory.

Four necessary conditions must be fulfilled in order to be able to move from laboratory experiments to larger scale reactor without employing (a) experimentally adjustable parameters and (b) costly successive physical changes of reactor sizes.

These conditions are (i) it is needed to have a validated kinetic scheme (a detailed mechanism or a precise empirical representation), (ii) it is needed to have a validated, intrinsic reaction kinetic expression as a function of position and time $[R(x, t)]$, (iii) it is needed to use in both reactors the same spectral radiation output power distribution [λ for monochromatic radiation and $f(\lambda)$ for the polychromatic cases], and (iv) it is needed to apply and correctly solve a rigorous mathematical model to both the laboratory and the large-scale reactor.

In other words, the scale-up of photochemical or photocatalytic reactors not only needs a good chemical knowledge of the reaction (which is indispensable), but the application of the fundamental principles of chemical reaction engineering that, in addition, almost always call for a decision to use without aversion, physics, mathematics, and numerical methods as well.

ACKNOWLEDGMENTS

The authors are grateful to Universidad Nacional del Litoral (UNL), Consejo Nacional de Investigaciones Científicas y Técnicas (CONICET), and Agencia Nacional de Promoción Científica y Tecnológica (ANPCyT) for financial support. We also thank Tec. Antonio C. Negro and Eng. Gerardo Rintoul for their participation in some parts of the experimental work and Eng. Claudia Romani for her technical assistance.

Table 7 has been reproduced from Helv. Chim. Acta; Reference: Labas et al., 2002, "Scaling-Up of a Photoreactor for Formic Acid Degradation Employing Hydrogen Peroxide and UV Radiation", Table; Copyright 2002, with permission from Helv. Chim. Acta.

Table 10 has been reproduced from Appl. Catal. B: Environ.; Reference: Satuf et al., 2008, Photocatalytic degradation of 4-chlorophenol: A kinetic Study, Table 4; Copyright 2008, with permission from Elsevier Ltd.

Table 11 has been reproduced from Appl. Catal. B: Environ.; Reference: Satuf et al., 2008, Photocatalytic degradation of 4-chlorophenol: A kinetic Study, Table 5; Copyright 2008, with permission from Elsevier Ltd.

Figures 1 and 2 have been reproduced from Int. J. Chem. React. Eng.; Reference: Alfano, O.M., Cassano, A.E., 2008, Photoreactor modeling. Applications to advanced oxidation processes, figures 1 and 2, Copyright 2008, with permission from The Berkeley Electronic Press.

Figures 3–8 and 10 have been reproduced from Chem. Eng. Sci.; Reference: Imoberdorf et al., 2007, Scaling-up from first principles of a photocatalytic reactor for air pollution remediation, figures 1–3, 5, 7, 8, and 10, Copyright 2006, with permission from Elsevier Ltd.

Figures 11–15 have been reproduced from Helv. Chim. Acta; Reference: Labas et al., 2002, Scaling-up of a photoreactor for formic acid degradation employing hydrogen peroxide and UV radiation, figures 1–7, Copyright 2002, with permission from Helv. Chim. Acta.

Figures 16–21 have been reproduced from Appl. Catal. B: Environ.; Reference: Satuf et al., 2008, Photocatalytic degradation of 4-chlorophenol: A kinetic Study, figures 1–4, 6, 10 and 11, Copyright 2008, with permission from Elsevier Ltd.

Figures 22 and 23 have been reproduced from Catal. Today, Vol. 129; Reference: Satuf et al., 2007b, Scaling up of slurry reactors for photocatalytic degradation of 4-chlorophenol, figures 3 and 4, Copyright 2007, with permission from Elsevier Ltd.

NOTATION

A	area (cm^2)
a_V	catalytic surface area per unit suspension volume (cm^{-1})
C	molar concentration $(mol\ cm^{-3})$
Cm	catalyst mass concentration $(g\ cm^{-3})$

c speed of light (m s^{-1})
D_i diffusion coefficient of species i in the mixture (cm^2 s^{-1})
E radiative energy (J)
e thickness (cm)
e^a local volumetric rate of photon absorption (Einstein cm^{-3} s^{-1})
$e^{a,s}$ local superficial rate of photon absorption (Einstein cm^{-2} s^{-1})
G incident radiation (Einstein cm^{-2} s^{-1})
g asymmetry factor (dimensionless)
I radiation intensity (Einstein cm^{-2} sr^{-1} s^{-1})
K equilibrium constant (m^3 mg^{-1})
L length (cm)
n number of times that a radiation beam has been attenuated by a glass or a film (dimensionless)
$\underline{n}$ unit normal vector (dimensionless)
P emission power (Einstein s^{-1})
p phase function (dimensionless)
Q volumetric flow rate (cm^3 s^{-1})
q net radiative flux (Einstein cm^{-2} s^{-1})
R radius (cm)
R reaction rate (mol cm^{-3} s^{-1})
r radial coordinate (cm); also, molar ratio C_P/C_F (dimensionless)
s linear coordinate along the direction $\underline{\Omega}$ (cm)
t time (s)
V volume (cm^3)
v_z axial velocity (cm s^{-1})
W gain or loss of energy (Einstein cm^{-3} sr^{-1} s^{-1})
X conversion (dimensionless)
x, y, z rectangular Cartesian coordinates (cm)
$\underline{x}$ spatial position vector (cm)
z axial coordinate (cm)

Greek letters

α_i kinetic parameter
β volumetric extinction coefficient (cm^{-1})
Γ global reflection coefficient (dimensionless)
ε_L liquid hold-up (dimensionless)
η directional cosine (dimensionless)
θ spherical coordinate (rad)
θ_n angle between the ray trajectory and the outwardly directed normal (rad)
κ volumetric absorption coefficient (cm^{-1})
λ radiation wavelength (nm)
μ directional cosine (dimensionless)

v stoichiometric coefficient (dimensionless)
σ volumetric scattering coefficient (cm^{-1})
ϕ spherical coordinate (rad)
Φ primary quantum yield (mol Einstein^{-1})
χ internal/external radius ratio (dimensionless)
Ω solid angle (sr)
$\underline{\Omega}$ unit vector in the direction of radiation propagation (dimensionless)

Subscripts

a	air property
c	relative to critical angle
exp	experimental
F	relative to formic acid
f	relative to film
g	relative to glass
HQ	relative to hydroquinone
i	relative to species i
L	relative to the lamp
mod	model
P	relative to hydrogen peroxide
PCE	relative to perchloroethylene
R	reactor property
T	total value
Tk	tank property
W	wall
λ	dependence on wavelength
$\underline{\Omega}$	relative to the direction of propagation
0	relative to the reactor window at $x = 0$; also initial condition
4-CC	relative to 4-chlorocatechol
4-CP	relative to 4-chlorophenol

Superscripts

abs	absorption
em	emission
in	inlet condition
in-scatt	in-scattering
ou	outlet condition
ou-scatt	out-scattering

Special symbol

$^{\circ}$	initial condition
$\langle . \rangle$	average value
$\overline{(.)}$	averaged value over the wavelength interval

REFERENCES

Alfano, O.M., Cabrera, M.I., and Cassano, A.E. "Photocatalytic reactions involving hydroxyl radical attack. I. Reaction kinetics formulation with explicit photon absorption effects". *J. Catal.* **172**, 370 (1997).

Alfano, O.M., and Cassano, A.E. "Photoreactor modeling. Applications to advanced oxidation processes". *Int. J. Chem. React. Eng.* **6**(P2), 1 (2008).

Alfano, O.M., Negro, A.C., Cabrera, M.I., and Cassano, A.E. "Scattering effects produced by inert particles in photochemical reactors. 1. Model and experimental verification". *Ind. Eng. Chem. Res.* **34** (2), 488 (1995).

Alfano, O.M., Romero, R.L., and Cassano, A.E. "A cylindrical photoreactor irradiated from the bottom. III. Measurement of absolute values of the local volumetric rate of energy absorption. Experiments with polychromatic radiation". *Chem. Eng. Sci.* **41**, 1163 (1986).

Almquist, C.B., and Biswas, P.A. "A mechanistic approach to modeling the effect of dissolved oxygen in photo-oxidation reactions on titanium dioxide in aqueous systems". *Chem. Eng. Sci.* **56**, 3421 (2001).

Blelski, B.H.J., Cabe III, D.E., Arudi, R.L., and Ross, A.B. "Reactivity of HO_2/O_2^- radicals in aqueous solutions". *J. Phys. Chem. Ref. Data.* **4**, 1041 (1985).

Brandi, R.J., Alfano, O.M., and Cassano, A.E. "Modeling of radiation absorption in a flat plate photocatalytic reactor". *Chem. Eng. Sci.* **51**, 3169 (1996).

Brandi, R.J., Alfano, O.M., and Cassano, A.E. "Rigorous model and experimental verification of the radiation field in a flat plate solar collector simulator employed for photocatalytic reactions". *Chem. Eng. Sci.* **54**, 2817 (1999).

Brandi, R.J., Rintoul, G., Alfano, O.M., and Cassano, A.E. "Photocatalytic reactors. Reaction kinetics in a flat plate solar simulator". *Catal. Today* **76**, 161 (2002).

Buxton, G.V., Greenstock, C.L., Helman, W.P., and Ross, A.B. "Critical review of rate constants for reactions of hydrated electrons, hydrogen atoms and hydroxyl radicals ($\cdot OH/\cdot O^-$) in aqueous solutions". *J. Phys. Chem. Ref. Data* **17**, 513 (1988).

Cabrera, M.I., Negro, A.C., Alfano O.M., and Cassano, A.E. "Photocatalytic reactions involving hydroxyl radical attack. II. Kinetic of the decomposition of trichloroethylene using titanium dioxide". *J. Catal.* **172**, 380 (1997).

Cassano, A.E., and Alfano, O.M. "Reaction engineering of suspended solid heterogeneous photocatalytic reactors". *Catal. Today* **58**, 167 (2000).

Cassano, A.E., Martín, C.A., Brandi, R.J., and Alfano, O.M. "Photoreactor analysis and design: Fundamentals and applications". *Ind. Eng. Chem. Res.* **34**, 2155 (1995).

Dijkstra, M.F.J., Panneman, H.L., Winkelman, J.G.N., and Kelly, J.J. *Chem. Eng. Sci.* **57**, 4895 (2002).

Duderstadt, J.J., and Martin, W.R. "Transport Theory". Wiley, New York, (1979).

Imoberdorf, G.E., Irazoqui, H.A., Alfano, O.M., and Cassano, A.E. "Scaling-up from first principles of a photocatalytic reactor for air pollution remediation". *Chem. Eng. Sci.* **62**, 793 (2007).

Imoberdorf, G.E., Irazoqui, H.A., Cassano, A.E., and Alfano, O.M. "Modeling of a multi-annular photocatalytic reactor for PCE degradation in air". *AIChE J.* **52**, 1814 (2006).

Imoberdorf, G.E., Irazoqui, H.A., Cassano, A.E., and Alfano, O.M. "Photocatalytic degradation of tetrachloroethylene in gas phase on TiO_2 films: A kinetic study". *Ind. Eng. Chem. Res.* **44**, 6075 (2005).

Irazoqui, H.A., Cerdá, J., and Cassano, A.E. "The radiation field for the point and line source approximations and the three-dimensional source models. Applications to photoreactions".*Chem. Eng. J.* **11**, 27 (1976).

Labas, M.D., Zalazar, C.S., Brandi, R.J., Martín, C.A., and Cassano, A.E. "Scaling-up of a photoreactor for formic acid degradation employing hydrogen peroxide and UV radiation". *Helv. Chim. Acta* **85**, 82 (2002).

Mills, A., and Morris, S. "Photomineralization of 4-chlorophenol sensitized by titanium dioxide: A study of the initial kinetics of carbon dioxide photogeneration". *J. Photochem. Photobiol. A.* **71**, 75 (1993).

Mills, A., Morris, S., and Davies, R. "Photomineralisation of 4-chlorophenol sensitised by titanium dioxide: A study of the intermediates". *J. Photochem. Photobiol. A.* **70**, 183 (1993).

Minero, C., Catozzo, F., and Pelizzetti, E., "Role of adsorption in photocatalyzed reactions of organic molecules in aqueous TiO$_2$ suspensions". *Langmuir.* **8**, 481 (1992).

Ozisik, M.N. "Radiative Transfer and Interactions with Conduction and Convection". Wiley, New York (1973).

Palmisano, G., Addamo, M., Augugliaro, V., Caronna, T., Di Paola, A., García López, E., Loddo, V., Marcí, G., Palmisano, L., and Schiavello, M. "Selectivity of hydroxyl radical in the partial oxidation of aromatic compounds in heterogeneous photocatalysis". *Catal. Today*, **122**, 118 (2007).

Salaices, M., Serrano, B., and de Lasa, H.I. "Photocatalytic conversion of phenolic compounds in slurry reactors". *Chem. Eng. Sci.* **59**, 3 (2004).

Satuf, M.L., Brandi, R.J., Cassano, A.E., and Alfano O.M. "Experimental method to evaluate the optical properties of aqueous titanium dioxide suspensions". *Ind. Eng. Chem. Res.* **44** (17), 6643 (2005).

Satuf, M.L., Brandi, R.J., Cassano, A.E., and Alfano O.M. "Photocatalytic degradation of 4-chlorophenol: A kinetic study". *Appl. Catal. B: Environ.* **82**, 37 (2008).

Satuf, M.L., Brandi, R.J., Cassano, A.E., and Alfano O.M. "Quantum efficiencies of 4-chlorophenol photocatalytic degradation and mineralization in a well-mixed slurry reactor". *Ind. Eng. Chem. Res.* **46**(1), 43 (2007a).

Satuf, M.L., Brandi, R.J., Cassano, A.E., and Alfano O.M., "Scaling-up of slurry reactors for the photocatalytic degradation of 4-chlorophenol". *Catal. Today*, **129**, 110 (2007b).

Siegel, R., and Howell, J.R., "Thermal Radiation Heat Transfer". 4th edition. Hemisphere Publishing Corp., Bristol, PA, 2002.

Stefan, M.I., and Bolton, J.R. "Mechanism of the degradation of 1,4-dioxane in dilute aqueous solution using the UV/hydrogen peroxide process". *Environ. Sci. Technol.* **32**, 1588 (1998).

Terzian, R., Serpone, N., Minero, C., Pelizzetti, E., and Hidaka, H., *J. Photochem. Photobiol. A.* **55**, 243 (1990).

Theurich, J., Linder, M., and Bahnemann, D.W. "Photocatalytic degradation of 4-chlorophenol in aerated aqueous titanium dioxide suspensions: A kinetic and mechanistic study". *Langmuir*, **12**, 6368 (1996).

Turchi, C.S., and Ollis, D.F., "Photocatalytic degradation of organic water contaminants: Mechanisms involving hydroxyl radical attack". *J. Catal.* **122**, 178 (1990).

Yamazaki, S., and Araki, K., "Photocatalytic degradation of tri- and tetrachloroethylene on porous TiO$_2$ pellets". *Electrochemistry*, **70**, 412 (2002).

Zalazar, C.S., Labas, M.D., Martín, C.A., Brandi, R.J., Alfano, O.M., and Cassano, A.E. "The extended use of actinometry in the interpretation of photochemical reaction engineering data". *Chem. Eng. J.* **109**, 67 (2005).

CHAPTER 8

Photocatalytic Treatment of Air: From Basic Aspects to Reactors

Yaron Paz

Contents

1. Introduction 290
2. Types of Air-Treatment Applications 293
   2.1 Indoor air treatment 294
   2.2 Outdoor air treatment 295
   2.3 Process gases 295
   2.4 Dissolved pollutants 296
3. Basic Aspects of Photocatalysis for Air Treatment 296
   3.1 Reaction kinetics 297
   3.2 Mass transport 300
   3.3 Light sources 301
4. Types of Target Pollutants 303
   4.1 BTEX 304
   4.2 Trichloroethylene 305
   4.3 NOx 308
   4.4 Mixtures 308
5. Photocatalytic Reactors for Air Treatment: Modes of Operation 310
   5.1 Batch reactors (including recirculation systems) 310
   5.2 Continuous, one-pass flow reactors 311
6. Types of Photocatalytic Reactors for Air Treatment 312
   6.1 Tubular reactors 312
   6.2 Annular reactors 322
   6.3 Flat plate reactors 326
   6.4 Generalized models and comparisons between reactors 327
   6.5 Combined adsorptive-photocatalytic reactors 328
7. Current Problems and Future Trends 329
   7.1 Visible light 329
   7.2 Mixtures 330
   7.3 Standardization 330
   7.4 Deactivation 331

Department of Chemical Engineering, Technion-Israel Institute of Technology, Haifa, Israel.
E-mail address: paz@tx.technion.ac.il

Advances in Chemical Engineering, Volume 36
ISSN 0065-2377, DOI: 10.1016/S0065-2377(09)00408-6

8. Concluding Remarks	331
Acknowledgment	331
List of Symbols	331
Abbreviations	332
References	333

1. INTRODUCTION

The growing awareness to our environment and to the conditions of living brings along an increasing scientific and commercial tendency to develop new, effective, and inexpensive means to improve the quality of water and air. Advanced oxidation processes (AOP) and in particular TiO_2 photocatalysis are now considered as true competitors to the classical techniques of purification. Accordingly, an increasing number of scientific manuscripts and patents can be found in the literature.

If one looks at the way photocatalysis has developed over the years, one finds out that the first applications to be considered following the pioneering paper of Fujishima and Honda (1972) were related to energy, that is, to water splitting, (Cunningham et al., 1981; Pelizzetti et al., 1981) as energy is something we all pay for, something whose shortage can be foreseen within, historicallywise, a very short time. It took awhile until the possibility of using the photocatalytic properties of titanium dioxide for environmental purposes was realized. From the chronicle point of view, the interest in water purification began prior to air treatment. In fact, this seniority, in terms of the number of scientific manuscripts, is still kept until these days. A quantified view of these statements is presented in Figure 1, which

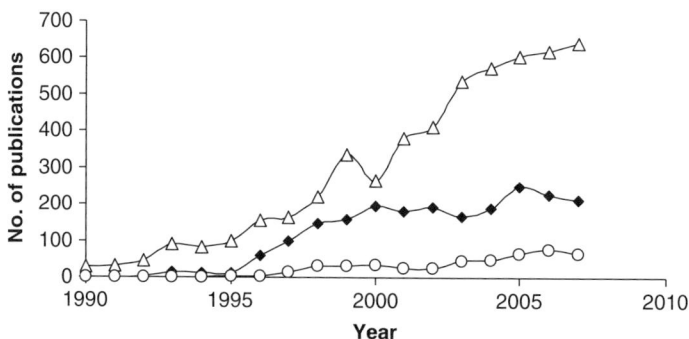

Figure 1 Estimated number of scientific manuscripts on titanium dioxide photocatalysis per year, categorized according to water treatment (empty triangles), air treatment (filled diamonds), self-cleaning surfaces (empty circles).

presents the estimated number of scientific manuscripts on titanium dioxide photocatalysis per year, categorized according to water treatment, air treatment, and self-cleaning surfaces.

The results presented in Figure 1 were acquired by counting hits upon using SciFinder Scholar™ (2007) as our search engine, having "titanium dioxide," "photocatalysis," and "air treatment/water treatment/self-cleaning" as keywords. It is obvious that the total number of publications on titanium dioxide photocatalysis is larger than the sum of papers in these three categories, as these cannot cover all the publications in the field, while the percentage of publications, which belong to more than one category, is (according to our impression) quite low. It is likely that other search engines might give different number of hits. It is further likely that a few of the relevant publications did not mention explicitly the medium in their abstracts. Yet, regardless of the exact numbers, it is reasonable to believe that the main features of the graphs, namely, the lag in the research on air treatment, the monotonically increase in the number of publications in each of the categories, and the fact that water treatment is still the subject of most of the TiO_2 photocatalysis manuscripts are genuine.

There could be several reasons for this situation: the common feeling that air (despite being essential for life) is not regarded as something that one (whether be a politician or the average person in the street) has to pay for, unlike water whose price is familiar to all (including grants agencies), in particular in the arid countries. It is also possible, that the fact that the first group of researchers came from the electrochemistry community played a role here. Anyway, whatever the reasons are, the higher number of publications on water treatment relative to air treatment cannot be denied. Interestingly enough, it seems that the scientific community was aware of this fact quite a few years ago (Alberici and Jardim, 1997). Nevertheless, it also seems that the scientific community was much less aware to the growing interest of the industrial and the commercial community, manifested, for example, by the number of patents in air – treatment.

Figure 2 presents the annual number of new patents on titanium dioxide photocatalysis categorized according to the three categories of Figure 1. From the figure, it is evident that the year 1995 was a turning point year with respect to the number of issued patents. The years 1995–2000 seem to be the "booming" years in which the annual number of patents grew fast and monotonically. Ever since year 2000, the annual number of patents remained constant more or less, probably indicating that the field enters its maturity period. The fact that as of 1997 the number of patents on air treatment surpasses that of water treatment is quite striking, especially if one takes into account that the number of scientific publications on water treatment is larger than that of air treatment (Figure 1).

The large annual number of patents on photocatalytic air treatment is portrayed again in Figure 3, which presents the distribution of patents

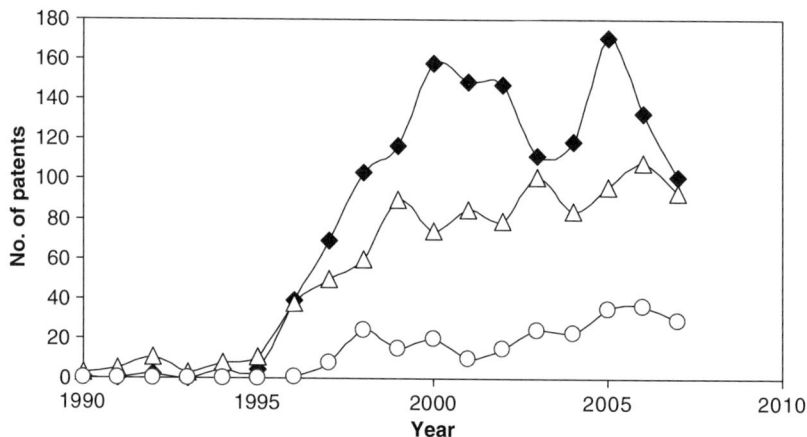

Figure 2 Estimated number of patents on titanium dioxide photocatalysis per year, categorized according to water treatment (empty triangles), air treatment (filled diamonds), self-cleaning surfaces (empty circles).

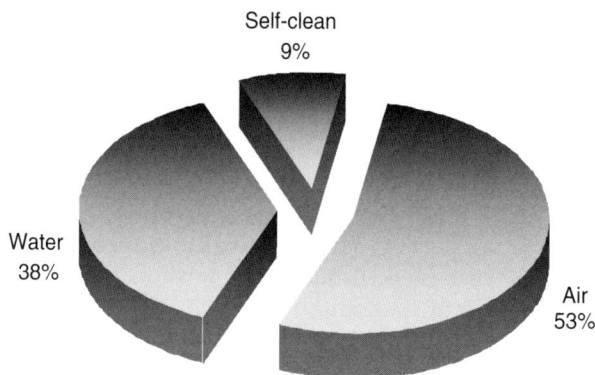

Figure 3 The distribution of patents between the three major categories (air treatment, water treatment, self-cleaning) on a cumulative basis.

between the above-mentioned three categories on a cumulative basis. Evidently, the number of patents on air treatment is larger than the sum of patents on water treatment and on self-cleaning surfaces. The fact that there are more patents on photocatalytic air treatment compared with the other two categories is not obvious, especially if one considers the fact that cleaned surfaces are in demand in the market, being tangible products that people are willing to pay for, and likewise are water, whose economical value is clear to anyone, even for those who do not use the so-called shadow cost of water.

This chapter tries to analyze the use of titanium dioxide for photocatalytic air treatment. For this, the various types of applications will be discussed, as well as the mechanisms and reaction kinetics of some of the most popular model contaminants. Care will be given to characterize the large variety of photocatalytic reactors in use for air treatment.

2. TYPES OF AIR-TREATMENT APPLICATIONS

In the context of gas-phase photocatalysis, it is possible to identify four major types of applications: indoor, outdoor, process gases, and dissolved pollutants. These four types have characteristic properties, which may (in certain cases may not) affect the effectiveness of photocatalysis relative to other means of treatment and the type of optimal photoreactor and material to be used.

Figure 4 presents the annual number of manuscripts dedicated to photo-catalytic indoor air treatment as well as the annual number of papers dedicated to photocatalytic outdoor treatment, as deduced from the number of SciFinder Scholar hits. Here, there was no point in trying to get information on the other two types, as the terms "process gases" and "dissolved pollutants" (or any other terms) are not associated strongly enough and specifically enough with these types of applications, hence the error, if such an attitude would have been taken, could have been quite large. The figure clearly demonstrates that the scientific interest in photocatalytic indoor air treatment is by far larger than that of photocatalytic outdoor air treatment.

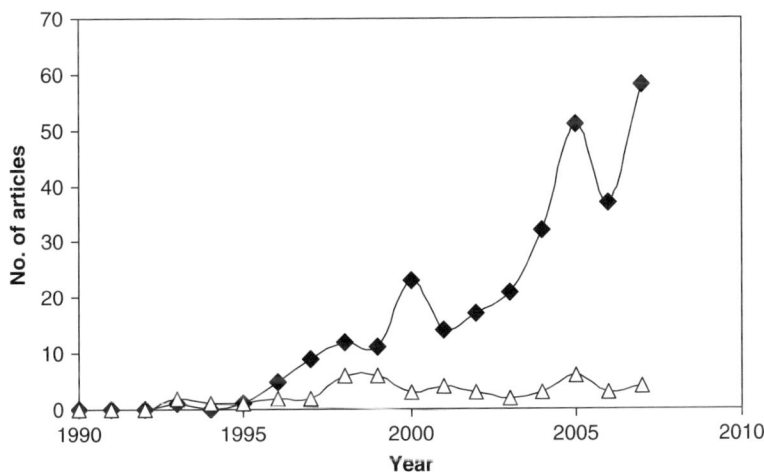

Figure 4   Estimated number of manuscripts per year on photocatalytic indoor air treatment (filled diamonds) and on photocatalytic outdoor air treatment (empty triangles).

Furthermore, the number of manuscripts on photocatalytic indoor air treatment seems to grow very fast over the last decade, in correlation with the trend of the overall number of manuscripts on photocatalytic air treatment (Figure 1).

2.1. Indoor air treatment

The modern human being lives in an urban environment and spends as much as 70–90% of his time indoors (Aguado et al., 2004), where ventilation, under the constrains of energy price, architecture, and lack of awareness of its importance, might be far less than optimal. The variety of chemicals emitted indoors is enormous. A simple ink-jet printer, one of so many potential sources for indoor pollutants, emits no less than 18 pollutants, among which are benzene, toluene, chloroform, methylene chloride, styrene, ozone, and many more (Lee et al., 2001). Even an innocent-looking device, such as the computer on which this manuscript is written at this moment, emits 100–200 µg of volatile organic compounds (VOCs) per hour (Destaillats et al., 2008). Investigations of gaseous contaminants in buildings, including buildings that suffer from the so-called sick building syndrome (SBS), have shown that the concentrations of individual species are as low as 0.1 parts per million by volume (ppmv) and that the total concentration of VOCs are between 0.5 and –2.0 ppmv (Obee and Brown, 1995). The indoor level of pollutants was found to be on the average higher than that at the nearby outdoor environment, indicating the importance of indoor sources (Saarela et al., 2003). No wonder that the Environmental Protection Agency considered indoor air pollution to carry a higher health risk than outdoor air pollution (EPA, 1987). It should be emphasized that within the context of this chapter, the term "indoor air," includes any confined places having levels of pollutants which are above the ambient concentrations outside of the confined place. Accordingly, the term is relevant not only to buildings but also to transportation vehicles, aircrafts, storehouses, and the like.

Photocatalysis seems to be well suited for the purification of indoor air, in particular if compared with purification of water. A very good analysis of the logics of this statement was given by Agrios and Pichat (2005) who pointed out that the low concentrations of pollutants found in air facilitates continuous operation without saturating the surface of the photocatalyst. There are also several drawbacks associated with using photocatalysis for indoor air treatment. These drawbacks are often related to the formation of by-products or end-products that block the active sites. Among the deactivating pollutants, one may mention benzene and other aromatic compounds, trichloroethylene (TCE), and, probably most important, organo-silicone compounds that become more and more in use as sealants.

2.2. Outdoor air treatment

Several researchers have claimed that roads painted with titanium dioxide or the use of buildings covered with cementitious materials containing TiO_2 may improve the quality of ambient air (Takeuchi, 1998), for example, by removing the aromatic group of compounds known as BTEX (Strini et al., 2005). Indeed, measurements under conditions that simulated canyon-type streets measured NOx concentration values that were 37–82% lower than the ones measured in the absence of TiO_2 coating (Maggos et al., 2008). Nevertheless, due to mass transport limitations (the need for the pollutants to reach the photocatalyst surface), practical use of photocatalysis for air decontamination is expected to be restricted to confined places such as canyon streets, as was justly pointed out by Agrios and Pichat (2005). It is noteworthy that canyon streets are more prone to collect pollutants, so in that sense, the photocatalytic solution may enter exactly at a point where it is needed. Also, photocatalysis can be specifically adequate for outdoor treatment of NOx, emitted at large by diesel vehicles, as the nitric acid formed during the photocatalytic oxidation can be washed away by rain.

Outdoor air treatment utilizes, in general, photocatalytic cementitious materials or sprayed photocatalytic coatings on walls, roads, and buildings. Quite often, these types of coatings are referred as "self-cleaning coatings," as photocatalytic coatings on construction materials act to prevent the adsorption of soot or dust that tend to stick to grimy surfaces. Although TiO_2-coated surfaces can serve, in principle, for the dual purpose of self-cleaning and outdoor air treatment, optimization for one type of application does not necessarily coincides with optimization for the second type of application. For example, for outdoor air treatment, a corrugated, high surface area film can do better than a smooth film, whereas for self-cleaning a smooth surface is preferred.

2.3. Process gases

Several works were aimed at treating gaseous pollutants, emitted as point sources at a defined location. This type of application is characterized by the fact that the pollutant is well defined in terms of compound and range of concentrations. For most cases, the concentration of the pollutant emitted from an industrial point source is higher than in indoor or in outdoor application, and the conditions are usually so that a true one-pass treatment is needed, unlike in indoor air treatment. Knowing the type of pollutant may assist in optimizing the design and operational conditions of the photoreactor. For example, some pollutants are photooxidized best on Degussa P-25 while other are degraded best on TiO_2 particles produced by Hombikat (Wang et al., 2002). One may even think on designing photocatalysts with high specificity for specific pollutant, for example, by the "Adsorb and Shuttle" (A&S) approach (see below, Ghosh–Mukerji et al., 2003; Paz, 2006; Sagatelian et al., 2005) or by the imprinting method (Sharabi, submitted). Likewise, the optimal level of humidity depends, to some extent, on the type of pollutant. Hence a proper design should include means to control humidity at its pollutant-specific level. Also, if the pollutant-emitting process is not continuous, one should consider the coupling of an adsorbing-desorbing device together with a continuously operating photocatalytic system, thus continuing to photooxide the pollutants during idle times.

Examples for treatment of process gases include treating low-concentration gases emitted from pulp and paper mills such as methanol by using an annular packed bed reactor (PBR) (Stokke et al., 2006) and the photocatalytic degradation of pyridine, used widely in the synthesis of vitamins, drugs, and rubber chemicals, by a photocatalytic reactor based on zeolite-supported TiO_2 (Sampath et al., 1994). To these examples one may add also the oxidation of NOx to HNO_3 by titanium dioxide, as well as the reduction of NOx to N_2 by a photocatalytic zeolite matrix hosting less than 1% of TiO_2 (Kitano et al., 2007).

2.4. Dissolved pollutants

Photoreactors for dissolved pollutants operate by bubbling a carrier gas (usually air) through a reservoir of liquids (usually water). The pollutant-loaded gas is then passed though the photocatalytic reactor and released. Several types of reactors have been connected with this application, including PBRs and coated wall reactors.

To some extent, such photoreactors can be considered as a subtype of the process-gas photoreactors, as they are used to reduce the emission of volatile compounds formed in chemical processes into the ambient air. It should be noted that the practical use of this application is limited by volatility of the liquid-phase pollutants, so it is inadequate for low vapor pressure pollutants.

3. BASIC ASPECTS OF PHOTOCATALYSIS FOR AIR TREATMENT

As mentioned before, the photocatalytic process was studied extensively for water treatment, air treatment, self-cleaning of surfaces, as well as in the context of medical applications and energy conversion. Throughout the years many insights on the fundamentals of photocatalysis were obtained. Some of the insights, closely related to air treatment, are summarized briefly hereby.

3.1. Reaction kinetics

There is an intensive literature showing that in the absence of mass transfer limitations, many photocatalytic reactions follow the Langmuir–Hinshelwood (LH) kinetics described in Equation (1) for a one-component system:

$$(-r_A) \equiv -\frac{dC_A}{dt} = k_{r,A}\vartheta_A = \frac{k_{r,A}K_A C_A}{1 + K_A C_A} \qquad (1)$$

Here, C_A is the concentration of pollutant, t is the time, $k_{r,A}$ is the reaction rate constant, K_A is the adsorption coefficient of A, and θ_A is the surface coverage of this specie. Gas-phase examples include the degradation of toluene (Bouzaza and Laplanche, 2002), methyl tert-butyl-ether (Boulamanti and Philippopoulos, 2008), dimethylamine (Kachina et al., 2007), ethanol (Vorontsov and Dubovitskaya, 2004), and other VOCs (Mills and Le Hunte, 1997). When $K_A C_A$ is small, this equation is degenerated into a typical first-order expression, where k_{app} the apparent first-order constant is approximately $k_{r,A}K_A$. Since the apparent rate constant depends on both $k_{r,A}$ and K_A, a lower adsorption constant does not always result in a lower degradation rate, as demonstrated by Bouzaza et al. (2006).

It is noteworthy that observation of LH kinetics does not necessarily imply a surface reaction (Turchi and Ollis, 1990), although in most cases this may be correct. In addition, one should not overlook a possible contribution from oxidizing species that leave the photocatalyst surface and operate by the so-called remote degradation effect (Haick and Paz, 2001; Lee and Choi, 2002; Murakami et al., 2007; Tatsuma et al., 2001; Zemel et al., 2002).

Presenting Equation (1) in its reciprocal form,

$$\frac{1}{(-r_A)} = \frac{1}{k_{r,A}} + \frac{1}{k_{r,A}K_A}\frac{1}{C_A} \qquad (2)$$

provides an easy way to estimate the involved parameters, by making rate measurements at different concentrations. To reduce uncertainties, it is common to apply the method of initial rates to this equation, that is, to plot the reciprocal of the initial rate $(1/-r_{A0})$ versus the reciprocal of the initial concentration $(1/C_{A0})$ for various initial concentrations of the pollutant. Then, $k_{r,A}$ and K_A are estimated from the intercept and slope of the graph. The initial rates at specific initial concentrations are usually obtained by extrapolating concentration–time runs back to time zero. It was argued that the initial rate method suffered from a serious drawback due to the subjectivity in estimating the initial rates and that the method was in particular inadequate for multicomponent systems. Accordingly, a technique based on a well-known method of nonlinear estimation [the Box–Draper technique (Box and Draper, 1965)] was adapted for photocatalysis and successfully demonstrated in the photocatalytic degradation of TCE (Mehrvar et al., 2000).

Real life may involve the participation of more than one species in the photocatalytic process, due to existence of several pollutants in the feed and/or due to formation of intermediate products. In the simplest case, this can be described by a competitive adsorption over one type of sites, such that the disappearance rate of specie i can be given by

$$(-r_i) \equiv -\frac{dC_i}{dt} = k_{r_i}\vartheta_i = \frac{k_{r_i}K_iC_i}{1 + \sum_j K_jC_j} \tag{3}$$

A common example of the use of this type of equation is the considering of water molecules as competitive adsorbates (Junio and Raupp, 1993; Peral and Ollis, 1992).

The photocatalytic degradation of gaseous ethanol involves the formation of acetaldehyde as the main gaseous intermediate product (Vorontsov et al., 1997). It was found that in this case, fitting the kinetic data to a one-site model gave a very poor fit. Instead, a three-site photocatalytic model was proposed (Vorontsov and Dubovitskaya, 2004), following a previous work (Muggli and Falconer, 1998), which found that there were sites available only for acetaldehyde, sites available for both acetaldehyde and ethanol, and sites available only for ethanol. According to this model

$$(-r_{ET}) = (-r_{ETEX}) + \frac{k_{ET}K_{ET}C_{ET}}{1 + K_{ET}C_{ET} + K_AC_A} \tag{4}$$

$$(-r_A) = \frac{k_A^A K_A^A C_A}{1 + K_A^A C_A} + \frac{k_A K_A C_A}{1 + K_{ET}C_{ET} + K_AC_A} \tag{5}$$

where $(-r_{ET})$ and $(-r_A)$ are the rate of disappearance of ethanol and acetaldehyde, respectively. K_{ET}, k_{ET} and K_A, k_A are the adsorption constants and the rate constants of ethanol and acetaldehyde, on sites that are available for both. $(-r_{ETEX})$ is the ethanol photocatalytic oxidation on sites not available for acetaldehyde, and K_A^A, k_A^A are the adsorption constant and the rate constants of acetaldehyde on sites that are available only for acetaldehyde. A clear asymmetry is noticed in Vorontsov's model between the expression for ethanol and the expression for acetaldehyde. This asymmetry was rationalized by a competition between ethanol and surface intermediate products, such as acetic acid (Muggli et al., 1998).

Water adsorbed on TiO_2 is the source for hydroxyl radicals needed to initiate photocatalytic reactions. On the other hand, water competes with the pollutants on adsorptive sites. It is for this reason that one finds out that the performance of photoreactors varies significantly with the humidity, having a maximal efficiency at some intermediate level, which depends on the adsorpticity (capability of adsorption) of the contaminants as well as on the reaction pathway. For example, the photodegradation rate of *m*-xylene was found to increase with humidity up to 7% RH. Further increasing of the relative humidity caused a gradual decrease in the degradation rate (Peral and Ollis, 1992). In general, this optimal point is obtained already at a very low relative humidity, such that within the practical range of use, water acts to reduce the photodegradation rates of most contaminants. Increasing the initial concentration of the pollutants may shift the optimal point toward higher RH values (Obee and Brown, 1995).

A major obstacle for large-scale implementation of photocatalysis stems from the fact that photodegradation kinetics depends to large extent on the adsorption coefficient of the contaminants on the photocatalyst surface, such that it is extremely difficult to pohotodegrade molecules that hardly adsorb on the polar TiO_2. Unfortunately, many hazardous contaminants, in particular contaminants of low polarity, belong to this category. Moreover, in practice, the effluents that are needed to be treated contain several contaminants that compete on the adsorptive sites of the photocatalyst, thus limiting the validity of laboratory scale, single-component experiments, for practical use. These problems can be aggravated by lack of control on the formation of harmful by-products and toxic intermediates.

Attaching photocatalyst particles to inert domains may assist in improving the efficiency of the photocatalytic process. The basic concept, termed as "Adsorb & Shuttle," is based on using the inert domains for adsorbing target compounds that otherwise hardly adsorb on the photocatalyst. That way, a reservoir of the contaminants within a small distance from the photocatalytic sites is formed. Once adsorbed in the vicinity of the photocatalyst, the target molecules may surface-diffuse to the photocatalytic sites as shown schematically in Figure 5.

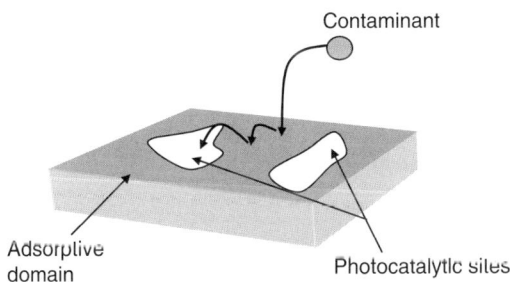

Figure 5 The concept of Adsorb and Shuttle.

Two typical gas-phase examples are the photocatalytic degradation of gaseous pyridine over a zeolite-supported titanium dioxide (Sampath et al., 1994) and the photodecomposition of propionaldehyde in air by a composite material made of titanium dioxide and activated carbon (AC) (Yoneyama and Torimoto, 2000). The photocatalytic activity of the composite photocatalysts was assessed by probing degradation kinetics and product distribution of the model contaminants. It was established that composite photocatalysts affect not only the degradation rates but also the distribution of intermediate products as well as the distribution of end-products. The extent by which these effects are manifested is governed by a large number of factors such as the surface area, the adsorpticity of the surface, the loading, the domain size, and even operational parameters such as humidity or temperature.

Too strong interaction between the adsorbate and the adsorbent might harm the diffusibility of the contaminant, thus canceling any synergistic effect. This was exactly what was found to happen at the photodegradation of dichloromethane on a composite photocatalyst consisting of titanium dioxide and AC (Torimoto et al., 1997). Here, the rate of CO_2 production was found to decrease monotonically with the increasing of the mass ratio of AC in the composite photocatalyst, even when the comparison was made on an equal amount of TiO_2 basis. In other words, in this case, not only that adding the inert adsorbent did not have any benevolent effect on the degradation rate, but, in fact, it competed against the photocatalyst, thus reducing the mineralization rate instead of enhancing it.

It is noteworthy that a large number of material-related parameters govern the activity of titanium dioxide in the gas phase. Among these parameters, one may outline chemical purity, phase (anatase or rutile), defect concentration, crystal size, porosity, surface area, and even aggregate size and particle–particle interaction (Ibrahim and de Lasa, 2002). In principle, these factors are basically the same as those in the aqueous phase and therefore will not be discussed at length within this chapter. We are not aware of any systematic work on a possible correlation between specific forms of titanium dioxide and specific types of use. However, it should be noted that, regardless of the experimental conditions, no shortage of oxygen can be expected for air treatment (unlike the situation with water containing high concentration of pollutants). At the same time, hydroxyl radicals are expected to be more abundant in the aqueous phase, hence the odds that a specific pollutant will be degraded by a direct mechanism of hole oxidation are larger in air treatment than in water treatment.

3.2. Mass transport

Mass transport, whether external (transfer from the bulk to the external surface of the photocatalyst) or internal (transfer within the pores of the photocatalyst), can, in principle, limit the overall degradation rate of pollutants. It seems that most experiments where photocatalysis was applied for the decontamination of air did not involve mass transfer limitations conditions and were dominated by reaction kinetics rather then mass transport. An example is the degradation of methanol and TCE in both a tube reactor and in an annular reactor, where the reaction rate was found to be independent of the gas flow rate (within the experimental range) (Doucet et al., 2006). One of the few cases where mass transfer did limit the overall degradation rate was presented by Sclafani et. al. (1993), who oxidized phenol in a fixed bed annular photoreactor and observed that the reaction rate was strongly dependent on the flow rate of the feed.

To check for internal mass transfer limitation, it is possible to use the nondimensional Weisz modulus, φ', which is a modified version of the Thiele modulus (Levenspiel, 1998):

$$\varphi' \equiv \frac{r_V \tau L^2}{D C_S \varepsilon} \tag{6}$$

Here, r_v is the experimental mean rate of reaction per unit volume of catalyst, L is a characteristic length of the porous photocatalyst (i.e., the film thickness), τ is the pore tortuosity (taken as three), D is the diffusion coefficient of the pollutant in air, C_s is the mean concentration at the external surface, and ε is the catalyst "grain" porosity (0.5 for Degussa's P25). Such a treatment was performed by Doucet et al. (2006) while taking D of the pollutants to be approximately 10^{-5} m^2 s^{-1}. The estimated Weisz modulus ranged between 10^{-3} and 10^{-5}, depending on the type of pollutant, that is, some three to five orders of magnitude smaller than the value of unity, which is often taken as a criterion for internal mass transport limitation.

3.3. Light sources

Light sources are among the most important parts of photocatalytic devices, based on the fact that photons are often regarded as the most expensive component of photocatalytic reactors (Nicolella and Rovatti, 1998). Hence, it is obvious that criteria for effective use of photons should be very important in the design and operation of photocatalytic devices. Unfortunately (or not), the odds that lamp manufacturers will produce UV lamps especially designed for photocatalysis for a competitive price are very slim. As a consequence, the design and even the size of a feasible reactor is very much constrained by the commercial availability of the radiation source (Imoberdorf et al., 2007).

Indoor air treatment is usually done with artificial light, whereas outdoor treatment is considered to be a solar light-induced process [maybe except for highway tunnels (Fujishima et al., 1999)]. Unlike water treatment, where light concentrators are sometimes in use (Monteagudo and Duran, 2006), air-treatment reactors rarely use concentrated light, following (sometimes unknowingly) the observation that above 1–2 mW cm^{-2} the quantum efficiency decreases with increasing UV flux. At any case, means for prevention of photon loss, such as mirrors, are quite common.

The most popular UV light sources are mercury plasma lamps, although sodium, zinc/cadmium, neon, and argon can be found in the market. The mercury lamps, emitting mostly in the UV-C range (200–280 nm), can be classified according to their pressure (Bolton et al., 1995). Low-pressure sources are characterized by short wavelength (80% of their output around 254 nm), long life time (> 5,000 h), 30% energy conversion, and very low energy density (~ 1 W cm^{-2}). Medium pressure lamps have shorter life time ($\sim 2,000$ h.), broad output range, moderate energy density (~ 125 W cm^{-1}) but low energy conversion ($\sim 15\%$). High-pressure mercury lamps have a life time of $\sim 3,000$ h, with strong emission below 250 nm, high energy density (~ 250 W cm^{-1}), and high energy conversion ($\sim 30\%$ into 200–300 nm light).

UV-A (315–400 nm) radiation is also used. In that case, the light is usually obtained by fluorescence caused by a mercury-lamp emission, using fluorescent media such as lead-doped barium silicate, europium-doped strontium borate, or europium-doped strontium fluoroborate. In addition, one may still find lamps that utilize the slightly broadened 365 nm spectral line of mercury from high pressure discharge. The UV-A lamps are often coated with a nickel-oxide-doped glass (Wood's glass), which blocks the visible light almost completely.

To large extent, the scientific literature is at least partially silent with respect to the effect of wavelength on photocatalysis. At most, there is some consideration to the fact that the absorption coefficient for 365 nm photons is significantly smaller than that for 254 nm photons, hence a thicker layer of photocatalyst is required with the former. Another effect of the shorter wavelength is the potential formation of active species that might play a role in the photooxidation reaction.

Accurate measurement of photon flux is essential in order to calculate quantum efficiency (whether based on absorbed photons' flux or the flux of photons impinging on the surface). The response of commercial sensors is wavelength dependence, hence, the readout, which is in energy flux units (W cm^{-2}), is usually calibrated according to either the 365 nm line of mercury or the 254 nm line, depending on the sensor. For more details on radiation sources see de Lasa et al. (2005). A different method for measuring photon flux is actinometry. This method was very popular in the past, however, very few researchers still use it due to its complexity and the time it consumes.

The efficiency of reactors is often given in terms of quantum yields (also called quantum efficiencies). One may find in the literature three types of quantum efficiencies, as outlined hereby. The primary quantum yield (PQY) is defined as the number of molecules degraded from a primary process over the number of absorbed photons (Cassano et al., 1995). The overall quantum yield (OQY) is defined as the ratio between the total number of pollutant molecules degraded via both primary and secondary processes over the total number of absorbed photons. Both PQY and OQY are often measured at initial conditions. Determining the exact number of absorbed photons is not easy, due to the diffused and specular reflectivity of the photocatalyst surface. For this reason, a third term [apparent quantum yield (AQY), also called global quantum yield] is often used. It is defined as the number of converted reactant molecules over the number of photons entering the reactor (Fox and Dulay, 1993). For more details about the definition of quantum yields see de Lasa et al. (2005). In many cases authors use the term "quantum efficiency" or "quantum yield" without defining whether they refer to PQY, OQY, or AQY. This is very unfortunate as the quantum efficiency is one of the major parameters when comparing different reactors.

As will be discussed later in this chapter, the geometrical attributes of photocatalytic reactors are closely related to the geometry and irradiance of existing UV sources. It is for this reason, that a large portion of the scientific efforts in developing models for photocatalysis was concentrated and is still concentrated in trying to model the distribution of light intensity at the surface of the photocatalyst, taking into account absorption and multiple scattering and in calculating the penetration depth of the light. Most of these models were constructed to meet specific geometrical designs (annular, tubular, etc.) or specific modes of introducing the catalyst (packed bed, coated film, etc.) and are discussed in the following parts according to the specific types of reactors.

4. TYPES OF TARGET POLLUTANTS

The number of gas-phase pollutants whose photocatalytic degradation was studied is quite large. Some of these pollutants such as aromatic compounds, chlorinated olefins, hydrocarbons (Obee and Brown, 1995), aldehydes (Chin et al., 2006), ethers (Araña et al., 2008) and alcohols (Tsuru et al., 2006) can be found indoor at sub-ppm level. Others, like NOx, are more typical for outdoor environment. The main groups are probably volatile aromatic compounds belonging to the BTEX family and short chlorinated hydrocarbons, like TCE. Figure 6 presents the estimated annual number of scientific publications on the photocatalytic air treatment of NOx, BTEX, and TCE based on SciFinder Scholar™ hits. These compounds were chosen based on the fact that they belonged to different chemical classes. NOx is representative for nonorganic compounds, BTEX

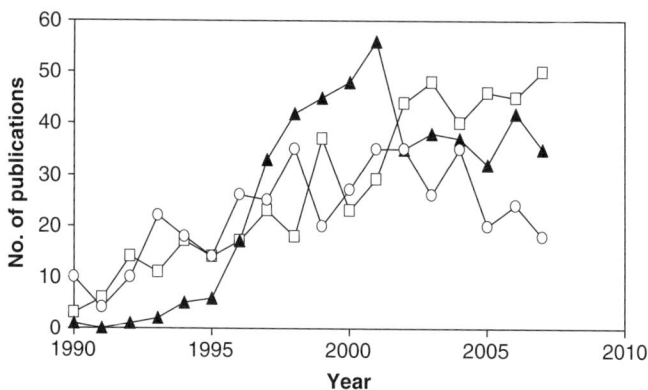

Figure 6 Estimated annual number of scientific publications, on the photocatalytic air treatment of NOx (filled triangles), BTEX (empty squares), and TCE (empty circles).

represents compounds that are degraded by hydroxyl mechanism, and TCE represents a specific case, where very high apparent quantum efficiency (AQE) is obtained due to radical chain reaction. In this context, it is noteworthy, that radical chain reaction leading to quantum efficiencies (OQY and even AQY), greater than 100%, are not limited to chlorinated compounds, as they were observed also for nonchlorinated molecules such as acetaldehyde (Ibrahim and de Lasa, 2003).

According to the figure, during the first years of research BTEX and TCE were more popular than NOx; however, over the years the importance of NOx in terms of publications was increased such that more and more scientific efforts were directed toward its degradation. As for TCE, it seems that it lost some of its early years popularity, most likely due to regulations that limited its use and consequently also its long-term potential threat. In the following part, the characteristics of these three gases as well as the main scientific findings regarding their photocatalytic degradation are discussed.

4.1. BTEX

Benzene, toluene, ethylbenzene, and xylenes (known as BTEX) are probably the most widely used aromatics, in abundant use in automotive fuel, as solvents or as feedstock for more complex compounds. The American Occupational and Safety Administration (OSHA) permissible exposure limit for benzene, for example, is as low as $3.26\,mg\,m^{-3}$ (1 ppm) due to its carcinogenic nature. This value should be compared with the $10–100\,\mu g\,m^{-3}$ of BTEX typically found in urban outdoor environment (Saarela et al., 2003). The photocatalytic degradation of BTEX might emit a variety of intermediate products and by-products. For example, the photocatalytic degradation of xylene was found to release benzaldehyde, 2,5-furandione, methyl-benzaldehydes, methyl benzoic acid, benzene dicarboxaldehyde, and 1,3-isobenzofurandione (Blanco et al., 1996). It is known that the concentration of pollutants might have a significant effect on the appearance and distribution of intermediate products as well as of end-products. This statement is in particular true for aromatic compounds, where low aromatic concentration has been shown to produce complete oxidation with little or no production of gas-phase intermediates (Ibsuki and Takeuchi, 1986; Obee and Brown, 1995), in contrast to degradation of 1.3 mol% in air, which led not only to the appearance of high concentrations of benzaldehyde but also to the production of benzene, benzyl alcohol, benzoic acid, and phenol (Martra et al., 1999).

As a consequence of the production of strongly adsorbed intermediates, the degradation of aromatics might not obey an apparent LH expression (Doucet et al., 2006). More important than the alteration of the kinetics expression is the gradual decrease in the degradation rate due to deactivation of the catalyst during continuous operation (Lewandowski and Ollis, 2003). The reason of deactivation is the formation of strongly bound intermediate products, which accumulate on the catalyst surface. The exact nature of these recalcitrant species is not always known. Even for toluene, whose degradation was studied intensively by numerous groups, there is still some debate on what are the chemical species responsible for deactivation, often accompanied by discoloration of the photocatalyst. Some researchers claimed that benzoic acid is the intermediate product to be blamed (Mendez-Roman and Cardona-Martinez, 1998), whereas others claimed that benzoic acid was not recalcitrant enough (Larson and Falconer, 1997). There is also a report on a mixture containing benzoic acid, benzyl alcohol, benzaldehyde, 4-hydroxybenzyl alcohol, 4-hydroxybenzaldehyde, and 3-hydroxybenzylaldehyde (d'Hennezel et al., 1998). In any case, whatever the actual recalcitrant species are, it is accepted that the aromatic ring structure was retained (Blount and Falconer, 2001).

Some regeneration was possible when the photocatalyst was exposed to UV in the presence of dry air (Luo and Ollis, 1996). However, this regeneration did not last for more than a few cycles, as the deactivation became irreversible, most likely due to accumulation of benzoic acid on the surface of the photocatalyst.

Overall, the BTEX compounds are among the most problematic pollutants, as they are both hazardous, quite abundant, and very difficult to handle.

4.2. Trichloroethylene

Trichlotoehtylene is a well-known pollutant. In aqueous systems this pollutant, like other chlorinated compounds such as chloroform, might be introduced into water following standard antibacterial treatment of drinking water by chlorine. In the gas phase, its main sources are related to its superb ability to dissolve organic compounds, hence is used in a variety of applications, from dry cleaning to the microelectronic industry.

The photocatalytic degradation of TCE was studied intensively, both in the liquid phase and in the gas phase. Unlike in the aqueous phase, where the quantum efficiency is no more than a few percents (Alberici and Jardim, 1997; Pruden and Ollis, 1983), the quantum efficiency in the gas phase can be higher than 100% (Upadhya and Ollis, 1998). This difference was attributed to the existence of two mechanisms. That the mechanism in the liquid phase was different than that in the gas phase could be deduced also by the fact that the intermediate products dichloroacetaldehyde (DCA) and dichloroacetic acid (DCAA) were identified only during liquid-phase photocatalysis (Pruden and Ollis, 1983).

The photocatalytic degradation of TCE was studied in the gas phase for the first time by Dibble and Raupp (1990, 1992). A positive order was found with respect to the molar ratio of oxygen and TCE, whereas for water a negative order of reaction was observed. No intermediate products were observed. It was clear that traces of water were required in order to preserve photocatalytic activity for long time, however, high relative humidity reduced the reaction rate. The rate could be described by a modified LH equation:

$$R = k' \left\{ \frac{K_1 y_{TCE}}{1 + K_1 y_{TCE} + K_3 y_{H_2O}} \right\} \left\{ \frac{K_2 y_{O_2}/y_{H_2O}}{1 + K_2 y_{O_2}/y_{H_2O} + K_4 y_{H_2O}} \right\}^2 \qquad (7)$$

where $K_1 - K_4$ are constants, and y_i represents the molar ration of component i. Phillips and Raupp (1992) studied the mechanism of the photocatalytic oxidation of TCE and concluded that UV-induced water desorption was a prerequisite for TCE adsorption and degradation. In fact, pre-exposure of the photocatalyst to UV light in the absence of any pollutants for some time may increase the photocatalytic degradation rate of TCE by a factor of 3.5 (Kim et al., 1999a, b).

The photocatalytic degradation of TCE could be performed very efficiently, with reported quantum efficiency values that could reach 40–90% in a fixed bed reactor (Yamazaki-Nishida et al., 1993). It is noteworthy that the quantum efficiency based on mineralization was quite smaller (4–17%) than the quantum efficiency based on degradation.

A thorough work on the photocatalytic oxidation of TCE was performed by Nimlos et al. (1993), who identified, apart from CO_2 and HCl, also Cl_2, dichloroacetyl chloride (DCAC ($CHCl_2COCl$)), CO, and phosgene ($COCl_2$), known to be much more toxic than TCE. The high quantum efficiency observed in this case (50–80%) was attributed to a chain reaction initiated by the formation of chlorine atoms, similar to the scheme proposed for the homogeneous case (Sanhueza et al., 1976). The chlorine atoms are formed following an attack of OH radicals on the TCE molecules:

$$CCl_2\!=\!CHCl + OH\bullet \rightarrow \bullet CCl_2CHClOH \qquad (8)$$

$$\bullet CCl_2CHClOH + O_2 \rightarrow \bullet OOCCl_2CHClOH \qquad (9)$$

$$2\bullet OOCCl_2CHClOH \rightarrow 2\bullet OCCl_2CHClOH + O_2 \qquad (10)$$

$$\bullet OCCl_2CHClOH \rightarrow CHClOHCCl(O) + Cl\bullet \qquad (11)$$

The chlorine atom attacks another TCE molecule to form an alkyl radical, which reacts with oxygen, thus forming a peroxo radical. Two peroxo radicals form the alkoxy radical, which reacts with a second peroxo radical yielding the alkoxy radical $CHCl_2CCl_2O\bullet$. This in turn may release chlorine to form DCAC or alternatively cleave a C—C bond to form phosgene (CCl_2O). The same group found, at a different study, that if the contact time is long enough, the phosgene may be hydrolyzed to yield CO_2 and HCl (Jacoby et al., 1994). It was further established that the product distribution is affected by the humidity. Increasing humidity increases the ratio between CO_2 and phosgene and reduces the mole fraction of DCAC in the product stream. However, this beneficial effect may come at the expense of reducing the overall rate of TCE degradation.

It should be pointed out that other groups (Fan and Yates, 1996) presented a contradictory explanation, based on using $^{18}O_2$ in the gas phase and monitoring the evolved CO_2, according to which the photocatalytic degradation of TCE was due to activated adsorbed oxygen and not due to hydroxyl radicals.

Nimlos et al. (1993) have found that the dependence of the gas-phase oxidation of TCE on the UV intensity, which was linear at high concentrations of TCE, changed its dependence into a square root law when the concentrations were low. This was explained (Upadhya and Ollis, 1998) by a change in the mechanism: at high concentrations the mechanism was a chain reaction mechanism induced by chlorine atoms, whereas at low concentrations the mechanism is the common holes/OH attack mechanism. This hypothesis correlated well with the fact that the quantum efficiency at high concentrations of TCE was 4–10 times higher than at low concentrations.

Treatment of an air stream containing high concentrations of TCE led to the deactivation of the photocatalyst. Several groups have reported that the surface of the photocatalyst became saturated, and as a consequence the amount of intermediate products that were released was increased (Driessen et al., 1998). The adsorbed specie was identified as dichloroacetate, which together with HCl is formed from DCAC reacting with hydroxyl radical (Hwang et al., 1998; Larson and Falconer, 1993). The activity of the photocatalyst was restored by flowing humid air in the dark for short time (Dibble and Raupp, 1992).

4.3. NOx

Nitrogen oxides (NOx) such as nitric oxide (NO), nitrous oxide (N_2O), and nitrogen dioxide (NO_2) make one of the most common groups of air pollutants. They are released mainly from internal combustion engines and furnaces and are considered to be harmful atmospheric pollutants, as they can cause acid rain, photochemical smog, and greenhouse effects (Zhou et al., 2007).

It has been established that photocatalysis may oxidize NO into NO_2 and eventually to HNO_3 (Hashimoto et al., 2000). The HNO_3 may remain on the surface of the catalyst; however, it can be washed easily by water. It is no wonder, therefore, that photocatalysis is considered to be one of the most promising techniques to handle this outdoor pollutant. It was found that the release of NO_2, an intermediate formed during the oxidation of nitric oxide, is reduced upon immobilizing the titanium dioxide on AC (Ao and Lee, 2005). It was claimed that the mechanism of NO oxidation involves two steps. In the first step, the NO is attacked by OOH to form NO_2 and hydroxyl radical, whereas in the second step the OH radical oxidizes the NO_2 thus forming HNO_3. The role of the AC is therefore to adsorb the NO_2 and to shuttle it to the photocatalytic domains. In this context, it is noteworthy that the competitive adsorption of water, which is the main cause for the adverse effect of humidity at high RH, was reduced upon using composite photocatalysts, where the pollutants are first adsorbed on the inert domains and then diffuse to the photocatalytic sites (Ao and Lee, 2004). This finding was not altered in the presence of other copollutants such as benzene, toluene, p-xylene, ethylbenzene, or SO_2.

An interesting effect of low loading of titanium dioxide within a zeolite matrix can be the formation of end-products that differ significantly from those produced at high loading. This effect was demonstrated with the use of zeolite MCM-41 matrix, hosting 0.6% by weight of TiO_2 (Kitano et al., 2007). Here, photocatalytic removal of NO led to the formation of the more friendly reductive products (N_2 and O_2) instead of the usual oxidative products (NO_2, HNO_3). Results were explained in terms of the formation of isolated titanium dioxide, having a lower coordination number.

4.4. Mixtures

As mentioned above, an indoor environment contains a large number of pollutants, each at a ppb level. It is reasonable to assume that the coexistence of these compounds may alter not only the rate by which they are degraded but also the type of by-products that are formed. The rate can be altered simply because of (the relatively easy to calculate) competitive adsorption, but also following interactions between intermediate active species. Despite the real need for the study of photocatalytic degradation of mixtures and the scientific challenge in understanding the related complex phenomena, the number of manuscripts on degradation of gas-phase mixtures is quite low. There are, however, several examples such as the study of mixtures of alcohols with their corresponding aldehydes (Araña et al., 2008) and a mixture of the four components of BTEX in equal amounts confirming that the activity order was benzene < toluene < ethylbenzene < o-xylene (Strini et al., 2005). It is for these reasons that the topic of mixtures was included explicitly in this chapter, thus conveying a message about the importance of studying mixtures in the context of photocatalytic air treatment.

The coexistence of various pollutants does not have to be deleterious, but, in certain cases, can be quite beneficial. The first evidence for this claim came probably from the work of Lichtin et al. (1994) who found that the edition of 0.03% by volume to an air-stream containing 0.1% iso-octane caused an enhancement in the photocatalytic oxidation of the latter. Likewise, a significant rate enhancement was recorded in the photocatalytic degradation of chloroform and dichloromethane in the presence of TCE. Similar effects were recorded also with other chlorinated olefins, such as perchloroethylene (PCE) and trichloropropene (TCP), which enhanced the photooxidation of toluene in a manner similar to that of TCE (Sauer et al., 1995).

The synergistic effect of TCE was attributed to a reaction with chlorine radicals formed during the decomposition of TCE. An extensive study of these observations, made by Luo and Ollis (1996), found that TCE concentrations higher than 225 mg m^{-3} promoted the oxidation rate of toluene (relative to its single component rate) as far as the toluene concentrations were lower than 160 mg m^{-3}. Higher concentrations of toluene acted to inhibit the decomposition rates of TCE. These findings were explained by an oxidation mechanism of TCE, which involved the TiO$_2$-aided formation of chlorine radicals. These radicals could induce a chain-reaction mechanism not only with TCE (thus preserving their concentration) but also caused a propagation reaction with molecules of toluene. When the toluene concentration was high, the consumption rate of chlorine radicals became so high that their concentration was reduced significantly and the TCE conversion fell accordingly.

The generality of the TCE-induced synergistic effect was studied later on in the co-oxidation of benzene, ethylbenzene, toluene, and m-xylene (d'Hennezel and Ollis, 1997). For all aromatic compounds, except for benzene, the rate of oxidation was enhanced in the presence of TCE; however, some decrease in the degradation rate of TCE was also noticed. As for benzene, no synergistic effect was observed; however, at the same time, the TCE degradation rate remained as it was in the absence of the copollutant. These results could be rationalized on the basis of a mechanism that operated through the abstraction of hydrogen atom from the alkyl group attached to the aromatic ring, so no wonder there was no synergistic effect with benzene. Some synergistic effect was also found with acetaldehyde, 2-butanone, although at the expense of some reduction in the rate of TCE. The degradation rate of other copollutants (acetone, chloroform, methylene chloride, and 1,1,1-trichloroethane) was found to be reduced when TCE competed with them for the precious hydroxyls. Overall, the higher the surface coverage of the copollutant was, the more the TCE conversion was reduced. Such effect could be due to overconsumption of the chlorine radicals (the so-called kinetic effect) or due to blocking of TCE adsorption sites (the so-called thermodynamic effect).

5. PHOTOCATALYTIC REACTORS FOR AIR TREATMENT: MODES OF OPERATION

Photoreactors for air treatment require proper design, aimed at achieving high quantum yield for long operation time at minimal cost. For that, it is necessary to choose the right reactor's configuration, the right UV source and its location, and the right design of photocatalyst. Two modes of operation can be found: batch reactors and continuous flow, one pass, reactors.

5.1. Batch reactors (including recirculation systems)

Batch reactors are often used at laboratory-scale experiments, usually for studying reaction kinetics (Vorontsov and Dubovitskaya, 2004) or for demonstrating the efficiency of recently developed photocatalyst. Although their potential for industrial-scale treatment of air is very limited, they can still be quite useful for obtaining kinetic rates expressions.

Most of the batch reactors operate in a recirculation mode, where gas flows into the "working" volume, treated, and then recirculated. Recirculation reactors provide a uniform concentration of reactants and products inside the system, thus enabling the obtaining of reliable kinetic measurements. For example, closed reirculation system was used to study the effect of photocatalysis for inactivating bacteria in the gas phase as a function of air stream conditions (Goswami et al., 1997). A complete inactivation was found to be possible, provided that the humidity is within the range of 50% RH and that the flow rate was low enough.

A specific type of recirculating reactor is the commercial indoor-air treatment device. The device is characterized by a small operating volume (a few liters at most), which is several orders of magnitude smaller than the reservoir (a room, vehicle, or aircraft, for example). The reactor itself can be either free standing or part of a larger system, usually an air-conditioning system, and is supposed to be able to take care of continuous emission of pollutants from construction materials as well as of short bursts of contaminants (Lewandowski and Ollis, 2003).

A word of caution needs to be added to this part on recirculating reactors, where the "working" volume is a very small part of the overall volume. As said before, recirculating reactors are, in essence, batch reactors (or semibatch in the case of inactivation of bacteria, if growth is allowed and nutrients are fed in). Therefore, one has to be very careful in using the terms of "contact time" or "retention time," while assigning them the same meaning as in continuous stirred tank reactor (CSTR) or plug flow reactor (PFR). Changing the flow rate, at a defined system, having a constant volume, has to do with changing the flow pattern and mass transport from the bulk to the film and should not be addressed as if the space time in a continuous reactor has been changed. It is unfortunate therefore that a small survey made by us revealed quite a few manuscripts suffering from this misconception.

Mixed flow reactors are quite rare in the context of gas-phase batch reactors. Nevertheless, they were used in several cases where it was essential to obtain uniform concentration of reactants at the photocatalyst's surface. Such cases include streams that contain very low level of pollutants (in the order of ppb), high conversion, and inadequacy of recirculating reactors due to parasitic adsorption of specific pollutants on the walls of the reservoir, the reactor, and the connecting tubes (Strini et al., 2005). Here, a cementitious photocatalyst was prepared by mixing P25 with a white cement powder, followed by the addition of water, hydration, humidity equilibration, and drying. The TiO_2 content in the cementitious material ranged between 0.5 and 6% on a dry cement basis. Linear dependence of the oxidation rate on pollutant's concentration was observed. A lower order was observed for the dependence of the catalytic activity on the TiO_2 loading, which was explained by formation of catalyst clusters.

5.2. Continuous, one-pass flow reactors

Continuous flow reactors are generally being in use whenever there is a continuous source of pollutants. These types of reactors are suitable in particular for handling process gases, thus reducing their release into the atmosphere. In such cases, it is expected to have at least some information on the composition and concentration of the pollutants in the inlet stream, which may assist significantly in designing the decontamination units.

Continuous flow reactors can also be found also at a laboratory scale, where the feed is constant and the output stream is constantly monitored by means such as GC-MS (Alberici and Jardim, 1997; Sun et al., 2007), GC-FID (Doucet et al., 2006), GC-TCD (Yamazaki–Nishida et al., 1996), or even FT-IR (Nimlos et al., 1993). In certain cases (for example: Doucet et al., 2006), continuous one-pass flow reactors are used for performing kinetic measurements later to be used for scaling up.

6. TYPES OF PHOTOCATALYTIC REACTORS FOR AIR TREATMENT

The photocatalytic reactors in use for air treatment can be categorized according to their gas flow geometry (tubular, annular, flat plate), according to the type of their light sources or according to the way by which the photocatalyst is introduced into the system. The preferred criterion is often a matter of personal taste and previous education. In order to summarize the various types a matrix was constructed, which represents the major types of photocatalytic reactors, care was taken in constructing this table, to assure that all the reactors mentioned in the Table 1 were in use for air treatment. For this reason, the fiber optic reactor (Peill and Hofffman, 1995), which was used for water treatment, was not introduced into the Table 1. The matrix is two dimensional (reactor geometry and catalyst form) since the third important dimension (light sources) is, to large extent, coupled to the other two dimensions. The fourth dimension, the photocatalyst itself and its manufacturing process, was not included here since a closed look at the data revealed a large collection of very specific production schemes and absence of data that could serve as a basis for true comparison or even generalization.

6.1. Tubular reactors

Tubular reactors are probably the most common photocatalytic reactors. Their popularity stems, most likely, from their simplicity. They are characterized by a gas flow along the axis of a tube, which contains the photocatalyst in many possible forms such as a thin coated film on its wall, fluidized particles, a coated monolith, or even as a free powder resting on an appropriate support. The light sources are located, in most cases, externally to the tube, in a parallel configuration relative to its axis. Reflecting surfaces encompass the lamps array, assuring that the only absorbance of photons would be that of the photocatalyst (Figure 7).

6.1.1. Powder layer tubular reactors

Powder layer reactors are based on an unattached layer of TiO_2 powder (usually Degussa's P25), deposited on a porous material such as glass. The flow of reactant vapors or contaminated air occurs in downward direction through the photocatalyst layer and the supporting frit. The catalyst load is usually adjusted to assure total absorption of photons, yet without limiting mass transport of contaminants. Illumination is preferably done from the front side of the photocatalyst-loaded frit to reduce the distance that the photoinduced charge carriers have to cross in order to reach the adsorbed pollutants. This type of reactor is so simple, and the replacement of used catalyst is so easy, that no wonder that this was one of the first types that were employed already in the seventies and the beginning of the eighties by

Table 1 Representative types of reactors, categorized by geometry and the way by which the photocatalyst is introduced into the reactor

	Tubular	Annular	Flat plate
Powder layer reactor	Lewandowski and Ollis, 2003; Peral & Ollis, 1992		
Fluidized bed	Dibble and Raupp, 1992		
Coated wall, parallel flow	Araña et al., 2008; Doucet et al., 2006	Bouzaza et al., 2006; Doucet et al., 2006; Lim and Kim, 2004; Sidheswaran and Taviarides, 2008	Salvado-Estivill et al., 2007a,b
		Lim et al., 2000; Lim and Kim, 2004; Matsuda and Hatano, 2005	
Packed bed	Araña et al., 2008; Tsoukleris, et al., 2007; Yamazaki-Nishida et al., 1996	Raupp et al., 1997; Stokke et al., 2006	
Plasma driven	Kim et al., 1999a, b; Sun et al., 2007; Thevenet et al., 2007		
Monolithic (honeycomb and foam)	Arabatzis et al., 2005; Blanco et al., 1996; Furman et al., 2007; Ibhadon et al., 2007; Nicolella, and Rovatti 1998; Nimlos et al., 1993; Obee and Brown, 1995		
Permeable layer	Tsuru et al., 2006	Ibrahim and de Lasa, 2002; Romero-Vargas Castrillon et al., 2006	

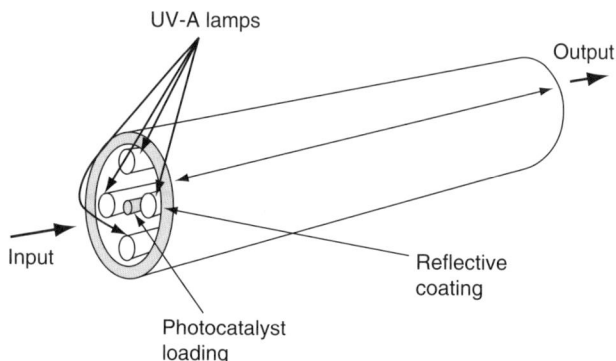

Figure 7 A schematic view of a tubular reactor.

Formenti et al. (1971) for partial oxidation of isobutane. This reactor was proved to be free of mass transfer problems and was regarded as an ideal reactor for laboratory kinetic studies.

As was pointed out justly by Lewandowski and Ollis (2003), powder layer reactors are not suitable for full-scale commercial application, as the high flow rate characteristic for commercial application might loosen the catalyst particles thus displacing the powder into the airstream or, in other cases depending on the configuration, might compress the bed, thus limiting the mass transport through the frit. Powder reactors are neither suitable for mobile or vibrating devices, as the vibrations might cause an uneven distribution of the powder, leading to channeling of contaminated air through the reactor with little contact with the catalyst powder, and in parallel, causing the formation of "dead volume" of photocatalyst at areas where accumulation of catalyst occurs.

6.1.2. Fluidized bed tubular reactors

Fluidized bed reactors (FBRs), characterized by an upward stream that is sufficient to lift and suspend the photocatalyst's particles, have several advantages. They enable high throughput of gas at a minor pressure drop, with very large contact surface between the pollutants and the photocatalyst. They are also very suitable for catalyst regeneration or replacement, an advantage that can be quite important for gas-phase pollutants that tend to deactivate the photocatalyst such as aromatic compounds or TCE (see Section 6.1, 6.2). No wonder that these reactors were among the first to be used (Dibble and Raupp, 1992), quite in parallel to the use of liquid–solid slurry reactors for water treatment. However, unlike the liquid-phase slurry reactors, they pose a real challenge in controlling the system – lifting the particles without carrying them out of the reactor. To succeed in this challenge, it is important that the particles have and maintain a proper size distribution. Typical dimensions of fluidized bed catalyst particles in the chemical industry are in the range of 10–300 microns (Hill, 1977), if reasonable velocities are to be employed. This is much larger than the size of typical TiO$_2$ powders, hence supporting particulates should be required. Agglomeration is a known problem in FBRs, as the higher velocities needed for lifting the agglomerates might carry out the nonagglomerated particles. For the combination of air as an effluent and titanium dioxide as the particles, this problem seems to be of less importance. On the other hand, attrition of the solid, forming very fine powder that might be blown out of the reactor, should be of great concern for long-term operation. A separation unit, for example a cyclone, as in a Riser Reactor, seems to be essential.

One of the first fluidized bed photocatalytic reactors was presented by Dibble and Raupp (1992), who used silica-supported titania catalysts in order to degrade TCE with an AQE of 13%. Here, the UV sources in this bench-scale reactor were located externally to the reactor. Catalyst loss was prevented in this laboratory-scale reactor by introducing a second glass frit located at the reactor outlet.

6.1.3. Coated wall tubular reactors

Coated wall reactors are made of hollow tubes, coated with the photocatalyst on their inner walls, where the carrier gas flows. Coating of the catalyst may be achieved by impregnation in a TiO$_2$ suspension (approximately $2\,g\,L^{-1}$) (Araña et al., 2008) or by spraying of either a TiO$_2$ suspension or a TiO$_2$ precursor solution. The impregnation (or spraying) is usually followed by thermal treatment at elevated temperatures to improve adherence, and in the case of in situ preparation of the photocatalyst also to evaporate the solvent and to form the TiO$_2$ at the required anatase phase. The thickness of the photocatalyst film has to be sufficient to assure that almost all nonreflected photons are absorbed. This amount was found to be around $1\,mg\,cm^{-2}$ (Doucet et al., 2006). Higher loadings are not expected to assist, and might act to reduce efficiency, as layers that are too thick might, in the back-illumination case, reduce the number of photogenerated charge carriers which reach the front, active, surface.

The light source is located externally, either in front of the entrance to the tubes (front-type illumination on an array of straight tubes, relatively short) or in parallel to the straight tubes, illuminating from the back of the photocatalyst film. Another configuration is at the center of a spiral made from the tube around a (usually elongated, fluorescent type) UV light source. In that case, the photocatalyst is back-illuminated. Figure 8 schematically presents such a spiral reactor, used for the degradation of methyl tert-butyl ether (MTBE) (Araña et al., 2008). In the first configuration, there is no restriction with respect to the tube's material, as far as the photocatalyst adheres well to it; however, care has to be taken that the tubes are not too long, otherwise a "dead volume" is formed. The second configuration requires the tubes to

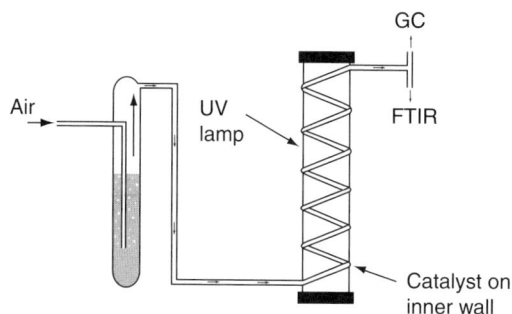

Figure 8 A coated wall tubular reactor in a spiral configuration (based on Araña et al., 2008; Vorontsov and Dubovitskaya, 2004).

be of a transparent material. If glass is used instead of silica and if the coating is made in situ by a sol-gel process, care has to be taken to prevent the migration of sodium ions from the glass into the nascent TiO_2 film (Paz et al., 1995; Paz and Heller, 1997). It is noteworthy that the spiral configuration, with its external light source, might suffer from inhomogeneous field of light impinging on the photocatalyst surface, the meaning of which might be less than optimal efficiency and difficulty in modeling the system.

A different configuration was presented by Doucet et al. (2006). Here, the catalyst (Degussa P25) was deposited on both sides of a pyrex glass plate, located inside a pyrex glass tube surrounded by six black fluorescence lamp. This front illumination reactor was used to study the degradation of TCE, methanol, and benzene.

6.1.4. Packed bed tubular reactors
Packed bed reactors are among the most used industrial reactors. However, they are of much less prospect for photocatalysis. PBRs are made of tanks or tubes, filled with photocatalyst pellets with reactants entering at one end and products leaving at the other. The gas flows in the void space around the pellets and reacts on the pellets. In the context of photocatalysis, the pellets can be made from inert materials coated with the photocatalyst or solely from titanium dioxide. This class may include, as many authors did, also columns packed with photocatalyst fine powder. A substantial complexity arises from the need to introduce the UV light into the packed bed. External light sources utilize only the outer part of the flow cross section, leaving a significant part of pellets in the dark. Assuming that some mixing occurs, the consequence is that large portion of the fluid might leave the reactor with much less "working" contact time that was considered by the designer. Three different PBRs, representing TiO_2 pellets, TiO_2 powder and spherules coated with TiO_2 are described herein.

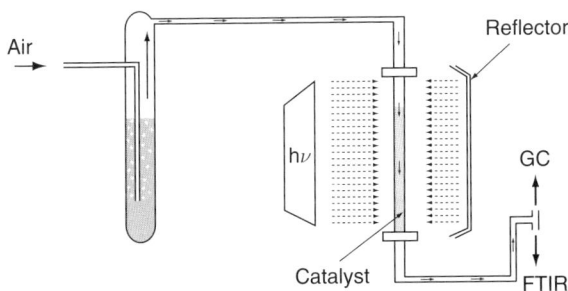

Figure 9 A scheme of the configuration of a tubular packed bed reactor (based on Araña et al., 2008).

Tubular reactor containing packed bed was used in a noncirculating mode to study the photocatalytic degradation of TCE and tetra-chloroethylene (Yamazaki-Nishida et al., 1996). The packed bed contained titanium dioxide pellets (1 mm in diameter), prepared by sol-gel and fired at 200–500°C. An inverse correlation between the firing temperature of the pellets and the formation of undesirable chlorinated compounds such as chloroform and carbon tetrachloride was found.

Packed bed reactors (Figure 9) with different lengths have been designed to study the photocatalytic degradation of MTBE, individual alcohols, and mixtures with their corresponding aldehydes. These studies have been performed with TiO_2 (Degussa P-25) and TiO_2 doped with Cu (Cu-TiO_2), the latter being more efficient than the former (Araña et al., 2008).

Another PBR introduced by the Tsoukleris et al. (2007) was based on glass spherules (0.5 cm in diameter) coated with a paste made of Degussa's P25, acetylacetone, and a binder (Triton X-100), which was then fired at 450°C to remove the organics and to improve the adherence. Photocatalytic degradation of BTEX with this reactor revealed remarkable stability, with practically no change in activity of the catalyst even after 30 consecutive tests.

6.1.5. Plasma-driven packed bed reactor

Plasma oxidation was suggested lately as a very promising way to treat VOCs in air (Oda et al., 1996, 2003). Efforts were made to study the effect of porous and various dielectric materials such as zirconates and titanates (Roland et al., 2002). It is no wonder, therefore, that the combination of titanium dioxide and plasma processes was tested as well.

A combination of packed bed titanium dioxide reactor and a plasma source was suggested as a means to obtain high efficiency and, more significantly, to reduce the appearance of undesired by-products. One of such designs (Kim et al., 1999b) utilized 5-mm pellets of TiO_2, packed in a quartz tube, into which a stainless wire held in its center served as a high-voltage electrode (Figure 10). The test gas contained, apart from N_2 as the

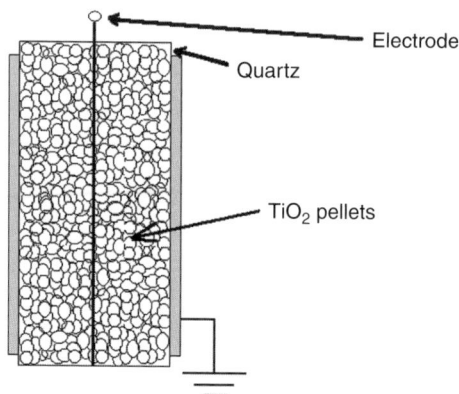

Figure 10   The plasma-driven reactor (based on Kim et al., 1999a).

carrier gas, also 400 ppm of NO, 10–15 ppm of NO_2, 10% O_2, and 1,000 ppm H_2O_2. The combination of discharge plasma with TiO_2 photocatalyst was found to be very effective for the removal of NOx by oxidizing it to HNO_3, in particular when H_2O_2 was added to the mixture. In addition, the emission of by-products, such as nitrous oxide (N_2O) or ozone, was significantly suppressed. It is noteworthy that the reactor did not contain a separate UV light source, so that photocatalytic contribution to the degradation was due to illumination by the plasma.

Plasma-driven photocatalysis was performed also for degrading gas-phase toluene (Sun et al., 2007), showing a significant increase in the removal of toluene, while decreasing the number of secondary by-products. It is noteworthy that the decrease in the emission of by-products could be also due to the loading of the photocatalyst on AC fibers, which could operate through the "Adsorb and Shuttle" mechanism (Paz, in press).

The synergistic characteristics of plasma-driven photocatalysis were studied also in the degradation of acetylene (Thevenet et al., 2007). Four modes of experiments were performed: photocatalytic, plasma degradation, plasma degradation + TiO_2 but without light, and a combined plasma and photocatalytic operation. It was found that plasma alone could not mineralize acetylene, as more than 50% of the carbon in the acetylene feed could not be accounted for, and, most likely, ended as strongly bound adsorbate by-products. Although the photocatalytic degradation per se was slower than that with plasma alone, the combination of the two was claimed to be very beneficial as it improved the mass balance of mineralized carbon (i.e., it reduced the formation rate of strongly bound by-products) and reduced the ratio between the production rate of carbon monoxide and that of carbon dioxide.

6.1.6. Monolith tubular reactor

Monolithic catalysts are solid structures pierced by parallel channels that facilitate the polluted air to flow through (Figure 11). The channels are usually wide enough to reduce the pressure drop by several orders of magnitude in comparison with powdered reactors. Moreover, if designed properly, in a manner that the light is guided into all the internal surfaces of the channels, monolithic reactors may provide a reasonable way to get quite a large area of exposure per source. The monolith itself may consist of a metal or oxide support, containing the photocatalyst either in the form of a coating layer or as a part of the matrix. Monolithic catalysts have been shown to be very suitable for gas contaminants treatment and are widely used in automotive emission control systems (Nicolella and Rovatti, 1998). Their large external surface and low pressure drop, together with a geometry that permits an increase in the lighted area, make them a very good candidate for commercial use. Monolith reactors are not affected by the problems that powder reactors have, with respect to separation between the photocatalyst fine-powdered particles and the gas stream. On the other hand, they are expected to be more expensive than the simple powdered reactors.

One of the first examples for the use of monolith reactors was presented by Blanco et al. (1996), who used a monolith made of titania dispersed in fibrous silicate to decontaminate air streams containing toluene or xylene, with a conversion up to 96% for toluene and 99% for xylene.

An alumina reticulate (4 channels cm^{-1}) was coated with Degussa P25 by simple wash-coat and used as a single-pass reactor (Obee and Brown, 1995) in order to study the effect of humidity on the degradation of formaldehyde, toluene, and 1,3-butadiene. The thickness of the reactor was 2 cm; however, the active depth was no more than 1 cm. For all three pollutants, the oxidation rate dependence on UV intensity within the range of 10–40 mw cm^{-2} followed a power law, with an exponent of 0.55. This was in agreement with previous works (Ollis et al., 1991; Peral and Ollis, 1992), which showed a linear dependence below 1–2 mW cm^{-2} and an exponent of 0.5 at higher intensities. Similar monolith, coated with P25, was used by Nimlos et al. (1993) to study TCE degradation. Parallel measurements were made by

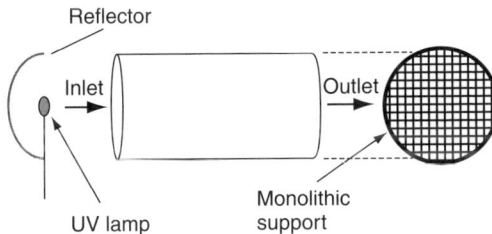

Figure 11   A schematic side view and front view of a monolithic photocatalytic reactor.

a film-on glass tube reactor. Both revealed the existence of by-products explained by radical chain mechanism induced by the formation of chlorine atoms.

The tubular monolithic reactor was a subject to quite a few models. One of the first models of honeycomb monolithic reactor was presented by Sauer and Ollis (1994). It assumes a CSTR reactor (the monolith) connected in series to a well-mixed recirculating loop, LH isothermal kinetics, and no mass transfer effects. It further assumed that the conversion per pass in the system was very low, that is, assumed a uniform concentration throughout the system. It should be mentioned that the model did not take into account the effect of radiation gradient along the pores and lumped all the radiation-related parameters. It did, however, make an estimation on the maximum active catalyst volume, taking into account the geometry of the pores, the geometry of the lamp, and its distance from the monolith. The model was validated for the degradation of acetone on a porous monolith (10 μm pore size) having 4-mm square channels, 15 cm in length, coated with P25 particles and operated under constant humidity conditions. The model showed very good predictive power at all conversions once the kinetic parameters were calculated based on initial rate measurements.

The influence of the geometry of the monolith on the reactor's efficiency was studied by Furman et al. (2007). A model, taking into account light absorption, hydrodynamic and transfer processes, and the reaction kinetics was developed. Three different geometries of the monolith reactor were tested: mixer, crossed channeled, and star. The monoliths were prepared by prefabricating an epoxy resin support using laser stereo lithography, followed by impregnation in P25 suspension. All three reactors had similar coated surface, of approximately 4×10^{-3} m^2. The model was analyzed assuming a very efficient external mass transfer, that is, that the surface concentration of the contaminant (methanol) was identical to the bulk concentration. LH kinetics were assumed, such that an apparent kinetic constant k could be calculated, taking K, the adsorption equilibrium constant as 129.7 m^3 mol^{-1} at 50°C, following published values. Under these assumptions, the apparent kinetic constant was 2.6×10^{-5}, 2.1×10^{-5}, and 1.7×10^{-5} (mol s^{-1}-m^{-2}) for the mixer, crossed channels, and star configurations, respectively. From this it was deduced that the mixer structure absorbs more photons per unit area of support than the two others. It is noteworthy that a good fit between the model and the experimental results was found only for flow rates higher than 0.2 L min^{-1}, indicating a mass transfer problem below this value.

A mathematical model of the operation of monolith reactors, focusing on the effect of light absorption on the heat and mass balance within the reactor and on the overall reaction kinetics, was presented by Nicolella and Rovatti (1998). The model assumed steady-state conditions, uniform pressure along the monolith channels, negligible axial diffusion, negligible conduction in the gas phase, and no homogeneous reaction. A circular dish lamp emitting monochromatic light was assumed. It was further assumed that the emitted rays were focused in front of the monolith entrance, that the monolith channels were placed axially relative to the lamp, and that reflection and scattering were negligible. Out of the many assumptions that were made, the last assumption is probably the most problematic as both the refractive index of titanium dioxide and the angle that the UV light impinge on the monolith surface suggest that reflection and scattering might play a dominant role. It is noteworthy that although the energy balance equations were taken into account in the model, the authors did acknowledge the fact that the energy effects are practically negligible due to the low concentration of contaminants in the air stream.

A different type of monolith reactor, the so-called foam monolith reactor, was designed and fabricated by Arabatzis and Falara (2003) and Arabatzis et al. (2005) for the treatment of VOCs. This type of photoreactor had the advantage that the photocatalyst could be formed in situ inside the reactor, thus could be obtained very easily in any desired geometry. The monolith was located within a centered glass tube located in parallel to the axis of four UV lamps, arranged in a symmetrical manner, forming a cross. The foam (Arabatzis and Falara, 2003) was prepared by adding H_2O_2 to a paste made of P25 and a surfactant in acetone. The H_2O_2 decomposed, releasing molecular oxygen bubbles that acted to gradually increase the overall volume within an hour. The foam possessed a highly porous structure, with an extended network of interconnected flakes forming irregular polygonal cavities of 200–500 nm in size. This structure was attributed both to H-bond interaction between the hexadecylamine (HDA) surfactant molecules and the nanocrystalline titanium dioxide and to hydrophobic interactions between adjacent HDA molecules responsible for the formation of a lamellar structure.

The modeling of this type of reactor was presented by Ibhadon et al. (2007) and validated by measuring the degradation of BTEX. It was found that this type of reactor could be quite efficient at low concentration of pollutants; however, its rate constant as well as the pollutants' conversion dropped considerably at high (millimolar) concentration of pollutants. A major obstacle for practical use was the fragility of the foam, which suggested that the reactor might suffer from severe stability problems under prolonged use, as there is no guarantee that the network architecture of the pores would not change over time.

6.1.7. Permeable layer tubular reactors

Permeable layer reactors consist of thin, porous metal or ceramic substrates onto which titanium dioxide is coated in a manner that allows for flowing through the porous substrates. The flow can be either perpendicular to the surface of the porous media or alternatively may combine perpendicular and parallel vectors. An example of such a reactor was presented by Tsuru et al. (2006). Here, a titania membrane (pore sizes 2.5–22 nm) was prepared by a sol-gel process, on the outer surface of an α-alumina microfiltration filter, using commercial anatase sol solution as an intermediate layer. The microfiltration filter was connected to a glass tube using glass frits, such that the air permeated through the membrane into the tube, from which it continued to a GC system equipped with thermal conductive detector (TCD). The membrane cell was positioned within a quartz tubular reactor, irradiated by eight black-light lamps. Photodeposition was used to dope the surface of the membrane with platinum using chloroplatinic acid as a precursor. An increase in the decomposition of methanol upon decreasing the pore diameter was observed and was attributed to better contact between the TiO_2 pore walls and the permeating reactants.

6.2. Annular reactors

Annular flow reactors are characterized by a cylindrical lamp surrounded by two concentric tubes such that the polluted air flows in the annulus between the inner and the outer tubes. That way, all emitted photons can be utilized without the use of expensive reflectors. In certain cases (Doucet et al., 2006), an optical (occasionally liquid) filter is introduced between the lamp and the flowing zone to control the wavelength and the power of the impinging light, as well as the temperature within the reactor.

6.2.1. Fluidized bed annular reactors

Fluidized bed reactor in an annular configuration was used for the study of NO (Lim et al., 2000) and TCE degradation (Lim and Kim, 2004). Here, P25 particles were attached to silica gel particles in order to improve their fluidization behavior. Light was introduced both from the center of the annulus and by affixing germicidal lamps outside of the reactor. The optimum phase holdup ratio of gas and solid phases for the TCE degradation was found to be 2.1, higher than values obtained for other photocatalytic FBR that used γ-alumina particles instead of silica gel, suggesting that silica gel is appropriate substrate for photocatalytic fluidization reactors.

A riser-type of FBR was presented by Matsuda and Hatano (2005) for the photocatalytic removal of NOx. Here, the FBR was coupled integrally to a cyclone-separating system that returned the carried-away titanium dioxide-loaded silica-gel particles.

6.2.2. Coated wall annular reactors

Coated wall annular reactors (Figure 12) are characterized by a thin film of photocatalyst located either on the outer side of inner tube or on the inner side of the outer tube. There are many reasons for preferring the inner side of the outer tube for the coating. The first reason has to do with the flux; it is known that at high flux of light (>1–$2\,mW\,cm^{-2}$) the quantum efficiency is proportional to 1 divided by the square root of the intensity. Hence, it is

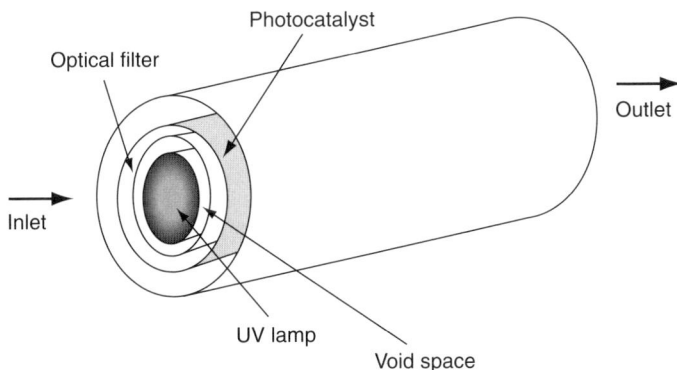

Figure 12 A typical annular flow reactor (after Doucet et al., 2006).

preferable to utilize the same number of photons per unit time on a larger area, as provided by the outer tube. The second reason has to do with the direction of illumination: for the inner tube the illumination is from the back side of the coating, whereas if the coating is the inner side of the outer tube, the illumination is at the front side of the photocatalyst. Still, many groups prefer to put the coating on the inner wall of the annulus.

It was shown that coated wall annular flow reactors can be modeled successfully as plug-flow reactors (Bouzaza et al., 2006). Here, the relative contributions of mass transfer and chemical reaction during the photocatalytic degradation were studied for two types of VOCs: TCE and toluene. A very simple steady-state model was used, following an approach developed previously (Ku et al., 2001; Lin, 2002). According to this model, which assumes LH kinetics, and under the assumption that the mass transfer in this PFR is not the limiting step, it is possible to describe the concentration of the pollutant along the reactor axis in the following very simple equation:

$$u\frac{dC}{dZ} + \frac{kKC}{1 + KC} = 0 \tag{12}$$

where $C(z)$ is the concentration in the bulk at an axial position Z, u is the flow velocity, k is the reaction rate constant ($mol\,L^{-1}\,min^{-1}$), and K is the Langmuir adsorption constant ($L\,mol^{-1}$). Integrating this equation provides (after some arrangements) a way to validate whether the system suffers from external mass transfer limitation or not, since in the absence of mass transfer limitation it is expected to have the following relation between the initial and the final concentration:

$$\frac{\ln(C_{in}/C_{out})}{C_{in} - C_{out}} = \frac{kKL}{u(C_{in} - C_{out})} - K \tag{13}$$

where C_{in} and C_{out} are the concentrations of pollutant at the reactor's inlet and outlet, respectively, and L is the reactor length. Accordingly, if all assumptions are correct, a plot of $[\ln(C_{in}/C_{out})]/(C_{in} - C_{out})$ versus $1/[u(C_{in} - C_{out})]$ should be linear, and it should be able to calculate kK from the slope kKL. This provides a way to estimate the effect of mass transfer resistance by comparing kK with k^*, the apparent first-order rate constant. k^* is calculated separately by measuring the concentration at the outlet, since the reactor obeys the simple first-order reaction PFR equation:

$$\frac{C_{out}}{C_{in}} = \exp\left[-\frac{k^*L}{u}\right] \qquad (14)$$

Estimating the effect of mass transfer resistance was done based on the connection between k^* and k:

$$\frac{1}{k^*} = \frac{1}{kK} + \frac{1}{k_m a_V} \qquad (15)$$

where k_m is the mass transfer coefficient from the gas to catalyst surface $(m\ min^{-1})$ and a_V is the total effective catalyst area per unit volume of the reactor $(m^2\ m^{-3})$. Under the conditions of their experiments, the apparent rate constant (k^*) with toluene was very close in its value to the product kK, over a wide range of velocities $(0.1-0.5\ m\ s^{-1})$, demonstrating that the resistance due to mass transfer was negligible.

It was already mentioned here that in back-illuminated film reactors (a class that includes annular film reactors), the performance as a function of film thickness is expected to go through a maximum. If the photocatalytic film is too thin, some of the photons might escape due to insufficient absorption and might not be utilized. If, however, the film is too thick, all photons will be absorbed, but lesser number of charge carriers will reach the surface due to the long distance between the location of charge carrier formation and the surface, which promotes charge recombination. For some reason, this point was somehow overlooked, and very few manuscripts give any attention to optimize the thickness of the photocatalyst film, such that one finds sometimes coatings having a thickness of 20 μm or so, at least 4–5 times thicker than the (wavelength dependent) optimal thickness.

6.2.3. Packed bed annular reactors

A two-flux radiation field model for an annular packed bed photocatalytic oxidation reactor was presented by Raupp et al. (1997). Similar to other annular flow reactors, the UV source was located at the center of the cylindrical reactor. Yet, the photocatalyst was not introduced in the form of a thin film but rather as spherules filling the annular space between the lamp and the housing. The principal assumptions made in this model included a steady state, isothermal operation, cylindrical symmetry, uniform packing shape and distribution, plug flow, negligible radial and angular diffusion, negligible light absorption by the gas and dilute reactants (i.e., negligible heat of reaction and constant density). These assumptions led, in essence, to a pseudohomogeneous model. The radiation balance was solved first, under the assumption that there was no absorption by the gas and that the total UV irradiance at a given point was the sum of the forward and backward irradiances (i.e., back-scattering in the radial direction). The model was constructed for reactor walls made of nonabsorbing, nonreflecting material. This might be not so realistic since making the outer walls from a reflecting material (thus reducing photon loss) is commonly, and justly, done. The intensity profile of the UV light, calculated by a fourth-order Runge-Kutta method, was then used to solve the mass balance equations by a central finite difference method. Very good agreement between the model's predictions and the conversion of acetone was obtained, using 1 mm OD silica beads, onto which Degussa P25 was attached. It is noteworthy that the model suggested that radial gradient in the local volumetric rate of energy absorption (defined as mole photons per unit time per unit volume) in this configuration may lead to significant gradients in concentration and reaction rate, which ultimately might limit the reactor performance.

6.2.4. Permeable layer annular reactors

Similar to permeable layer tubular reactors, the permeable layer annular reactors are characterized by thin, porous substrates onto which titanium dioxide is coated in a manner that allows for flowing through the porous substrates. The flow can be either perpendicular to the surface of the porous media or alternatively may combine perpendicular and parallel vectors. The venturi reactor (also called the photo-CREC-air reactor) developed by de Lasa's group (de Lasa and Ibrahim, 2004; Ibrahim and de Lasa, 2002) followed this definition. This reactor was made of a venturi-like tube, containing a square basket covered with woven fiberglass screen, impregnated with TiO_2 on its side walls. The air flew through the venturi throat and around the basket base, and then contacted the impregnated mesh, where oxidation occurred, and continued its flow through the mesh. In a more advanced model (Romero-Vargas Castrillon et. al., 2006), several changes were introduced in the design: perforated plates replaced the wire-mesh basket, the front base of the basket received an aerodynamic shape, and cross section of the basket was altered from a rectangular shape into a cylindrical one. The flow took place between the TiO_2-impregnated screen and an outer house, thus forming an annular (or close to annular in the original scheme) flow cross section. A major difference between the annular flow reactors described above and the venturi reactor was the location of the UV lamps. In the annular reactors described above, the lamps were located at the center of the cylinder, along its axis, whereas at the venturi reactor the UV sources were mounted outside of the venturi divergent section, housed inside parabolic reflectors.

A commercial computational fluid dynamics (CFD) package (CFX-10) was used to solve the Reynolds-averaged mass, momentum, and contaminant transport equations under isothermal conditions (Romero-Vargas Castrillon et al., 2007). The process took into account the pressure drop, assumed a turbulent flow in the vicinity of the photocatalyst support, and a LH-type kinetics. The model contaminant was acetone, as this pollutant degrades photocatalytically without any observable gas-phase species except for carbon dioxide and water (Ibrahim and de Lasa, 2003). Significant differences were found between the two designs with respect to single-pass conversion values, which were attributed to both higher UV flux and available surface area in the advanced design.

6.3. Flat plate reactors

Flat plate reactors are characterized by a thin layer of photocatalyst-coated on a flat (or zigzagged) support, made of glass or metal, such that the flow of contaminated air is in parallel to the photocatalyst panel (Figure 13). The UV light sources can be either located within the embodiment of the reactor, or otherwise can be located externally to the reactor, introducing their light through a glass window. Care has to be taken to design the inlet and outlet in a manner that achieves uniform and fully developed flow over the photocatalytic plate (Salvado-Estivill et al., 2007a, b).

Flat plate reactors, being very simple for construction and analysis, are often used as a tool to obtain kinetic data, later to be used in the modeling of more complex system, for example, that of an annular reactor (Mohseni and Taghipour, 2004).

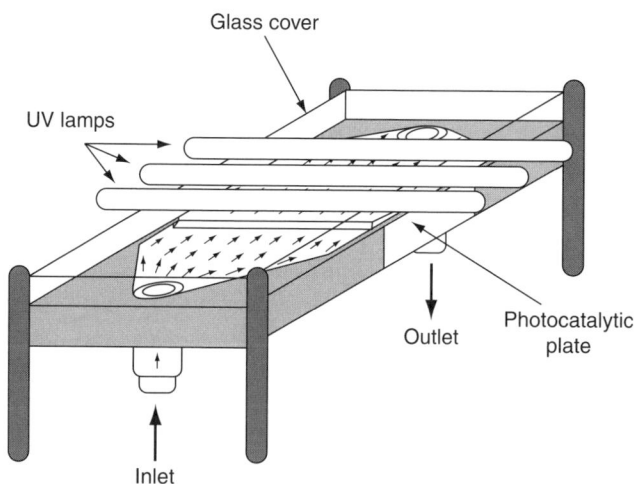

Figure 13 Schematics of a flat plate reactor (after Salvado-Estivill et al., 2007a, b).

6.4. Generalized models and comparisons between reactors

In the last years, there is a growing tendency to develop tools that will facilitate to simulate results for one type of a reactor by performing measurements on a different type of a reactor having a smaller size. Along this line, a first-principles approach that enables scaling-up was developed by Cassano's group (Imoberdorf et al., 2007) for reactors consisting of catalytic walls coated with a thin layer of titanium dioxide. The approach, which did not use any adjustable parameters, was demonstrated in the photocatalytic degradation of PCE. By utilizing kinetic information obtained in a laboratory scale, flat plate reactor operating at a steady state in a continuous, well-mixed recirculation reactor, it was possible to make very good (validated) predictions for a pilot-scale continuous, multiannular reactor having a catalytic surface 60 times larger than that of the flat plate reactor. The methodology was based on few principles: using fundamental chemistry research and detailed degradation mechanism, employing the same catalyst preparation protocol and morphology in both reactors, using a laboratory device as simple as possible to be operated under isothermal conditions, and most important, carrying the laboratory-scale experiments under the kinetic control regime (i.e., under conditions free of mass transfer limitations).

To large extent, the above-described method (as well as other scale-up methods) points into the importance of obtaining the complete kinetic network. Such an understanding involves measuring the time-dependent concentration of all species involved, that is, not only the reactant and the most common end-product (CO_2) but also the concentration of all intermediate products, as well as that of end-products other then carbon dioxide. Such a measurement can be quite problematic, taking into account that it is very rare to find a single experimental technique that would be suitable. In that case, choosing relatively simple model pollutants such as acetone or acetaldehyde (Ibrahim and de Lasa, 2004) might assist.

Computational fluid dynamics approach was utilized in the study of photocatalytic destruction of gas-phase vinyl chloride in an annular flow reactor (Mohseni and Taghipour, 2004). The kinetic data for the model was obtained from a differential glass plate reactor. The modeling results indicated significant gradient of vinyl chloride in the radial direction and nonuniform flow distributions, which resulted in reduced efficiency over the entire range of inlet concentrations.

Three-dimensional CFD-coupled with radiation field modeling and photocatalytic reaction dynamics was employed by Salvado-Estivill et al. (2007b) to model the decomposition of TCE in a flat-plate, single-pass photocatalytic reactor containing immobilized P25. The outcome was pollutant-specific kinetic rate parameters, which were independent of the reactor geometry, radiation field, and fluid dynamics. This was followed by a two-dimensional model, which assumed fully developed laminar flow, uniform irradiation, large width to thickness ratio, negligible axial diffusion. The model coupled the photocatalytic reaction kinetics (prededuced from the 3D model) with unsteady state continuity equation and with a radiation field to give the transient and the steady-state behavior of the reactor (Salvado-Estivill, 2007a). Application of the model to the photocatalytic oxidation of TCE in humidified air streams closely approximated the experimental results, demonstrating the possibility to use a simple, less time-consuming 2D approach for computational modeling of such reactors.

This tendency for generalization goes to large extent hand in hand with the works of researchers, who are seeking the optimal configuration and design for air-treatment reactors. An example for this statement is a comparison between two PBRs and a coated wall reactor, both having a diameter of 4 mm (Araña et al., 2008). The two PBRs had a working length of 15 and 75 cm, whereas the length of the coated wall reactor was 140 cm. Each of the reactors utilized the same amount of photocatalyst (0.15 g), with approximately the same contact time. APQs defined as the ratio between the moles of MTBE that were degraded per s cm^2 and the moles of incident photons per s cm^2 were calculated. The APQ values were 0.009 with the coated wall reactor, 0.021 with the long PBR, and 0.03 with the short PBR. The differences were attributed to variations in the light penetration, since all other parameters had equal values. The importance of comparing between reactors is very well demonstrated in the case of TCE. Here, it was shown that the type of reactor may have a crucial effect on the product distribution as it affects the actual contact time. A coated wall annular reactor, used by Jacoby et al. (1994) yielded significant amounts of phosgene but no DCAC. In contrast, a PBR did not yield any phosgene (Yamazaki-Nishida et al., 1993). It was claimed that the different results were due to "dark" reaction taking place in the PBR, where the phosgene is hydrolyzed to HCl and CO_2. The fact that in both cases same relative amount of CO was obtained brought the researchers to claim that the released CO originated directly from the photocatalytic degradation of the TCE and not from the consecutive hydrolysis of the phosgene.

6.5. Combined adsorptive-photocatalytic reactors

One way to improve the performance of photocatalytic air-treatment reactors is to decouple between cleaning the air and degrading the pollutants. Figure 14 shows schematically such a system (Chin et al., 2006; Shiraishi et al., 2003). The system includes two independent continuous flow systems interconnected by a rotating, cylindrical ceramic honeycomb. The rotation cycles the honeycomb rotor through a low temperature zone, where adsorption takes place, and a high temperature zone where desorption takes place, thus regenerating the monolith. The contaminated air flows

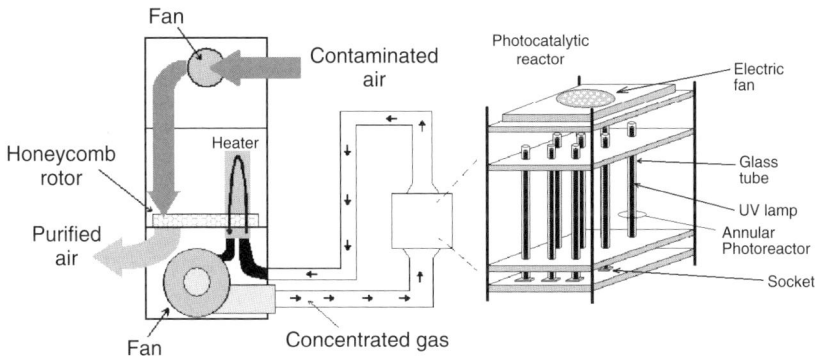

Figure 14 Schematics of a combined adsorptive-photocatalytic system (after Shiraishi et al., 2003).

through the monolith and leaves the device as a purified air. The pollutant (formaldehyde) that has been desorbed from the monolith is pumped in a recirculation manner to the so-called small box reactor, which comprises of nine coated- wall annular reactors.

The reactor was modeled by Chin et al. (2006) to determine adsorption, desorption, and pseudo-first-order rate constant. A photocatalytic rate constant, calculated based on the experimental data of Shiraishi and the model, was found to be in good agreement with previously published value.

7. CURRENT PROBLEMS AND FUTURE TRENDS

Taken the large number of groups and the diversity of topics, it is quite difficult to foresee the progress that will be made in this field in the next years to come. Yet, it is believed that it is possible, based on analyzing the current status, to pinpoint the current major bottlenecks in the photocatalytic treatment of air, while hoping that if these are indeed the major obstacles for large scale implementation and if these obstacles are recognized and defined, then a systematic research on these issues will eventually solve them. To some extent, this is an optimistic way of looking at scientific endeavors. We do hope that this optimism will justify itself.

7.1. Visible light

Constant efforts are being taken in the last years to shift the activity of titanium dioxide toward the visible by introducing dopants such as carbon, nitrogen, or iron. Although some literature exists on air treatment (Sidheswaran and Taviarides, 2008; Yin et al., 2008; Zhou et al., 2007), the overwhelming majority of the experiments was done in the context of water treatment. This is simply due to the fact that water seems to be the natural candidate for the use of solar energy and therefore for such an improvement. Nevertheless, red shifting of the activity could greatly assist in the area of outdoor air cleaning. In the context of indoor air treatment, red shifting is unlikely to make revolutionary changes, although it may push forward indoor air treatment simply due to the promotion of the general field of photocatalysis. Direct contribution to indoor air treatment is, in our eyes, less likely to occur, although it should not be ruled out.

7.2. Mixtures

Taking into account the large number of manuscripts dedicated to photocatalytic air treatment, it is amazing how small is the number of scientific papers on the properties of mixtures of gases. This is despite the fact that indoor air contains always a mixture of pollutants, which might interact between themselves during photocatalysis. The partial data that is already in hand point toward the existence of synergistic mechanisms that could, in principle, be utilized to enhance degradation rates, if not in the context of indoor air (since the concentration in indoor air might be too low to get a significant change), then at least in the context of treating process gases.

7.3. Standardization

Intensive efforts have been made to develop new, modified photocatalysts. Likewise, a lot of skills and imagination were put into new designs of reactors. Nevertheless, the ability of the community to utilize these immense efforts is very much limited due to the fact that there are hardly any tools that enable the reader to compare between different sets of experiments done by different groups around the globe. Moreover, in many cases the kinetic data is published without mentioning the photocatalyst's irradiated area, thus preventing any comparison. Recent years have witnessed the development of models that facilitate some comparison between experiments that were carried out in reactors of different configurations. This includes also the developing of evaluation tools based on energy considerations, such as the photocatalytic thermodynamic efficiency factor (PTEF) (Serrano and de Lasa, 1997) used in water treatment. Nevertheless, this is not enough. There should be ways to facilitate a simple comparison between catalysts and reactor configurations. An intensive effort in the direction of standardization is taking place these days, initiated by Japan and the European Community. It is hoped that this effort will be materialized, as this may have a huge impact toward successful products.

7.4. Deactivation

The problem of deactivation was known to most of the researchers in the community for many years, as discussed above. Over the years, some means to fight this phenomenon have been proposed (Alberici and Jardim, 1997; Ameen and Raupp, 1999; Cao et al., 2000). Yet, research, focusing on trying to solve this problematic issue, is quite scarce. Whenever deactivation and regeneration are mentioned, they always appear as an unintended spin-off of the research. To the best of our knowledge and understanding, for this reason there are in the market quite a few products that might not meet the life span declared by the manufacturer. Our feeling is that deactivation is not just an issue for manufacturers to take care of, as it may require extensive basic research.

8. CONCLUDING REMARKS

This chapter tried to summarize a few of the aspects related to photocatalytic air treatment. Particular efforts were made to identify the large variety of photocatalytic reactors for indoor air treatment and to point toward the prospects and obstacles in utilizing titanium dioxide for this purpose.

In a manuscript published in 1997, Gregory Raupp has stated that "although it is relatively straightforward to design a laboratory scale reactor that simultaneously contacts a uniformly irradiated catalyst and air effectively, the issues of UV distribution and utilization (energy cost) make reactor design a difficult problem at the commercial level" (Raupp et al., 1997). It seems that the statement made by Raupp et al. in 1997 is still valid and will continue to inspire researchers in their continuing efforts. That way or another, what comes clear from this review is that there cannot be one single solution to be crowned as the ultimate photocatalytic reactor design.

Nevertheless, the increasing number of patents and, more important, the increasing number of products in the market send a clear message that although this aim was not achieved yet, it is not beyond the abilities of science.

ACKNOWLEDGMENT

The assistance of Mr. Yuval Gamliel in the literature search and in drawing the figures is gratefully acknowledged.

LIST OF SYMBOLS

Latin letters

a_V The total effective catalyst area per unit volume of the reactor ($m^2\,m^{-3}$)
C Bulk gas-phase concentration
C_A The concentration of pollutant A

C_{A0} The initial concentration of pollutant A
C_{in} The concentration of pollutant at the reactor's inlet
C_{out} The concentration of pollutant at the reactor's outlet
C_s The mean concentration at the external surface
Cs   Concentration at the surface of the photocatalyst
D The diffusion coefficient of the pollutant in air
k True rate constant in LH expression (mol L^{-1} min^{-1})
k^* Apparent rate constant in LH expression
K The Langmuir adsorption constant (L mol^{-1})
L The reactor length
u The flow velocity
Z Axial position along the reactor
K_A The adsorption coefficient of A
k_m The mass transfer coefficient from the gas to catalyst surface (m min^{-1})
$k_{r,A}$ The reaction rate constant
L The characteristic length of the porous photocatalyst
m Meter
r_A The disappearance of A
r_{A0} The initial disappearance of A
r_v The experimental mean rate of reaction per unit volume of catalyst
s   Second
t Time
y_i Represents the molar ration of component i

Greek letters

ε The catalyst "grain" porosity
θ_A The surface coverage of this specie
τ The pore skewness

ABBREVIATIONS

AC   Activated carbon
A&S   Adsorb and Shuttle
AOP   Advanced oxidation processes
APQ   Apparent quantum efficiency
app.   Approximately
BTEX   An acronym for a group of aromatic compounds consisting of benzene, toluene, ethylbenzene, and xylenes
CFD   Computational fluid dynamics
CSTR   Continuous stirred tank reactor
DCA   Dichloroacetaldehyde (CCl_2HCOH)
DCAA   Dichloroacetic acid (CCl_2HCOOH)

DCAC	Dichloroacetyl chloride ($CHCl_2COCl$)
EPA	Environmental Protection Agency
AQY	Apparent quantum yield
FBR	Fluidized bed reactor
FT-IR	Fourier-transformed infra red spectroscopy
GC-FID	Gas chromatograph, flame ionization detector
GC-MS	Gas chromatograph, mass spectroscopy
HAD	Hexadecylamine
LH	Langmuir–Hinshelwood kinetics expression
MTBE	methyl tert-butyl ether
NO_X	Nitrogen oxides
OQY	Overall quantum yield
OSHA	The Occupational and Safety Administration (US)
PBR	Packed bed reactor
PCE	Perchloroethylene
PFR	Plug flow reactor
ppb	Parts per billion (usually by volume)
ppmv	Parts per million by volume
PQY	Primary quantum yield
PTEF	Photocatalytic thermodynamic efficiency factor
RH	Relative humidity
SBS	Sick building syndrome
TCD	Thermal conductive detector
TCE	Trichloroethylene
TCP	trichloropropene
UV	Ultraviolet
VOCs	Volatile organic compounds
W	Watt

REFERENCES

Agrios, A.G., and Pichat, P. *J. Appl. Electrochem.* **35**, 655 (2005).

Aguado, S., Polo, A., Bernal, M., Coronas, J., and Santamaria, J. *J. Membrane Sci.* **240**, 159 (2004).

Alberici, R.M., and Jardim, W.F. *Appl. Catal. B* **14**, 55 (1997).

Ameen, M.M., and Raupp, G.B. *J. Catal* **184**, 112 (1999).

Ao, C.H., and Lee, S.C. *Chem. Eng. Sci.* **60**, 103 (2005).

Ao, C.H., and Lee, S.C. *J. Photochem. Photobiol. A Chem.* **161**, 131 (2004).

Arabatzis, I.M., and Falara, P. *Nanoletters* **3**, 249 (2003).

Arabatzis, I.M., Spyrellis, N., Loizos, Z., and Falaras, P. *J. Mater. Process. Technol.* **161**, 224 (2005).

Araña, J., Peña Alonso, A., Doña Rodriguez, J.M., Herrera Melian, J.A., Gonzales Diaz, O., and Perez Peña, J. *Appl. Catal. B* **78**, 355 (2008).

Blanco, J., Avila, P., Bahamonde, A., Alvarez, E., Sanchez, B., and Romero, M. *Catal. Today* **29**, 437 (1996).

Blount, M.C., and Falconer, J.L. *J. Catal.* **200**, 21 (2001).

Bolton, J.R., Safarzadeh-amiri, A., and Cater, S.R. The detoxification of waste water streams using solar and artificial UV light sources. *in* F.S. Sterret (Ed.), "Alternative Fuels and the Environment". Lewish Publishers, Boca Raton, FL (1995), p. 187.

Boulamanti, A.K., and Philippopoulos, C.J. *J. Hazard. Mater.* **160**, 83 (2008).

Bouzaza, A., and Laplanche, A. *J. Photochem. Photobiol. A Chem.* **150**, 207 (2002).

Bouzaza, A., Vallet, C., and Laplanche, A. *J. Photochem. Photobiol. A Chem.* **177**, 212 (2006).

Box, G.E.P., and Draper, N.R. *Biometrica* **52**, 355 (1965).

Cao, L.X., Gao, Z., Suib, S.L., Obee, T.N., Hay, S.O., and Freihaut, J.D. *J. Catal.* **196**, 253 (2000).

Cassano, A., Martin, C., Brandi, R., and Alfano, O. *Ind. Eng. Chem. Res.* **34**, 2155 (1995).

Chin, P., Yang, L.P., and Ollis, D.F. *J. Catal.* **237**, 29 (2006).

Cunningham, J., Tobin, J.P.J., and Meriaudeau, P. *Surf. Sci.* **108**, L465 (1981).

de Lasa, H., and Ibrahim, H.U.S. Patent, 6,752,957 (2004).

de Lasa, H., Serrano, B., and Salaices, M. "Photocatalytic Reaction Engineering". Springer, New York (2005).

Destaillats, H., Maddalena, R.L., Singer, B.C., Hodgson, A.T., and McKone, T.E. *Atmos. Environ.* **42**, 1371 (2008).

d'Hennezel, O., and Ollis, D.F. *J Catal* **167**, 118 (1997).

d'Hennezel, O., Pichat, P., and Ollis, D.F. *J. Photochem. Photobiol. A* **118**, 197 (1998).

Dibble, L.A., and Raupp, G.B. *Catal. Lett.* **4**, 345 (1990).

Dibble, L.A., and Raupp, G.B. *Environ. Sci. Technol.* **26**, 492 (1992).

Doucet, N., Bocquillon, F., Zahraa, O., and Bouchy, M. *Chemosphere*, **65**, 1188 (2006).

Driessen, M.D., Goodman, A.L., Miller, T.M., Zaharias, G.A., and Grassian, V.H. *J. Phys. Chem. B* **102**, 549 (1998).

Environmental Protection Agency (EPA). "Total Exposure Assessment Methodology(TEAM) study". Report 600/6-87/002a, Washington, DC (1987).

Fan, J., and Yates, Jr. J.T. *J. Am. Chem. Soc.* **118**, 4686 (1996).

Formenti, M., Juillet, F., Meriaudeau, P., and Teichner, S. *Chem. Technol.* **1**, 680 (1971).

Fox, M.A., and Dulay, M.T. *Chem Rev.* **93**, 341 (1993).

Fujishima, A., Hashimoto, K., Watanabe, T., "TiO_2 Photocatalysis: Fundamentals and Applications". BKC Inc., Tokyo (1999), p. 59.

Fujishima, A., and Honda, K. *Nature* **238**, 37 (1972).

Furman, M., Corbel, S., Le Gall, H., Zahraa, O., and Bouchy, M. *Chem. Eng. Sci.* **62**, 5312 (2007).

Ghosh-Mukerji, S., Haick, H., and Paz, Y. *J. Photochem. Photobiol.* **160**, 77 (2003).

Goswami, D.Y., Trivedi D.M., and Block, S.S. *J. Solar Energy Eng.* **119**, 92 (1997).

Haick, H., and Paz, Y. *J. Phys. Chem. B* **105**, 3045 (2001).

Hashimoto, K., Wasada, K., Toukai, N., Kominami, H., and Kera, Y. *J. Photochem. Photobiol. A Chem.* **136**, 103 (2000).

Hill, C.G. "An Introduction to Chemical Engineering Kinetics & Reactor Design". John Wiley & Sons, NY (1977), p. 429.

Hwang, S.-J., Petucci, C., and Raftery, D. *J. Am. Chem. Soc.* **120**, 4388 (1998).

Ibhadon, A.O., Arabatzis, I.M., Falaras, P., and Tsoukleris, D. *Chem. Eng. J.* **133**, 317 (2007).

Ibrahim, H., and de Lasa, H. *AIChE J.* **50**, 1017 (2004).

Ibrahim, H., and de Lasa, H. *Appl. Catal.* **38**, 201 (2002).

Ibrahim, H., and de Lasa, H. *Chem. Eng. Sci.* **58**, 943 (2003).

Ibsuki, T., and Takeuchi, K. *Atmos. Environ.* **20**, 1711 (1986).

Imoberdorf, G.E., Irazoqui, H.A., Alfano, O.M., and Cassano, A.E. *Chem. Eng. Sci.* **62**, 793 (2007).

Jacoby, W.A., Nimlos, M.R., Blake, D.M., Noble, R.D., and Koval, C.A. *Environ. Sci. Technol.* **28**, 1661 (1994).

Junio, C.T., and Raupp, G.B. *Appl. Surf. Sci.* **72**, 321 (1993).

Kachina, A., Preis, S., and Kallas, J. *Int. J. Photoenergy* **2**, 79847/1 (2007).

Kim, J.S., Itoh, K., Murabayashi, M., and Kim B.A. *Chemosphere* **38**, 2969 (1999a).

Kim, H.H., Tsunoda, K., Katsura, S., and Mizuno, A. *IEEE Trans. Ind. Appl.* **35**, 1306 (1999b).

Kitano, M, Matsuoka, M, Ueshima, M., and Anpo, M. *Appl. Catal. A Gen.* **325**, 1 (2007).

Ku, Y., Ma, C.-M., and Shen, Y.-S. *Appl. Catal. B Environ.* **34**, 181 (2001).

Larson, S.A., and Falconer, J.L. *Catal. Lett.* **44**, 57 (1997).

Larson, S.A., and Falconer, J.L. Characterization of TiO2 Used in Liquid Phase and Gas Phase Photooxidation of Trichloroethylene. *in* D.F. Ollis, and H. Al-Ekabi (Eds.), "Photocatalytic Purification of and Treatment of Water and Air". Elsevier Science Publishers, Amsterdam (1993), p. 473.

Lee, M.C, and Choi, W. *J. Phys. Chem. B* **106**, 11818 (2002).

Lee, S.C., Lam, S., and Fai, H.K. *Build. Environ.* **36**, 837 (2001).

Levenspiel, O. Solid Catalysed Reactions. in "Chemical Reaction Engineering, 3rd Edn.". John Wiley & Sons, New York (1998), p. 388.

Lewandowski, M., and Ollis, D.F. Photocatalytic Oxidation of Gas- Phase Aromatic Contaminants. in V. Ramamurthy, and K.S. Schanze (Eds.), "Semiconductor Photochemistry and Photophysics". Marcel Dekker, New York (2003), p. 249.

Lichtin, N.N., Avudaithai, M., Berman, E., and Dong, J. *Res. Chem. Intermed.* **20**, 755 (1994).

Lim, T.H., Jeong, S.M., Kim, S.D., and Gyenis, J. *J. Photochem. Photobiol. A Chem.* **134**, 209 (2000).

Lim, T.H., and Kim, S.D. *Chemospere* **54**, 305 (2004).

Lin, H.F., Ravikrishna, R., Valsaraj, K.T., *Sep. Purif. Technol.* **28**, 87 (2002).

Luo, Y., and Ollis, D.F. *J. Catal.* **163**, 1 (1996).

Maggos, Th., Plassais, A., Bartzis, J.G., Vasilakos, Ch., Moussiopoulos, N., and Bonafous, L. *Environ. Monit. Assess.* **136**, 35 (2008).

Martra, G, Coluccia, S., Marchese, L., Auguliaro, V., Loddo, V., Palmisano, L., and Schiavello, M. *Catal. Today* **53**, 695 (1999).

Matsuda, S., and Hatano, H. *Powder Technol.* **151**, 61 (2005).

Mehrvar, M., Anderson, W.A., Moo-Young, M., and Reilly, P.M. *Chem. Eng. Sci.* **55**, 4885 (2000).

Mendez-Roman, R., and Cardona-Martinez, N. *Catal. Today* **40**, 353 (1998).

Mills, A., and Le Hunte, S. *J. Photochem. Photobiol. A Chem.* **108**, 1 (1997).

Mohseni, G.M., and Taghipour, F. *Chem. Eng. Sci.* **59**, 1601 (2004).

Monteagudo, J.M., and Duran, A. *Chemosphere* **65**, 1242, (2006).

Muggli, D.S., and Falconer, J.L. *J. Catal.* **175**, 213 (1998).

Muggli, D.S., Lowery, K.H., and Falconer, J.L. *J. Catal.* **180**, 111 (1998).

Murakami, Y., Endo, K., Ohta, I., Nosaka, A.Y., and Nosaka, Y. *J. Phys. Chem. C* **111**, 11339 (2007).

Nicolella C., and Rovatti M. *Chem. Eng. J.* **69**, 119 (1998).

Nimlos, M.R., Jacoby, W.A., Blake, D.M., and Milne, T.A. *Environ. Sci. Technol.* **27**, 732 (1993).

Obee, T.N., and Brown, R.T. *Environ. Sci. Technol.* **29**, 1223 (1995).

Oda, T., Yamashita, R., Haga, I., Takahushi, T., and Masuda, S. *J. Electrostat.* **57**, 293 (2003).

Oda, T., Yamashita, R., Takahushi, T., and Masuda, S. *IEEE Trans. Ind. Appl.* **32**, 118 (1996).

Ollis, D.F., Pelizzetti, E., and Serpone, N. *Environ. Sci. Technol.* **25**, 1523 (1991).

Paz, Y. *Comptes Rendus Chimie* **9**, 774 (2006).

Paz, Y. "Composite Titanium Dioxide Photocatalysts and the "Adsorb & Shuttle" Approach: A review", in "Solid- State Chemistry and Photocatalysis of Titanium Dioxide". Nowotny J. (Editor), Trans Tech Publications, UK [in press].

Paz, Y., and Heller, A. *J. Mater. Res.* **12**, 2759 (1997).

Paz, Y., Luo, Z., Rabenberg, L., and Heller, A. *J. Mater. Res.* **10**, 2842 (1995).

Peill, N., and Hoffmann, M. *Environ. Sci. Technol.* **29**, 2974 (1995).

Pelizzetti, E., Visca, M., Borgarello, E., Pramauro, E., and Palmas, A. *Chimica e l'Industria* **63**, 805 (1981).

Peral, J., and Ollis, D.F. *J. Catal.* **136**, 554 (1992).

Phillips, L.A., and Raupp, G.B. *J. Mol. Catal.* **77**, 297 (1992).

Pruden, A.L, and Ollis, D.F. *J. Catal.* **82**, 404, (1983).

Raupp, G.R., Nico, J.A., Annangi, S., Changrani, R., and Annapragada, R. *AIChE J.* **43**, 792 (1997).

Roland, U., Holzer, F., and Kopinke, F.-D. *Catal. Today* **73**, 315 (2002).

Romero-Vargas Castrillon, S., and De-Lasa, H.I. *Ind. Eng. Chem. Res.* **46**, 5867 (2007).

Romero-Vargas Castrillon, S., Ibrahim, H., and De-Lasa, H.I. *Chem. Eng. Sci.* **61**, 3343 (2006).

Saarela, K., Tirkkonen, T., Laine-Ylijoki, J., Jurvelin, M.J., Nieuwenhuijsen, M.J., and Jantunen, M. *Atmos. Environ.* **37**, 5563 (2003).

Sagatelian, Y., Sharabi, D., and Paz, Y. *J. Photochem. Photobiol. A* **174**, 253 (2005).

Salvado-Estivill, I., Brucato, A., and Li Puma, G. *Ind. Eng. Chem. Res.* **46**, 7489 (2007a).

Salvado-Estivill, I., Hargreaves, D.M., and Li Puma, G. *Environ. Sci. Technol.* **41**, 2028 (2007b).

Sampath, S., Uchida, H., and Yoneyama, H. *J. Catal.* **149**, 189 (1994).

Sanhueza, E., Hisatsune, I.C., and Heicklen, J. *Chem. Rev.* **76**, 801 (1976).

Sauer, M.L., Hale, M.A., and Ollis, D.F. *J. Photochem. Photobiol. A* **88**, 169 (1995).

Sauer, M.L., and Ollis, D.F. *J. Catal.* **149**, 81 (1994).

Sclafani, A., Brucato, A., and Rizzuti, L. Mass transfer limitations in a packed bed photoreactor used for phenol removal. in D.F. Ollis, H. Al-Ekabi (Eds.), "Photocatalytic Purification and Treatment of Water and Air", vol. 3. Elsevier, (1993), p. 511.

Serrano, B., and de Lasa, H. *Ind. Chem. Eng. Res.* **36**, 4705 (1997).

Sharabi, D., "Preferential Photodegradation of Contaminants by Molecular Imprinting on a Photocatalytic Substrate" Msc. thesis, Technion, Israel (2007).

Shiraishi, F., Yamaguchi, S., and Ohbuuchi, Y. *Chem. Eng. Sci.* **58**, 929 (2003).

Sidheswaran, M., and Taviarides, L.L. *Ind. Eng. Chem. Res.* **47**, 3346 (2008).

Stokke, J.M., Mazyck, D.W., Wu, C.Y., and Sheahan, R. *Environ. Prog.* **25**, 312 (2006).

Strini, A., Cassese, S., and Schiavi, L. *Appl. Catal. B Environ.* **61**, 90 (2005).

Sun, R.-B., Xi, Z.-G., Chao, F.-H., Zhang, W., Zhang, H.-S., and Yang, D.-F. *Atmos. Environ.* **41**, 6853 (2007).

Takeuchi, K. *J. Jpn. Soc. Atmos. Environ.* **33**, 139 (1998).

Tatsuma, T., Tachibana, S., and Fujishima, A. *J. Phys. Chem. B* **105**, 6987 (2001).

Thevenet, F., Guaiella, O., Puzenat, E., Herrmann, J.-M., Rousseau, A., and Guillard, C. *Catal. Today* **122**, 186 (2007).

Torimoto, T., Okawa, Y., Takeda, N., and Yoneyama, H. *J. Photocem. Photobiol. A Chem.* **103**, 153 (1997).

Tsoukleris, D.S., Maggos, T., Vassilakos, C., and Falaras, P. *Catal. Today* **129**, 96 (2007).

Tsuru, T., Kan-no, T., Yoshioka, T., and Asaeda, M. *J. Memb. Sci.* **280**, 156 (2006).

Turchi, C.S., and Ollis, D.F. *J. Catal.* **122**, 178 (1990).

Upadhya, S., and Ollis, D.F. *J. Adv. Oxid. Technol.* **3**, 199 (1998).

Vorontsov, A.V., and Dubovitskaya, V.P. *J. Catal.* **221**, 102 (2004).

Vorontsov, A.V., Savinov, E.N., Barannik, G.B., Troitsky, V.N., and Parmon, V.N., *Catal. Today* **39**, 207 (1997).

Wang, C.-Y., Rabani, J., Bahnemann, D.W., and Dohrmann, J.K. *J. Photochem. Photobiol. A Chem.* **148**, 169 (2002).

Yamazaki–Nishida, S., Fu, X., Anderson, M.A., and Hori, K. *J. Photochem. Photobiol. A Chem.* **97**, 175 (1996).

Yamazaki-Nishida, S., Nagano, K.J., Phillips, L.A., Cervera-March, S., and Anderson, M.A. *J. Photochem. Photobiol. A Chem.* **70**, 95 (1993).

Yin, S., Liu, B., Zhang, P., Morikawa, T., Yamanaka, K.-I., and Sato, T. *J. Phys. Chem. C* **112**, 12425 (2008).

Yoneyama, H., and Torimoto, T. *Catal. Today* **58**, 133 (2000).

Zemel, E., Haick, H., and Paz, Y. *J. Adv. Oxid. Technol.* **5**, 27 (2002).

Zhou, L., Tan, X., Zhao, L., and Sun, M. *Korean J. Chem. Eng.* **24**, 1017 (2007).

A

Acetaldehyde, 76, 298
Acetic acid, 75
Acetonitrile, 7
Acid Orange 7, 47, 77
Acid Orange 24, 209
Acrylic acid, 75
Acrylic windows, 238f
Actinometry, 260, 263, 284
Activated carbon, 57, 304
Activation step, 231
Adsorb & Shuttle, 299
Adsorption-desorption equilibrium, 8
Advanced oxidation processes (AOP),
    290
Agglomeration, 315
Air humidifier, 238f
Air mass (AM), 119
Air treatment
    annular reactors, 322–326
    applications, 293–296
    batch reactors, 310–311
    BTEX in, 304–305
    coated wall annular reactors, 322–324
    coated wall tubular reactors, 315–316
    combined adsorptive-photocatalytic
        reactors, 328–329
    continuous, one-pass flow reactors,
        311
    current problems, 329–331
    deactivation, 331
    dissolved pollutants and, 293, 296
    flat plate reactors, 326
    fluidized bed annular reactors, 322
    fluidized bed tubular reactors,
        314–315
    generalized models, 327–328
    indoor, 294
    light sources, 301–303
    manuscripts on, 293f
    mass transport, 300–301
    mixtures, 308–310, 330
    modes of, 310–311
    monolith tubular reactor, 319–321
    NOx, 308
    outdoor, 295
    packed bed annular reactors, 324–325
    packed bed reactors, 317–318
    patents, 291–292
    permeable layer annular reactors,
        325–326
    permeable layer tubular reactors,
        321–322
    photocatalysis for, 296–303
    plasma-driven packed bed reactor,
        317–318
    powder layer tubular reactors,
        312–314
    process gases and, 295–296
    reaction kinetics, 297–300
    reactors for, 312–329
    recirculation systems, 310–311
    standardization, 330
    target pollutants in, 303–310
    trichloroethylene in, 305–308
    visible light and, 329–330
Aldehyde, 20, 44
Alizarin red, 48
AlTCPc. See Hydroxoaluminiumtri-
    carboxymonoamide
    phthalocyanine
Alumina reticulate, 319–320
Aluminum hydroxide, 59
AM. See Air mass
Angular variable, 216
Anion doping, 126–127

band structure, 127f
Annular lamp reactor, 218–220
Annular reactors, 322–326
   coated wall, 322–324
   fluidized bed, 322
   packed bed, 324–325
   permeable layer, 325–326
AOP. *See* Advanced oxidation processes
Apparent quantum efficiency
   (AQE), 304
AQ-1000, 177
AQE. *See* Apparent quantum efficiency
AQ-Red dye, 177
AQ-RF, 177
Aqua regia, 53
Argon lamps, 302
Arsenic
   hetereogeneous photocatalysis, 58–61
   oxidation, 60
   pollution, 58
   titanium dioxide and, 59–60
   toxicity of, 58
Avogadro's number, 196

B
Band structure
   anion doping, 127f
   cation doping, 127f
   of photocatalyst composite,
      127f, 129f
Band-gap energy, 113
   of semiconductors, 124f
Band-gap engineering, 126–129
   anion doping, 126–127
   cation doping, 126–127
Batch reactors, 310–311
Beam solar radiation, 217f
Benzaldehyde, 305
Benzene, 294, 309–310
Benzene dicarboxaldehyde, 305
Benzoic acid, 305
   degradation rate, 155f, 157f
Benzoquinone (BQ), 269
1,4-benzoquinone (1,4-BQ), 76
   oxidation of, 90, 96t
Benzyl alcohol, 4, 5, 6
   concentration, 24f

experimental values of, 18f, 19f
   irradiation, 19f
   photoadsorption, 22–25
   photodegradation of, 18
   unit mass, 24f
Bessel functions, 216
Best fitting procedure, 32
Borosilicate glass tubes, 244f, 266
Boundary conditions, 241
   Marshak, 215, 220
   RTE, 232
Box-Draper technique, 298
BQ. *See* Benzoquinone
1,4-BQ. *See* 1,4-benzoquinone
BTEX, 295, 309, 321
   as target pollutant, 304–305
Buchi Rotovapor M, 5
Bulk recombination, 27
1-butanol, 56
Byphenyls, 77

C
Calcination, 76
Calomel, 52
   formation, 50
CAR. *See* Classical annular reactor
Carbaryl, 196
   in CPC reactors, 201f
   evolution of, 205f
Carbon dioxide, 20, 21, 300
   kinetic modeling and, 92–94
Carbon tetrachloride, 146
Carboxylic acid, 56, 85, 99
Cat, 2
Catalyst elemental analysis, 80–81
Catechol, 20
Cathodic pathway, 39–40
Cation doping, 126–127
   band structure of, 127f
CB. *See* Conduction band
4-CC. *See* 4-chlorocatechol
CdS, 138–139
   photoactivity of, 139
Cellulose acetate filter, 7
Centrifugal pump, 258f
CFD, 166–167, 170, 262, 326,
   327–328

Charge separation, 26
Chemical decay, 3
Chemical reactions, initiating, 2
Chemical reactors, design, 2
Chemisorption, 15
Chlorine, 59, 235, 306–307
4-chlorocatechol (4-CC), 263, 269, 271f
    concentrations, 275
    experimental concentrations, 276f
    in heterogeneous photocatalysis,
        281–282
    predicted concentrations, 276f
Chloroform, 146
4-chlorophenol (4-CP), 48, 196, 263,
        269, 271f
    concentrations, 275
    conversions, 282
    degradation of, 264f
    experimental concentrations, 276f
    experimental conversions, 281f
    in heterogeneous photocatalysis,
        281–282
    kinetic parameters, 275t
    predicted concentrations, 276f
    predicted conversions, 281f
    reaction pathway, 264f
Chromium, 39
    heterogeneous photocatalysis and,
        44–49
    net reaction for, 45
    photocatalytic reaction, 45–46
    photocatalytic reduction, 45, 47
    photoreduction of, 73
    titanium dioxide and, 48
Circulation rates, in water
        purification, 148
Citrate, 52
Citric acid, 43, 56
Classical annular reactor (CAR),
        161, 179
Coated wall annular reactors,
        322–326
Coated wall tubular reactors, 315–316
Cocatalysts, deposition of, 129–130
Combined adsorptive-photocatalytic
        reactors, 328–329
    schematics of, 333f
Composite semiconductors, 127–129

Compound parabolic collectors
        (CPC), 186, 190
    carbaryl in, 205f
    comparison of, 201–202
    concentrating, 196, 206–210
    cross section, 203f
    design, 195
    geometries, 201f
    geometry of, 194f
    multiple, 204f
    nonconcentrating, 194–195
    optical systems, 193f
    photoreactor, 191f, 192–195
    reflectors, 195f
    test bed, 204f
    truncation of suns, 195f
    tubular receiver, 193–194
Conduction band (CB), 113
    electrons, 38
Conductivity, 214
Constants
    intrinsic kinetic, 93
    photoadsorption equilibrium, 13
    pseudo-first-order rate, 12
    Temkin equilibrium absorption, 32
    wavelength-dependent, 214
Constrained relationships, 94–95
Continuous, one-pass flow
        reactors, 311
Continuous stirred tank (CST), 80, 311
Copper, 39
4-CP. See 4-chlorophenol
CPC. See Compound parabolic
        collectors
CREC
    advances in, 78–79
    catalyst elemental analysis, 80–81
    EDX and, 80–81
    experimental methods used in,
        79–81
    iron analysis, 80
    model pollutant analysis, 80
    reactants, 80
    reaction setup, 79–81
    substrate analysis, 80–81
    XPS and, 80–81
CST. See Continuous stirred tank
CTAB surfactant, 51

D

DCA. *See* Dichloroacetaldehyde
Deactivation, 331
Decay
  chemical, 3
  physical, 3
Degradation rate
  of 4-CP, 264f
  of benzoic acid, 155f, 157f
  of methanol, 301
  PCE, 236t
Degussa, 38
Dichloroacetaldehyde (DCA), 306
Dichloroacetic acid, 199, 306
Dichloroacetyl acid, 306
Dichloromethane, 146, 300
Dichromate, 45
Dicyanomercury, 51
Differential equations, 97
  Helmholtz, 214
  integro, 231
  ordinary, 94
  second-order partial, 214
Diffusion approximation, 214
Dimensionless asymmetry factor, 272
1,2-dimethoxybenzene, 77
Dimethylamine, 297
Direct reduction, 42
  lead, 56
Discrete ordinates method (DOM), 207,
  211–212, 273
Dissolved pollutants, 293, 296
Distribution heads, 244f
Distributive type, 162
  comparison of, 162t
DOM. *See* Discrete ordinates method
Doping techniques, 71
Doubly periodic wavy flow
  (DPWF), 171
DPWF. *See* Doubly periodic wavy flow
Duran glass tubes, 202

E

EDTA, 43
  lead and, 53
  mercury and, 50
  uranium and, 58

EDX. *See* Energy dispersive X-ray
  spectroscopy
Eigenfunctions, 220
Eigenvalues, 220
Electrolysis, 112
Emission, 233
  lamp model, 260–261
  superficial model, 241
  volumetric lamp model, 261f
  volumetric model, 259
Energy dispersive X-ray
  spectroscopy (EDX), CREC and,
  80–81
Enthalpy, 123
Environmental Protection Agency
  (EPA), 294
EPA. *See* Environmental Protection
  Agency
Epoxy resin, 320
Equations. *See also* Radiative transfer
  equation
  differential, 94, 97, 214, 231
  Henry's law, 16
  integro-differential, 231
  Langmuir, 16
  mass conservation, 241
  model, 259
  ordinary differential, 94
  pseudo-first-order rate, 12
  radiation transfer, 210–211
  second-order partial differential,
  214
  substituting, 237
Escherichia coli, 202
Ethanol, 44, 56, 301
  mercury and, 50
Ethylbenzene, 309
Euler-type method, 209–210
Europium-doped strontium
  borate, 302
Europium-doped strontium
  fluoroborate, 302
Experimental apparatus, 5–8
  set up, 6f
Experimental run procedure, 7f
External type, 161
  comparison of, 162t

F

FBR. *See* Fluidized bed tubular reactors
Feed tank, 258f
Ferric ions, 74
 on oxidation rate of phenol, 84f
Ferrihydrite, 57
Ferrous ions, 74
 on oxidation rate of phenol, 84f
Ferrous sulfate, 45
Fixed-bed flow-through reactor, 60
Flat plate photoreactor, 238f
Flat plate reactors, 326
Floccules, 85
Flow homogenizer, 238f
Flow instability, 172
Flow pattern, 153
 slurry reactor, 154
 turbulent vortex, 179f
 TVF, 179f
 WTWVF, 179f
FLUENT, 153, 166, 170
Fluidized bed annular reactors, 322
Fluidized bed tubular reactors (FBR),
 314–315
Fluoride, 59
Formic acids, 44, 46, 56, 75, 86, 257, 259
 mineralization, 250t
 photolysis of, 255
Four phase system, 148
Fourier expansion, 217
Fourier series, 218
Freundlich expression, 15
Freundlich isotherm, 14–16, 22, 25, 28
Freundlich maximum adsorption
 capacity, 15
Freundlich model
 linear form of, 25f
 parameters, 26f
Freundlich photoadsorption, 20f
Freundlich relationship, 15
Fumaric acid, 86
2,5-furandione, 305

G

GAMBIT, 170
Gas inlet, 238f
Gas outlet, 238f

Gas scrubber, 238f
Gaussian distribution, 21
Germicidal lamps, 257
Gibbs free energy change, 112
 for water splitting, 117f
Global energy loss, 123
Global reaction, for metallic mercury
 deposition, 50
Goethite, 57
Gold, 39
Graham condenser, 5
Granular ferric hydroxide, 57

H

HDA. *See* Hexadecylamine
Heat exchanger, 242f
Helmholtz differential equation, 214
Hematite, 57, 74
Henry's law equation, 16
Henyey-Greenstein phase function,
 207–208, 272
Heterogeneous photocatalysis, 38
 4-CC in, 281–282
 4-CP in, 281–282
 arsenic, 58–61
 chromium, 44–49
 defined, 2–3
 experimental setup, 271f
 hydroquinone in, 281–282
 kinetic results, 273–277
 laboratory reactor, 266–273
 lead, 53–57
 mechanistic pathways, 41–44
 mercury, 49–53
 mobile windows mechanism, 270f
 phenol, 69–106
 pilot scale reactor, 277
 of PMA, 52
 of PMC, 52
 radiation absorption and, 263–282
 radiation model, 279–280
 reaction mechanisms, 264t
 reaction scheme, 263–266
 reactor model, 277–279
 scaling up, 263–282
 scattering and, 263–282
 schematic representation of, 269f

simplified diagram of, 39f
thermodynamical considerations, 41–44
of titanium dioxide, 39f
treatment of metals in water with,
    37–62
uranium, 57–58
variation, 279–282
Hexadecylamine (HDA), 321
High-performance liquid
    chromatograph (HPLC), 7–8, 20, 80
Homogeneous photochemical reactors
    experiments, 262t
    kinetic model, 250–251
    kinetic results, 256–257
    laboratory, 251–256
    model equation, 259
    pilot scale, 257–258
    predictions, 262t
    radiation model, 259–262
    reaction scheme, 250–251
    reactor model, 258–259
    scaling up of, 250–263
    validation, 262–263
HPLC. *See* High-performance liquid
    chromatograph
Humic acid, 59
hv, 2
Hydrogen, 114
    evolved, 123
    peroxide, 74, 250–251, 259, 263
    in water splitting, 132f
Hydroquinone, 20, 263, 271
    concentrations, 275
    experimental concentrations, 276f
    in heterogeneous photocatalysis,
        278–279
    predicted concentrations, 276f
Hydroxoaluminiumtri-
    carboxymonoamide
    phthalocyanine (AlTCPc), 48
3-hydroxybenzylaldehyde, 305
4-hydroxybenzaldehyde, 305
4-hydroxybenzyl alcohol, 305
Hydroxyl ions, 75
Hydroxyl radicals, 9, 38, 44, 72, 75, 234
    abundance of, 300
    cathodic pathway and, 39–40

I
Imaging optical systems, 193f
Immersion type, 161
    comparison of, 162t
    lamps, 169
Impregnation, 76
Incident Radiation, 260
Indirect reduction, 42, 43–44
    lead, 56
Indoor air treatment, 295
Inequalities, 30
Initial rates, 297
Innovative type, 163
Inorganic reduction
    in heterogeneous photocatalysis, 72
    organic oxidation and, 73
In-scattering, 233
Integro-differential equations, 231
Intrinsic kinetic constants, 93
Ion-implantation technique, 127
Iron
    analysis, 80
    assisted photocatalytic mineralization
        of phenol, 81–104
    chloride, 59
    disappearance rates of phenol and,
        83f
    kinetic modeling, 92–104
    oxidation of phenol and, 82–89
    oxide, 57, 59
    in photocatalytic process, 74–78
    titanium dioxide and, 75
Iron hydroxide, 59
    floccules, 85
Irradiation, 2
    absorption, 3
    benzyl alcohol, 19f
    phenol, 20f, 21f
    titanium dioxide surface
        modifications under, 8–10
1,3-isobenzofurandione, 305
Isopropyl alcohol, 46
Isotherms, 12
    Freundlich, 14–16, 22, 25, 28
    Langmuir, 13–14, 15, 22, 25, 28
    Redlich-Peterson, 16–18, 22, 25, 28
    Temkin, 31–32

J

Jetter's method, 217

K

Ketones, 44
Kinematic viscosity, 170
Kinetic modeling, 78
  aromatics, 96, 99
  carbon dioxide and, 98–104
  of iron-assisted phenol oxidation,
    90, 104
  lumped acids, 98–104
  overall, 92–94
  parameter estimation, 94–104
  parameters, 24f, 26f
  of photocatalysis, 92–104
  reaction network for, 96f
  reaction schemes, 101f
  scaling up and, 235–237
  schematic representation of, 101f
  series-parallel, 77, 275
  of unpromoted photocatalysis
    oxidation, 92–104
Kinetic reactors, in water purification,
  151–152
Kinetic regimes, 152–154
Kinetic scheme, for PCE degradation,
  236t

L

Laboratory reactors
  coordinate system for, 242f
  description, 251–256, 266–273
  experimental conversions, 244f
  heterogeneous photocatalysis,
    266–273
  kinetics results, 243
  model, 243–246, 266–273
  operating conditions, 239t–240t,
    253t–254t, 267t–268t
  pilot scale, 243
  predicted outlet conversions, 244f
  scaling up, 237–242, 251–256
  schematic representation of, 238f, 252f
Lambert-Beer type law, 206, 211
Laminar Taylor vortex flow (LTVF), 171
  time-dependent, 178

Lamps
  annular, 218–220
  argon, 302
  contour, 261f
  emission model, 260–261
  germicidal, 257
  immersion type, 169
  mercury medium-pressure, 5
  mercury plasma, 302
  neon, 302
  sodium, 302
  tube light reactor, 174
  UV, 174, 238f, 244f
  volumetric, 261f
  zinc/cadmium, 302
Langmuir equation, 16
Langmuir isotherm theory, 13–14, 15, 22,
  25, 28
  asymptotic cases of, 29–31
Langmuir model
  linear form of, 23f
  parameters, 24f
Langmuir photoadsorption, 18f, 19f
Langmuir relationship, 13
Langmuir-Hinshelwood (LH) model,
  10–11, 78
Langmuir-type dependence, 151
Lead, 39
  direct reduction, 56
  EDTA and, 53
  heterogeneous photocatalysis, 53–57
  indirect reduction, 56
  oxidative route to, 54
  ozone and, 54–55
  titanium dioxide and, 54
Lead-doped barium silicate, 302
Levenberg-Marquardt method, 273
LH model. *See* Langmuir-Hinshelwood
  model
Light intensity, 198
Light sources, 301–303
Limiting efficiency, 121
  for single-band gap photocatalyst,
    122f
Limiting values, 242
Limonite, 71
Linear coefficients, 23

Liquid-solid catalytic reactions, 4
Local incident radiation, 210
  distribution, 221f
Local superficial rate of photon
  absorption (LSRPA), 230, 237, 246
  values, 248
Local volumetric rate of photon
  absorption (LVRPA), 196,
  230, 252, 255, 265, 276
  homogeneous value for, 197
  for polychromatic radiation, 273
  profiles, 274f
  titanium dioxide and, 207
LSRPA. *See* Local superficial rate of
  photon absorption
LTVF. *See* Laminar Taylor vortex flow
Lumped acids, kinetic modeling and,
  98–104
LVRPA. *See* Local volumetric rate of
  photon absorption

M
Macrokinetic studies
  important factors in, 149t–150t
  water purification, 148–159
Magnetic stirrer, 5
Magnetite, 57, 74
Maleic acid, 75, 76, 86
  concentration, 87f
  phenols and, 77
Manganese, 39
Marquardt's percent standard deviation
  (MPSD), 5
Marshak boundary condition,
  215, 220
Mass conservation equation, 241
Mass flowmeter, 238f
Mass transfer, 168, 236
Mass transport
  air treatment, 296–303
  external, 300
  internal, 300
MCM. *See* Monte Carlo method
Mechanistic pathways, heterogeneous
  photocatalysis, 41–44
Merbromin, 52
Mercuric chloride, 50

Mercurochrome, titanium dioxide
  and, 52
Mercury, 39, 47
  in agricultural pesticides, 49
  EDTA and, 50
  ethanol and, 50
  global reaction for, 50
  heterogeneous photocatalysis, 49–53
  salts, 50
  SDS and, 50–51
  time profiles of, 50
  titanium dioxide and, 49, 51
  toxicity of, 51
Mercury medium-pressure lamp, 5
Mercury plasma lamps, 302
Metalloids, treatment of, in water by
  heterogeneous photocatalysis,
  37–62
Metals
  nitrides, 137
  oxynitrides, 137
  reduction potentials, 41f, 73f
  treatment of, in water by
  heterogeneous photocatalysis,
  37–62
Methanol, 7, 44, 56, 135, 320
  degradation of, 301
Method of initial rates, 297
Methyl benzoic acid, 305
Methyl tert-butyl-ether, 297, 315, 317
Methylbenzaldehydes, 305
Methylmercury, 51
Micro steady-state approximation
  (MSSA), 250
Microphotoelectrodes, 116
Mobile windows mechanism, 270f
Model pollutant analysis, 80
Modeling dye degradation, 208–210
Mole balance, 11
Monochlorobenzene, 168
Monochromatic incident radiation, 255
Monochromatic radiation, 273
Monochromatic specific intensities, 255
Monolith tubular reactor, 319
Monte Carlo method (MCM), 211–213
MPSD. *See* Marquardt's percent
  standard deviation

MSSA. *See* Micro steady-state
 approximation
Muconic acid, 77
Multiannular photocatalytic reactor,
 238f
Multiple tube reactor, 164–168, 178, 181
 mixing inside, 167f
 schematic diagram of, 173f
 titanium dioxide in, 164

N

Neon lamps, 302
Nickel-oxide-doped glass, 302
Niobates, in water splitting, 135–136
Nitric oxides (NOx), 295, 296, 318, 322
 as target pollutant, 303
Nitrides, in water splitting, 137
Nitrogen, 136
Nonimaging optical systems, 195f
Nonuniform coating, 241
Novel kinetic reactor, 152–153
NOx. *See* Nitric oxides

O

OC. *See* Organic carbon
Occupational and Safety Administration
 (OSHA), 304
ODE. *See* Ordinary Differential Equation
o-DHB. *See* Ortho-dihydroxybenzene
Optical properties, 206–208
Optical systems
 CPC, 195f
 imaging, 193f
 nonimaging, 193f
Optimal catalyst loading, 156, 158
OQY. *See* Overall quantum yield
Orange II, 176, 179
 decomposition, 180t
Ordinary Differential Equation
 (ODE), 94
Organic carbon (OC), 99
 oxidation and, 100f
Organic oxidation
 in heterogeneous photocatalysis, 72
 inorganic reduction and, 73
Organo-silicone compounds, 294
Ortho-dihydroxybenzene (o-DHB), 77

concentration profiles, 87f
oxidation of, 89–90, 95t
predicted profiles of, 97f
OSHA. *See* Occupational and Safety
 Administration
Outdoor air treatment, 295
Overall quantum yield (OQY), 303
Oxalic acids, 44, 86
 degradation of, 201f
 in solar reactors, 201f
Oxidation, 27
 of 1,4-benzoquinone, 90, 96t
 arsenic, 60
 estimated parameters, 103t
 kinetic modeling, 92–104
 of o-DHB, 89–90, 95t
 organic carbon and, 100f
 of p-DHB, 90
 phenol, 83f
 of phenol with ferric ions, 83f
 of phenol with ferrous ions, 84f
 of phenol with iron, 82–89
 predicted profiles for, 102f
 rates of phenol, 84f
 reaction, 42
Oxidative removal, 42, 44–45
Oxygen
 coverage, 11–12
 supply, 271f
 titanium dioxide and, 39
 in water splitting, 130f
Oxynitrides, in water splitting, 139
Ozone
 lead and, 54–55
 ROS and, 55

P

P1 approximation, 213–222
 applicability of, 221–222
Packed bed annular reactors, 324–325
Packed bed reactors (PBR), 296, 316–317
 configuration of, 317f
Palladium, 39
Parabolic trough concentrators (PTC),
 188, 189f, 190
 coordinate systems, 218f
 flux concentration distribution in, 219f

geometries, 201f
    reaction rate optical factor and, 199f
    solutions for, 217–218
Para-dihydroxybenzene (p-DHB), 76
    oxidation of, 90
    predicted profiles of, 97f
Parameter estimation, 94–104
    constrained relationships for, 94–95
Particulate systems, 116–117
Patents
    air treatment, 291–292
    distribution of, 292f
    per year, 292f
    titanium dioxide, 291
PBR. See Packed bed reactors
PCE. See Perchloroethylene
p-DHB. See Para-dihydroxybenzene
Perchloroethylene (PCE), 309
Perhydroxyl radicals, 75
Permanganate, 59
Permeable layer annular reactors,
    325–326
Permeable layer tubular reactors,
    321–322
Perspex, 174
Persulfate, 118
PFR. See Plug flow reactor
pH, 151
Phenol, 6
    CREC and, 78
    disappearance rates of, 83f
    experimental values, 20f, 21f
    ferric ions and oxidation rate of, 84f
    ferrous ions and oxidation rate of, 84f
    heterogeneous photocatalysis, 69–106
    iron and oxidation of, 82–85
    iron assisted photocatalytic
        mineralization of, 81–104
    irradiation time, 20f, 21f
    kinetic modeling, 78, 92–104
    mineralization of, 69–106
    oxidation rates, 83f
    photoadsorption, 25–26
    predicted profiles of, 97f
Phenolic intermediates, 77
Phenylmercury chloride (PMC), 51
    photocatalysis of, 52
Philips Lighting, 174

Photoadsorption, 3
    center, 3
    determination, 4, 10–18
    discussion, 21–28
    equilibrium constant, 13
    Freundlich, 20f
    Langmuir, 18f, 19f
    phenol, 25–26
    of polycrystalline semiconductor
        oxide, 4–5
    reaction mechanism, 26–28
Photo-Cat, 148
Photocatalysis, 5
    air treatment, 296–303
    band structure of, 128f, 129f
    band-gap engineering, 126–129
    batch reactors, 310–311
    composite, 128f
    composite semiconductors, 127–129
    continuous, one-pass flow reactors,
        311
    conversion rates, 151
    cost of, 70–71
    design challenges in, 159–181
    developing, 125–140
    doping, 71
    dual, 119f
    estimated parameters, 198t
    improving, 71–72
    inorganic reduction, 72
    iron assisted, of phenol, 81–104
    iron in, 74–78
    kinetic modeling, 92–104
    large scale, 159–181, 299
    light sources, 301–303
    mass transport, 300–301
    metal nitrides in, 137
    metal oxynitrides in, 137
    modes of operation, 310–311
    multiple tube reactors, 164–168
    organic oxidation, 72
    phenol, 69–106
    reaction kinetics, 297–300
    reactor comparison, 162t
    reactors for air treatment, 312–329
    recirculation systems, 310–311
    scaling up, 230–282
    semiconductor alloys, 129

series-parallel reaction mechanism for, 90–92
surface modification by deposition of cocatalysts, 129–130
surface modification schematic diagram, 130f
Taylor vortex reactor, 169–172
tube light reactor, 168–169
unpromoted, 96t
for water splitting, 124–140
Photocatalytic plate, 238f
Photocatalytic reaction, chromium, 45–46
Photocatalytic reduction, chromium, 45, 47
Photocatalytic thermodynamic efficiency factor (PTEF), 330
Photocatalytic wall reactor
laboratory reactor, 237–242
radiation absorption and, 234–249
radiation reflection and, 234–249
scaling up of, 234–249
Photoconversion efficiency, 123
Photocorrosion, 138
Photo-CREC Water-II reactor, 79f, 82, 220
experiments in, 81
in water purification, 163
Photo-CREC-air reactor, 325
Photodegradation, of benzyl alcohol, 18
Photoelectrochemistry
cell diagram, 115f
of water splitting, 113–123
Photo-fenton, 186, 211
Photon absorption
effects, 3
rate, 231f
reaction rate and, 277t
Photon distribution, characterization of, 232f
Photon mol, 196
Photoreactivity, 30
Photoredox, 58
Photoreduction, 73
Physical decay, 3
Pilot scale reactor, 243
coordinate system for, 247t

experimental conversion for, 249f
experimental device, 278f
heterogeneous photocatalysis, 277
homogeneous, 259
operating conditions, 253t–254t, 267t–268t
predicted outlet conversions for, 249f
scaling up, 243
schematic representation of, 244f, 278f
Planck's constant, 196
Plasma-driven packed bed reactor, 317–318
Platinization, 58
Platinum, 39
Plug flow reactor (PFR), 311, 323
PMA. See Polymercury acetate
PMC. See Phenylmercury chloride
Pn approximation, 213
Polychromatic radiation, 273
Polycrystalline semiconductor oxide, photoadsorption of, 4–5
Polymercury acetate (PMA), 51
photocatalysis of, 52
Polyometalates, 59
Potassium ferrioxalate reaction, 256
Powder layer tubular reactors, 312–314
PQY. See Primary quantum yield
Primary quantum yield (PQY), 303
Process gasses, 293, 295–296
2-propanol, 44, 56
Pseudo-first-order kinetics, 15
Pseudo-first-order rate constant, 12
Pseudo-first-order rate equation, 12
PTC. See Parabolic trough concentrators
PTEF. See Photocatalytic thermodynamic efficiency factor
Purifics Environmental Canada Technologies, 148
Pyrex, 5, 173, 258f

Q
Quantum yields (QY), 121, 123, 302–303
overall, 303
primary, 303
QY. See Quantum yields

R

Radiation
  emitting system, 269, 271f
  field, 246–248
  intensity, 196–199
  monochromatic, 273
  polychromatic, 273
  properties, 232
Radiation absorption, 186
  heterogeneous photocatalytic reactors
    and, 263–282
  scaling up with, 234–250, 263–282
Radiation model, homogeneous reactor,
    259–260
Radiation reflection, scaling up with,
    234–250
Radiation transfer
  equation, 210–211
  in solar reactors, 206–213
Radiative transfer equation
    (RTE), 187, 210–211, 214,
    231, 272
  boundary conditions for, 232
  radiation propagation and, 232
  solutions of, 211–212, 280
Radiometers, 260
Ray tracing technique, 241, 259
Reaction kinetics, 297–300
Reaction mechanism
  heterogeneous photocatalyst reactors,
    264f
  photoadsorption, 26–28
Reaction rate optical factor (RROF), 198
  for parabolic trough solar
    photocatalytic reactor, 199f
Reaction schemes
  kinetic models and, 101f
  scaling up and, 234–237
Reaction setup, CREC, 79–81
Reaction-specific parameters, 160
Reactive oxygen species (ROS), 44, 49
  ozone and, 55
Reactors. *See specific types*
Reactor-specific parameters, 160
Recirculation systems, 310–311
Recombination centers, 85
Recycle pump, 238f

Redlich-Peterson isotherm, 16–18, 22, 25,
    28
Redox reagents, 118
Redox shuttle, 119f
Reduction, 27
Reduction potentials
  of metal ions, 73f
  metallic couples, 41f
  titanium dioxide and, 73f
Refractive index, of titanium dioxide,
    165
Reversible redox couple, 118
Reynolds number, 179f, 277
Rhodium, 39
Riser Reactor, 315
RMSE. *See* Root mean square error
Root mean square error (RMSE), 249,
    282
ROS. *See* Reactive oxygen species
RROF. *See* Reaction rate optical factor
RTE. *See* Radiative transfer equation
Runge-Kutta method, 325

S

Salicylic acid, 43, 73
Sampling device, 238f
Sampling valve, 271f
SBS. *See* Sick-building syndrome
Scaling up
  experiments, 262t
  heterogeneous photocatalytic
    reactors, 263–282
  of homogeneous photochemical
    reactors, 250–263
  kinetic model for, 235–237
  kinetic results, 256
  kinetics results, 243
  laboratory reactor, 237–240, 251–256
  methodology, 235f
  photocatalytic reactors, 230–282
  photocatalytic wall reactor, 234–250
  pilot scale reactor, 243, 257, 277
  predictions, 262t
  with radiation absorption, 234–250
  radiation model, 259–262
  with radiation reflection, 234–250
  reaction scheme and, 235–237

reactor model, 243–246, 258–259, 277–279
  validation, 249, 262–263, 280–282
Scattering/absorbing medium, 206
  heterogeneous photocatalytic reactors and, 263–282
SciFinder Scholar, 291
  target pollutants in, 303–310
SDS, mercury and, 50–51
Sealants, 294
Second-order partial differential equations, 214
Self-cleaning surfaces, 290f
Semiconductors, 118
  alloys, 129
  band-gap energy of, 124f
  composite, 127–129
  sulfide, 138
Series-parallel kinetic model, 77, 275
Series-parallel reaction mechanism
  detailed, 91f
  for photocatalysis, 90–92
Shielding effect, 151
Sick building syndrome (SBS), 294
Siderite, 74
Silver, 39
Singly periodic wavy vortex flow (SPWVF), 171
Sink terms, 233
Slurry reactor (SR), 148
  comparison of, 162t
  distributive, 161, 162t
  external, 161, 162t
  flow pattern inside, 154f
  immersion, 161, 162t
  innovative, 163
  regimes for, 154–159
  solar, 187, 191
  type, 161
Sodium lamps, 302
Sodium meta-bisulfite, 45
Sodium sulfite, 46
Sodium thiosulfate, 45
Solar photocatalytic reactors, 187–206
  annular lamp reactor, 218–220
  comparisons, 199–202
  concentrating, 187

coordinate system, 242f
  fixed, 187
  geometries, 201f
  modeling dye degradation, 208–210
  Monte Carlo method and, 211–213
  nonconcentrating, 186
  optical properties, 206–208
  oxalic acid in, 201f
  P1 approximation, 213–222
  parabolic trough, 199f
  radiation intensity, 196–199
  radiation transfer equation in, 210–211
  radiation transfer in, 206–213
  reaction rates, 189, 196–199
  RTE solutions, 211–213
  slurry, 187, 191
  thin film fixed bed, 190f
  types, 187–192
Solar radiation, 106
  beam, 219f
  characteristics of, 187
  conversion, 112
  spectral, 120f
  titanium dioxide and, 187
Solar spectrum
  energy distribution in, 120t
  irradiance, 120f
  water splitting and, 119–123
Sol-gel preparation, 76, 192, 203f
Solid phase dispersed (SPD), 159
Solid phase stationary (SPS), 159–160
Source terms, 233
SPD. See Solid phase dispersed
Spectral specific intensity, 233
Spherical coordinate system, 260
SPS. See Solid phase stationary
SPWVF. See Singly periodic wavy vortex flow
SR. See Slurry reactor
ST-1000, 179
Steady-state conditions, 4
Substituting equation, 237
Substrate, 30
Substrate analysis, 80–81
Sulfides, 118
  semiconductors, 138
  in water splitting, 138–140

Sulfites, 118
Sulfur anions, 135
Sulfur dioxide, 45
Sunlight, 111–112
Suns, 188
Superficial emission model, 241
Superhydrophilicity, 9
Superoxide radicals, 85
Surface modifications, 8–10
    by deposition of cocatalysts, 129–130
Surface reactions, 297
Surface recombination, 27
Surface roughening, 166

T
Tantalates, in water splitting, 135–136
Target pollutants
    in air purification, 303–310
    BTEX, 304–305
    mixtures, 308–310
    NOx, 308
    trichloroethylene, 305–308
Taylor number, 172
Taylor series expansion, 276
Taylor vortex flow, 178f
    flow pattern, 179f
Taylor Vortex photocatalytic reactor
        (TVR), 172
    analysis of, 177
    experimental set-up of, 175f
    schematic view of, 174
Taylor vortex reactor, 169–172
    flow configuration of, 171f
Taylor-Couette flow, 169, 170, 177
    instability, 172
Taylor-Couette geometry, 170
Taylor-Couette vortices, 170
t-butanol, 56
TC. See Tubular collectors
TCD. See Thermal conductive detector
TDSVE model, 259
Temkin equilibrium absorption
        constant, 32
Temkin isotherm, 31–32
Temkin model, 16
Tetracyanomercurate, 51
Thallium, 39

1,2,3-THB. See 1,2,3-trihydroxybenzene
Thermal conductive detector (TCD), 322
Thermal plasma, 76
Thermodynamical considerations
    heterogeneous photocatalysis, 41–44
    of titanium dioxide, 42
Thermohygrometer, 238f
Thermostatic bath, 238f, 271f
Thiele modulus, 301
Thin film fixed bed reactor, 190f
Time accurate solutions, 170
Time-dependent Taylor vortex flow,
        178f
TiO_2. See Titanium dioxide
Titanates, 317
    in water splitting, 134–135
Titanium dioxide (TiO_2), 2, 5, 20, 25, 28
    anodically biased, 43–44
    arsenic and, 59–60
    chromium and, 48
    coating, 295
    doped, 106
    Fe-doped, 85
    heterogeneous photocatalysis at, 39f
    immobilized on glass beads, 60
    iron and, 75
    lead and, 54
    LVRPA and, 196
    mercurochrome and, 52
    mercury and, 49, 51
    in multiple tube reactors, 164
    nanocrystalline, 46
    oxygen and, 39
    patents on, 291
    photoadsorption determination,
        10–18
    platinization of, 58
    polar, 299
    reduction potentials and, 73f
    refractive index of, 165
    scientific manuscripts on, 290f
    in SciFinder Scholar, 291
    solar radiation and, 189
    superhydrophilicity of, 9
    surface modifications under
        irradiation, 8–10
    suspension, 315

thermodynamic ability of, 42
in water splitting, 134–135
zeolite-supported, 300
TLR. *See* Tube light reactor
TOC profiles, 88, 281
TOC Shimadzu analyzer, 8
Toluene, 297, 309, 324
Total internal reflection, 166
Toxicity
of arsenic, 58
of mercury, 51
Transition metal oxides, 136
Trichloroethylene, 146, 263, 294, 305–308
in air purification, 304–308
Trichloropropene (TCP), 309
1,2,3-trihydroxybenzene (1,2,3-THB), 76, 92
Tube light reactor (TLR), 168–169
lamps in, 174
schematic diagram of, 174f
Tubular collectors (TC), 191
Tubular reactors, 215
for air treatment, 312–326
coated wall, 315–316
fluidized bed, 314–315
illumination of, 219f
monolith, 319
permeable layer, 321–322
powder layer, 312–314
schematic view of, 314f
spectral volumetric absorbed power in, 219f
Tubular receivers, CPC, 194–197
Turbulent vortex flow (TVF), 171
flow pattern, 179f
TVF. *See* Turbulent vortex flow
TVR. *See* Taylor Vortex photocatalytic reactor

U
Ultrafine catalysts, 160
Unit mass, 23, 26
benzyl alcohol, 24f
Uranium
EDTA and, 58
heterogeneous photocatalysis, 57–58
UV illumination, 8–9

UV lamps, 174, 238f, 244f
UV light-electronic phase, 148
UV radiometer, 200

V
Valence band holes, 38
Velocity profile, 245
Velocity vectors, 167f
Ventilation, 294
Visible light, 329–330
VOCs. *See* Volatile organic compounds
Volatile organic compounds (VOCs), 294
Volumetric emission model, 259
Volumetric flow rate, 160
Volumetric lamps, 261f
Vorontsov's model, 298
VTC. *See* V-trough collector
V-trough collector (VTC), 201
geometries, 205f

W
Waste tank, 258f
Water, 10
heterogeneous photocalysis, 37–62
oxidation reaction, 42
tank, 258f
treatment of metals in, 37–62
treatment patents, 292f
Water purification, 145–148
catalyst immobilization in, 175–176
catalyst in, 175
circulation rates in, 151
conversion rates in, 151
experimental details, 172–181
experimental procedure, 176–177
experimental setup, 176
kinetic reactors in, 152–153
large scale, 159–181
macrokinetic studies, 148–159
model analysis, 176
model component, 176
multiple tube reactor, 164–168
photo-CREC Water-II reactor in, 163
reactor comparison, 162t
reactors, 172–175
slurry system regimes, 154–159

Taylor vortex reactor, 169–172
tube light reactor, 168–169
Water splitting, 111–113
  configurations, 114–117
  energy requirements, 117–119
  Gibbs free energy change for, 115f
  hydrogen in, 130f
  material requirements for, 124–125
  metal nitrides in, 137
  metal oxynitrides in, 137
  niobates in, 135–136
  oxygen in, 130f
  photocatalysts for, 124–140
  photoelectrochemistry of, 113–123
  principle of, 114f
  schematic diagram, 118f
  solar spectrum and, 119–123
  tantalates in, 135–136
  titanates in, 134–135
  titanium dioxide in, 134–135
  transition metal oxides in, 136
Wavelength-dependent asymmetry
    parameters, 207
Wavelength-dependent constants, 214
Weakly turbulent wavy vortex flow
    (WTWVF), 171
  flow pattern, 181f

Weisz modulus, 301
WHO. See World Health Organization
Wood's glass, 302
World Health Organization (WHO), 44,
    49, 57
WTWVF. See Weakly turbulent wavy
    vortex flow

X
XPS. See X-ray photoelectron
    spectroscopy
X-ray diffraction (XRD), 56
X-ray photoelectron spectroscopy (XPS),
    56
  CREC and, 80–81
XRD. See X-ray diffraction

Z
Zeolite, 300
Zero-order reaction rates, 88
Zinc/cadmium lamps, 302
Zirconates, 317
$ZnCl_2$, 51
ZnO, 139
ZnS, 139

CONTENTS OF VOLUMES IN THIS SERIAL

Volume 1 (1956)

J. W. Westwater, *Boiling of Liquids*
A. B. Metzner, *Non-Newtonian Technology: Fluid Mechanics, Mixing, and Heat Transfer*
R. Byron Bird, *Theory of Diffusion*
J. B. Opfell and B. H. Sage, *Turbulence in Thermal and Material Transport*
Robert E. Treybal, *Mechanically Aided Liquid Extraction*
Robert W. Schrage, *The Automatic Computer in the Control and Planning of Manufacturing Operations*
Ernest J. Henley and Nathaniel F. Barr, *Ionizing Radiation Applied to Chemical Processes and to Food and Drug Processing*

Volume 2 (1958)

J. W. Westwater, *Boiling of Liquids*
Ernest F. Johnson, *Automatic Process Control*
Bernard Manowitz, *Treatment and Disposal of Wastes in Nuclear Chemical Technology*
George A. Sofer and Harold C. Weingartner, *High Vacuum Technology*
Theodore Vermeulen, *Separation by Adsorption Methods*
Sherman S. Weidenbaum, *Mixing of Solids*

Volume 3 (1962)

C. S. Grove, Jr., Robert V. Jelinek, and Herbert M. Schoen, *Crystallization from Solution*
F. Alan Ferguson and Russell C. Phillips, *High Temperature Technology*
Daniel Hyman, *Mixing and Agitation*
John Beck, *Design of Packed Catalytic Reactors*
Douglass J. Wilde, *Optimization Methods*

Volume 4 (1964)

J. T. Davies, *Mass-Transfer and Inierfacial Phenomena*
R. C. Kintner, *Drop Phenomena Affecting Liquid Extraction*
Octave Levenspiel and Kenneth B. Bischoff, *Patterns of Flow in Chemical Process Vessels*
Donald S. Scott, *Properties of Concurrent Gas–Liquid Flow*
D. N. Hanson and G. F. Somerville, *A General Program for Computing Multistage Vapor–Liquid Processes*

Volume 5 (1964)

J. F. Wehner, *Flame Processes–Theoretical and Experimental*

J. H. Sinfelt, *Bifunctional Catalysts*

S. G. Bankoff, *Heat Conduction or Diffusion with Change of Phase*

George D. Fulford, *The Flow of Lktuids in Thin Films*

K. Rietema, *Segregation in Liquid–Liquid Dispersions and its Effects on Chemical Reactions*

Volume 6 (1966)

S. G. Bankoff, *Diffusion-Controlled Bubble Growth*

John C. Berg, Andreas Acrivos, and Michel Boudart, *Evaporation Convection*

H. M. Tsuchiya, A. G. Fredrickson, and R. Aris, *Dynamics of Microbial Cell Populations*

Samuel Sideman, *Direct Contact Heat Transfer between Immiscible Liquids*

Howard Brenner, *Hydrodynamic Resistance of Particles at Small Reynolds Numbers*

Volume 7 (1968)

Robert S. Brown, Ralph Anderson, and Larry J. Shannon, *Ignition and Combustion of Solid Rocket Propellants*

Knud Østergaard, *Gas–Liquid–Particle Operations in Chemical Reaction Engineering*

J. M. Prausnilz, *Thermodynamics of Fluid–Phase Equilibria at High Pressures*

Robert V. Macbeth, *The Burn–Out Phenomenon in Forced-Convection Boiling*

William Resnick and Benjamin Gal–Or, *Gas-Liquid Dispersions*

Volume 8 (1970)

C. E. Lapple, *Electrostatic Phenomena with Particulates*

J. R. Kittrell, *Mathematical Modeling of Chemical Reactions*

W. P. Ledet and D. M. Himmelblau, *Decomposition Procedures foe the Solving of Large Scale Systems*

R. Kumar and N. R. Kuloor, *The Formation of Bubbles and Drops*

Volume 9 (1974)

Renato G. Bautista, *Hydrometallurgy*

Kishan B. Mathur and Norman Epstein, *Dynamics of Spouted Beds*

W. C. Reynolds, *Recent Advances in the Computation of Turbulent Flows*

R. E. Peck and D. T. Wasan, *Drying of Solid Particles and Sheets*

Volume 10 (1978)

G. E. O'Connor and T. W. F. Russell, *Heat Transfer in Tubular Fluid–Fluid Systems*

P. C. Kapur, *Balling and Granulation*

Richard S. H. Mah and Mordechai Shacham, *Pipeline Network Design and Synthesis*

J. Robert Selman and Charles W. Tobias, *Mass-Transfer Measurements by the Limiting-Current Technique*

Volume 11 (1981)

Jean-Claude Charpentier, *Mass-Transfer Rates in Gas–Liquid Absorbers and Reactors*
Dee H. Barker and C. R. Mitra, *The Indian Chemical Industry—Its Development and Needs*
Lawrence L. Tavlarides and Michael Stamatoudis, *The Analysis of Interphase Reactions and Mass Transfer in Liquid–Liquid Dispersions*
Terukatsu Miyauchi, Shintaro Furusaki, Shigeharu Morooka, and Yoneichi Ikeda, *Transport Phenomena and Reaction in Fluidized Catalyst Beds*

Volume 12 (1983)

C. D. Prater, J, Wei, V. W. Weekman, Jr., and B. Gross, *A Reaction Engineering Case History: Coke Burning in Thermofor Catalytic Cracking Regenerators*
Costel D. Denson, *Stripping Operations in Polymer Processing*
Robert C. Reid, *Rapid Phase Transitions from Liquid to Vapor*
John H. Seinfeld, *Atmospheric Diffusion Theory*

Volume 13 (1987)

Edward G. Jefferson, *Future Opportunities in Chemical Engineering*
Eli Ruckenstein, *Analysis of Transport Phenomena Using Scaling and Physical Models*
Rohit Khanna and John H. Seinfeld, *Mathematical Modeling of Packed Bed Reactors: Numerical Solutions and Control Model Development*
Michael P. Ramage, Kenneth R. Graziano, Paul H. Schipper, Frederick J. Krambeck, and Byung C. Choi, *KINPTR (Mobil's Kinetic Reforming Model): A Review of Mobil's Industrial Process Modeling Philosophy*

Volume 14 (1988)

Richard D. Colberg and Manfred Morari, *Analysis and Synthesis of Resilient Heat Exchange Networks*
Richard J. Quann, Robert A. Ware, Chi-Wen Hung, and James Wei, *Catalytic Hydrometallation of Petroleum*
Kent David, *The Safety Matrix: People Applying Technology to Yield Safe Chemical Plants and Products*

Volume 15 (1990)

Pierre M. Adler, Ali Nadim, and Howard Brenner, *Rheological Models of Suspenions*
Stanley M. Englund, *Opportunities in the Design of Inherently Safer Chemical Plants*
H. J. Ploehn and W. B. Russel, *Interations between Colloidal Particles and Soluble Polymers*

Volume 16 (1991)

Perspectives in Chemical Engineering: Research and Education

Clark K. Colton, *Editor*

Historical Perspective and Overview

L. E. Scriven, *On the Emergence and Evolution of Chemical Engineering*
Ralph Landau, *Academic–industrial Interaction in the Early Development of Chemical Engineering*
James Wei, *Future Directions of Chemical Engineering*

Fluid Mechanics and Transport

L. G. Leal, *Challenges and Opportunities in Fluid Mechanics and Transport Phenomena*
William B. Russel, *Fluid Mechanics and Transport Research in Chemical Engineering*
J. R. A. Pearson, *Fluid Mechanics and Transport Phenomena*

Thermodynamics

Keith E. Gubbins, *Thermodynamics*
J. M. Prausnitz, *Chemical Engineering Thermodynamics: Continuity and Expanding Frontiers*
H. Ted Davis, *Future Opportunities in Thermodynamics*

Kinetics, Catalysis, and Reactor Engineering

Alexis T. Bell, *Reflections on the Current Status and Future Directions of Chemical Reaction Engineering*
James R. Katzer and S. S. Wong, *Frontiers in Chemical Reaction Engineering*
L. Louis Hegedus, *Catalyst Design*

Environmental Protection and Energy

John H. Seinfeld, *Environmental Chemical Engineering*
T. W. F. Russell, *Energy and Environmental Concerns*
Janos M. Beer, Jack B. Howard, John P. Longwell, and Adel F. Sarofim, *The Role of Chemical Engineering in Fuel Manufacture and Use of Fuels*

Polymers

Matthew Tirrell, *Polymer Science in Chemical Engineering*
Richard A. Register and Stuart L. Cooper, *Chemical Engineers in Polymer Science: The Need for an Interdisciplinary Approach*

Microelectronic and Optical Material

Larry F. Thompson, *Chemical Engineering Research Opportunities in Electronic and Optical Materials Research*
Klavs F. Jensen, *Chemical Engineering in the Processing of Electronic and Optical Materials: A Discussion Bioengineering*

Bioengineering

James E. Bailey, *Bioprocess Engineering*
Arthur E. Humphrey, *Some Unsolved Problems of Biotechnology*
Channing Robertson, *Chemical Engineering: Its Role in the Medical and Health Sciences*

Process Engineering

Arthur W. Westerberg, *Process Engineering*
Manfred Morari, *Process Control Theory: Reflections on the Past Decade and Goals for the Next*
James M. Douglas, *The Paradigm After Next*
George Stephanopoulos, *Symbolic Computing and Artificial Intelligence in Chemical Engineering: A New Challenge*

The Identity of Our Profession

Morton M. Denn, *The Identity of Our Profession*

Volume 17 (1991)

Y. T. Shah, *Design Parameters for Mechanically Agitated Reactors*
Mooson Kwauk, *Particulate Fluidization: An Overview*

Volume 18 (1992)

E. James Davis, *Microchemical Engineering: The Physics and Chemistry of the Microparticle*
Selim M. Senkan, *Detailed Chemical Kinetic Modeling: Chemical Reaction Engineering of the Future*
Lorenz T. Biegler, *Optimization Strategies for Complex Process Models*

Volume 19 (1994)

Robert Langer, *Polymer Systems for Controlled Release of Macromolecules, Immobilized Enzyme Medical Bioreactors, and Tissue Engineering*
J. J. Linderman, P. A. Mahama, K. E. Forsten, and D. A. Lauffenburger, *Diffusion and Probability in Receptor Binding and Signaling*
Rakesh K. Jain, *Transport Phenomena in Tumors*
R. Krishna, *A Systems Approach to Multiphase Reactor Selection*
David T. Allen, *Pollution Prevention: Engineering Design at Macro-, Meso-, and Microscales*
John H. Seinfeld, Jean M. Andino, Frank M. Bowman, Hali J. L. Forstner, and Spyros Pandis, *Tropospheric Chemistry*

Volume 20 (1994)

Arthur M. Squires, *Origins of the Fast Fluid Bed*
Yu Zhiqing, *Application Collocation*
Youchu Li, *Hydrodynamics*
Li Jinghai, *Modeling*
Yu Zhiqing and Jin Yong, *Heat and Mass Transfer*
Mooson Kwauk, *Powder Assessment*
Li Hongzhong, *Hardware Development*
Youchu Li and Xuyi Zhang, *Circulating Fluidized Bed Combustion*
Chen Junwu, Cao Hanchang, and Liu Taiji, *Catalyst Regeneration in Fluid Catalytic Cracking*

Volume 21 (1995)

Christopher J. Nagel, Chonghum Han, and George Stephanopoulos, *Modeling Languages: Declarative and Imperative Descriptions of Chemical Reactions and Processing Systems*
Chonghun Han, George Stephanopoulos, and James M. Douglas, *Automation in Design: The Conceptual Synthesis of Chemical Processing Schemes*
Michael L. Mavrovouniotis, *Symbolic and Quantitative Reasoning: Design of Reaction Pathways through Recursive Satisfaction of Constraints*
Christopher Nagel and George Stephanopoulos, *Inductive and Deductive Reasoning: The Case of Identifying Potential Hazards in Chemical Processes*
Keven G. Joback and George Stephanopoulos, *Searching Spaces of Discrete Soloutions: The Design of Molecules Processing Desired Physical Properties*

Volume 22 (1995)

Chonghun Han, Ramachandran Lakshmanan, Bhavik Bakshi, and George Stephanopoulos, *Nonmonotonic Reasoning: The Synthesis of Operating Procedures in Chemical Plants*
Pedro M. Saraiva, *Inductive and Analogical Learning: Data-Driven Improvement of Process Operations*
Alexandros Koulouris, Bhavik R. Bakshi and George Stephanopoulos, *Empirical Learning through Neural Networks: The Wave-Net Solution*
Bhavik R. Bakshi and George Stephanopoulos, *Reasoning in Time: Modeling, Analysis, and Pattern Recognition of Temporal Process Trends*
Matthew J. Realff, *Intelligence in Numerical Computing: Improving Batch Scheduling Algorithms through Explanation-Based Learning*

Volume 23 (1996)

Jeffrey J. Siirola, *Industrial Applications of Chemical Process Synthesis*
Arthur W. Westerberg and Oliver Wahnschafft, *The Synthesis of Distillation-Based Separation Systems*
Ignacio E. Grossmann, *Mixed-Integer Optimization Techniques for Algorithmic Process Synthesis*
Subash Balakrishna and Lorenz T. Biegler, *Chemical Reactor Network Targeting and Integration: An Optimization Approach*
Steve Walsh and John Perkins, *Operability and Control inn Process Synthesis and Design*

Volume 24 (1998)

Raffaella Ocone and Gianni Astarita, *Kinetics and Thermodynamics in Multicomponent Mixtures*
Arvind Varma, Alexander S. Rogachev, Alexandra S. Mukasyan, and Stephen Hwang, *Combustion Synthesis of Advanced Materials: Principles and Applications*
J. A. M. Kuipers and W. P. Mo, van Swaaij, *Computational Fluid Dynamics Applied to Chemical Reaction Engineering*
Ronald E. Schmitt, Howard Klee, Debora M. Sparks, and Mahesh K. Podar, *Using Relative Risk Analysis to Set Priorities for Pollution Prevention at a Petroleum Refinery*

Volume 25 (1999)

J. F. Davis, M. J. Piovoso, K. A. Hoo, and B. R. Bakshi, *Process Data Analysis and Interpretation*
J. M. Ottino, P. DeRoussel, S., Hansen, and D. V. Khakhar, *Mixing and Dispersion of Viscous Liquids and Powdered Solids*
Peter L. Silverston, Li Chengyue, Yuan Wei-Kang, *Application of Periodic Operation to Sulfur Dioxide Oxidation*

Volume 26 (2001)

J. B. Joshi, N. S. Deshpande, M. Dinkar, and D. V. Phanikumar, *Hydrodynamic Stability of Multiphase Reactors*
Michael Nikolaou, *Model Predictive Controllers: A Critical Synthesis of Theory and Industrial Needs*

Volume 27 (2001)

William R. Moser, Josef Find, Sean C. Emerson, and Ivo M, Krausz, *Engineered Synthesis of Nanostructure Materials and Catalysts*

Bruce C. Gates, *Supported Nanostructured Catalysts: Metal Complexes and Metal Clusters*

Ralph T. Yang, *Nanostructured Absorbents*

Thomas J. Webster, *Nanophase Ceramics: The Future Orthopedic and Dental Implant Material*

Yu-Ming Lin, Mildred S. Dresselhaus, and Jackie Y. Ying, *Fabrication, Structure, and Transport Properties of Nanowires*

Volume 28 (2001)

Qiliang Yan and Juan J. DePablo, *Hyper-Parallel Tempering Monte Carlo and Its Applications*

Pablo G. Debenedetti, Frank H. Stillinger, Thomas M. Truskett, and Catherine P. Lewis, *Theory of Supercooled Liquids and Glasses: Energy Landscape and Statistical Geometry Perspectives*

Michael W. Deem, *A Statistical Mechanical Approach to Combinatorial Chemistry*

Venkat Ganesan and Glenn H. Fredrickson, *Fluctuation Effects in Microemulsion Reaction Media*

David B. Graves and Cameron F. Abrams, *Molecular Dynamics Simulations of Ion–Surface Interactions with Applications to Plasma Processing*

Christian M. Lastoskie and Keith E, Gubbins, *Characterization of Porous Materials Using Molecular Theory and Simulation*

Dimitrios Maroudas, *Modeling of Radical-Surface Interactions in the Plasma-Enhanced Chemical Vapor Deposition of Silicon Thin Films*

Sanat Kumar, M. Antonio Floriano, and Athanassiors Z. Panagiotopoulos, *Nanostructured Formation and Phase Separation in Surfactant Solutions*

Stanley I. Sandler, Amadeu K. Sum, and Shiang-Tai Lin, *Some Chemical Engineering Applications of Quantum Chemical Calculations*

Bernhardt L. Trout, *Car-Parrinello Methods in Chemical Engineering: Their Scope and potential*

R. A. van Santen and X. Rozanska, *Theory of Zeolite Catalysis*

Zhen-Gang Wang, *Morphology, Fluctuation, Metastability and Kinetics in Ordered Block Copolymers*

Volume 29 (2004)

Michael V. Sefton, *The New Biomaterials*

Kristi S. Anseth and Kristyn S. Masters, *Cell–Material Interactions*

Surya K. Mallapragada and Jennifer B. Recknor, *Polymeric Biomaterias for Nerve Regeneration*

Anthony M. Lowman, Thomas D. Dziubla, Petr Bures, and Nicholas A. Peppas, *Structural and Dynamic Response of Neutral and Intelligent Networks in Biomedical Environments*

F. Kurtis Kasper and Antonios G. Mikos, *Biomaterials and Gene Therapy*

Balaji Narasimhan and Matt J. Kipper, *Surface-Erodible Biomaterials for Drug Delivery*

Volume 30 (2005)

Dionisio Vlachos, *A Review of Multiscale Analysis: Examples from System Biology, Materials Engineering,and Other Fluids-Surface Interacting Systems*

Lynn F. Gladden, M.D. Mantle and A.J. Sederman, *Quantifying Physics and Chemistry at Multiple Length-Scales using Magnetic Resonance Techniques*

Juraj Kosek, Frantisek Steĕpánek, and Miloš Marek, *Modelling of Transport and Transformation Processes in Porous and Multiphase Bodies*

Vemuri Balakotaiah and Saikat Chakraborty, *Spatially Averaged Multiscale Models for Chemical Reactors*

Volume 31 (2006)

Yang Ge and Liang-Shih Fan, *3-D Direct Numerical Simulation of Gas–Liquid and Gas–Liquid–Solid Flow Systems Using the Level-Set and Immersed-Boundary Methods*
M.A. van der Hoef, M. Ye, M. van Sint Annaland, A.T. Andrews IV, S. Sundaresan, and J.A.M. Kuipers, *Multiscale Modeling of Gas-Fluidized Beds*
Harry E.A. Van den Akker, *The Details of Turbulent Mixing Process and their Simulation*
Rodney O. Fox, *CFD Models for Analysis and Design of Chemical Reactors*
Anthony G. Dixon, Michiel Nijemeisland, and E. Hugh Stitt, *Packed Tubular Reactor Modeling and Catalyst Design Using Computational Fluid Dynamics*

Volume 32 (2007)

William H. Green, Jr., *Predictive Kinetics: A New Approach for the 21st Century*
Mario Dente, Giulia Bozzano, Tiziano Faravelli, Alessandro Marongiu, Sauro Pierucci and Eliseo Ranzi, *Kinetic Modelling of Pyrolysis Processes in Gas and Condensed Phase*
Mikhail Sinev, Vladimir Arutyunov and Andrey Romanets, *Kinetic Models of C1–C4 Alkane Oxidation as Applied to Processing of Hydrocarbon Gases: Principles, Approaches and Developments*
Pierre Galtier, *Kinetic Methods in Petroleum Process Engineering*

Volume 33 (2007)

Shinichi Matsumoto and Hirofumi Shinjoh, *Dynamic Behavior and Characterization of Automobile Catalysts*
Mehrdad Ahmadinejad, Maya R. Desai, Timothy C. Watling and Andrew P.E. York, *Simulation of Automotive Emission Control Systems*
Anke Güthenke, Daniel Chatterjee, Michel Weibel, Bernd Krutzsch, Petr Kočí, Miloš Marek, Isabella Nova and Enrico Tronconi, *Current Status of Modeling Lean Exhaust Gas Aftertreatment Catalysts*
Athanasios G. Konstandopoulos, Margaritis Kostoglou, Nickolas Vlachos and Evdoxia Kladopoulou, *Advances in the Science and Technology of Diesel Particulate Filter Simulation*

Volume 34 (2008)

C.J. van Duijn, Andro Mikelić, I.S. Pop, and Carole Rosier, *Effective Dispersion Equations for Reactive Flows with Dominant Péclet and Damkohler Numbers*
Mark Z. Lazman and Gregory S. Yablonsky, *Overall Reaction Rate Equation of Single-Route Complex Catalytic Reaction in Terms of Hypergeometric Series*
A.N. Gorban and O. Radulescu, *Dynamic and Static Limitation in Multiscale Reaction Networks, Revisited*
Liqiu Wang, Mingtian Xu, and Xiaohao Wei, *Multiscale Theorems*

Volume 35 (2009)

Rudy J. Koopmans and Anton P.J. Middelberg, *Engineering Materials from the Bottom Up – Overview*
Robert P.W. Davies, Amalia Aggeli, Neville Boden, Tom C.B. McLeish, Irena A. Nyrkova, and Alexander N. Semenov, *Mechanisms and Principles of 1D Self-Assembly of Peptides into β-Sheet Tapes*

Paul van der Schoot, *Nucleation and Co-Operativity in Supramolecular Polymers*
Michael J. McPherson, Kier James, Stuart Kyle, Stephen Parsons, and Jessica Riley, *Recombinant Production of Self-Assembling Peptides*
Boxun Leng, Lei Huang, and Zhengzhong Shao, *Inspiration from Natural Silks and Their Proteins*
Sally L. Gras, *Surface- and Solution-Based Assembly of Amyloid Fibrils for Biomedical and Nanotechnology Applications*
Conan J. Fee, *Hybrid Systems Engineering: Polymer–Peptide Conjugates*

Volume 36 (2009)

Vincenzo Augugliaro, Sedat Yurdakal, Vittorio Loddo, Giovanni Palmisano, and Leonardo Palmisano, *Determination of Photoadsorption Capacity of Polychrystalline TiO$_2$ Catalyst in Irradiated Slurry*
Marta I. Litter, *Treatment of Chromium, Mercury, Lead, Uranium, and Arsenic in Water by Heterogeneous Photocatalysis*
Aaron Ortiz-Gomez, Benito Serrano-Rosales, Jesus Moreira-del-Rio, and Hugo de-Lasa, *Mineralization of Phenol in an Improved Photocatalytic Process Assisted with Ferric Ions: Reaction Network and Kinetic Modeling*
R.M. Navarro, F. del Valle, J.A. Villoria de la Mano, M.C. Álvarez-Galván, and J.L.G. Fierro, *Photocatalytic Water Splitting Under Visible Light: Concept and Catalysts Development*
Ajay K. Ray, *Photocatalytic Reactor Configurations for Water Purification: Experimentation and Modeling*
Camilo A. Arancibia-Bulnes, Antonio E. Jiménez, and Claudio A. Estrada, *Development and Modeling of Solar Photocatalytic Reactors*
Orlando M. Alfano and Alberto E. Cassano, *Scaling-Up of Photoreactors: Applications to Advanced Oxidation Processes*
Yaron Paz, *Photocatalytic Treatment of Air: From Basic Aspects to Reactors*